Product Design and Development Handbook

An Innovative, Entrepreneurial, and Structured Approach for Engineering Capstone and Industry Projects

FIRST EDITION

Steven W. Trimble and Abdelrahman N. Shuaib

Arizona State University

SAN DIEGO

Bassim Hamadeh, CEO and Publisher
Amy Smith, Senior Project Editor
Alia Bales, Production Editor
Jess Estrella, Senior Graphic Designer
Kylie Bartolome, Licensing Associate
Natalie Piccotti, Director of Marketing
Kassie Graves, Senior Vice President, Editorial
Jamie Giganti, Director of Academic Publishing

Copyright © 2022 by Steven W. Trimble and Abdelrahman N. Shuaib. All rights reserved. No part of this publication may be reprinted, reproduced, transmitted, or utilized in any form or by any electronic, mechanical, or other means, now known or hereafter invented, including photocopying, microfilming, and recording, or in any information retrieval system without the written permission of Cognella, Inc. For inquiries regarding permissions, translations, foreign rights, audio rights, and any other forms of reproduction, please contact the Cognella Licensing Department at rights@cognella.com.

Trademark Notice: Product or corporate names may be trademarks or registered trademarks and are used only for identification and explanation without intent to infringe.

Cover image copyright © 2019 iStockphoto LP/lvcandy.

Printed in the United States of America.

Contents

List of Figures xv

List of Tables xxi

List of Worksheets xxv

Foreword xxvii
    Kyle Squires, PhD

Preface xxix

Acknowledgments xxxi

SECTION A Handbook Overview 1

- A.1 Purpose of the Handbook 2
- A.2 Organization of the Handbook 2

MODULE 1
Introduction to the Handbook and Course 4

- A.3 How to Use the Handbook 5
- A.4 Phased Product Development (PPD) Process 5
  - A.4.1 Overall PPD Process 5
  - A.4.2 PPD Process Tailored for the Capstone Project 6
  - A.4.3 PPD Process Phase Descriptions 8
- A.5 Organization of a Typical Capstone Course 10
  - A.5.1 Learning Modules 10
  - A.5.2 Course Schedules 10
  - A.5.3 Capstone Course Deliverables 10
- A.6 ABET Accreditation Outcomes 13
- A.7 Entrepreneurial Mindset 14
- A.8 Being Curious Worksheet 17
- A.9 Commercialization Process 19
- A.10 Transitioning from Student to Practicing Engineer 21
- A.11 Sustainability 22
- A.12 Responsibilities and Expectations of the Capstone Course Student 22
- A.13 Case Study of the PPD Process in Industry 23
- A.14 Review Quiz for Section A 25

APPENDIX AA1
Worksheet 2 Answers 28

SECTION B Phased Product Development (PPD) Process — 29

SECTION B: PHASE 1: PRECONCEPT DESIGN — 31

MODULE 2
Starting Work on the Proposal (Phase 1) — 32

- B.1 Phase 1: Preconcept Design — 33
  - B.1.1 Overview — 33
  - B.1.2 Teamwork — 35
    - B.1.2.1 Stages of Team Development — 35
    - B.1.2.2 Team Roles — 36
    - B.1.2.3 Team Charter — 36
    - B.1.2.4 Team Minutes — 37
    - B.1.2.5 Sources of Team Conflict — 37
    - B.1.2.6 Holding Teammates Accountable — 38
  - B.1.3 Selecting the Capstone Project — 38
  - B.1.4 Understanding the Basic Physics Involved — 41
  - B.1.5 Creating the Problem Statement — 42
  - B.1.6 Initial Product Requirements — 42

MODULE 3
Project Selection — 43

- B.1.7 Selecting the Team Project — 44
- B.1.8 Example Travel Iron Team Report for Week 2 — 45

MODULE 4
Selecting the Project Preconcept — 46

- B.1.9 Preconcept Design — 47
- B.1.10 Value Proposition — 47

MODULE 5
Preparing the Project Proposal — 49

- B.1.11 Preparing the Project Proposal—Part 1 — 50
  - B.1.11.1 Introduction — 50
  - B.1.11.2 Customer Need and Engineering Requirements — 50
  - B.1.11.3 Preconcept Design — 50
  - B.1.11.4 Value Proposition — 50
  - B.1.11.5 Strategies for Addressing Key Issues — 51
  - B.1.11.6 Technical Approach — 51
  - B.1.11.7 Management Approach — 51
  - B.1.11.8 Risk Management — 51
  - B.1.11.9 Work Breakdown Structure (WBS) and Task Dictionary — 51
- B.1.12 Example Travel Iron Team Report for Week 3 — 54

MODULE 6
Project Schedules and Budgets — 55

- B.1.13 Preparing the Project Proposal—Part 2 — 56
  - B.1.13.1 Project Schedules — 56
  - B.1.13.2 Project Labor Budget — 56
  - B.1.13.3 Project Monetary Budget — 59
  - B.1.13.4 Five Key Success Factors — 59
- B.1.14 Using a Labor Chart and Schedule to Manage the Project — 59
  - B.1.14.1 Example: Managing Team XYZ — 59

MODULE 7
Finalizing the Proposal — 62

- B.1.15 Five Key Success Factors — 63
- B.1.16 Phase 1 Project Plan — 63
- B.1.17 Commercialization Aspects of Phase 1 — 65
- B.1.18 Final Steps in Completing the Proposal — 65
- B.1.19 Example Travel Iron Team Report for Week 4 — 65
  - B.1.19.1 Commercialization Plan—Market Analysis — 66
  - B.1.19.2 Commercialization Plan—Business Case — 66

MODULE 8
Completing Phase 1 — 67

- B.1.20 Phase 1 Preconcept Design Review (DR1) — 68
- B.1.21 Example Travel Iron Team Report for Week 4.5 — 68
- B.1.22 Phase 1 Exit Criteria Checklist — 75

SECTION B: PHASE 2: CONCEPTUAL DESIGN — 81

MODULE 9
Introduction to Phase 2: Conceptual Design — 82

- B.2 Phase 2: Conceptual Design — 83
  - B.2.1 Overview — 83
  - B.2.2 Building Team Commitment — 85
  - B.2.3 Refining the Phase 2 Project Plan — 85
  - B.2.4 Example Travel Iron Team Report for Week 5 — 87

MODULE 10
Capturing the Voice of the Customer — 88

- B.2.5 Capturing the Voice of the Customer (VOC) — 89

MODULE 11
Updating Requirements with QFD — 91

- B.2.6 Establishing Design Requirements — 92

	B.2.6.1	Quality Function Deployment (QFD)	92
	B.2.6.2	Goal Function	92
	B.2.6.3	Engineering Requirements	94
	B.2.6.4	Requirements Validation Matrix	95
B.2.7	A Simple Example: Requirements for a Travel Iron		95
B.2.8	Requirements Matrix and Goal Function		103
B.2.9	Preparing the Draft of Section 4 of the Final Report		105
B.2.10	Example Travel Iron Team Report for Week 6		105

MODULE 12
Exploring the Design Space — 106

B.2.11	Updating the Preconcept Function Block Diagram	107
B.2.12	Exploring the Design Space	107
B.2.13	Using Morphological Analysis to Identify Candidate Conceptual Designs	108
B.2.14	Selecting Candidate Conceptual Designs	111
B.2.15	Defining the Candidate Conceptual Designs	111
	B.2.15.1 Configuration Block Diagram	111
	B.2.15.2 Product Conceptual Sketch	111
	B.2.15.3 Concept Narrative	112

MODULE 13
Selecting and Analyzing the Conceptual Design — 113

B.2.16	Selecting the Final Conceptual Design	114
B.2.17	Conceptual Design Analyses and Testing	115
B.2.18	General Guidance on Conducting Design Analyses	116
B.2.19	General Guidance on Conducting Testing	117
B.2.20	Example Travel Iron Team Report for Week 7	117

MODULE 14
Completing Phase 2: Conceptual Design — 120

B.2.21	Defining the Final Conceptual Design	121
B.2.22	Commercialization	121
B.2.23	Marketing Basics	122
	B.2.23.1 Marketing Definitions	122
	B.2.23.2 Marketing Types	122
	B.2.23.3 Marketing Mix	123
	B.2.23.4 Marketing Strategy	123
B.2.24	Business Opportunity Development	123
B.2.25	Simple Business Model	124
B.2.26	Business Case	125
B.2.27	Return-on-Investment Calculations	125
B.2.28	Phase 2 Conceptual Design Review (DR2)	126
B.2.29	Documenting the Conceptual Design	128
B.2.30	Example Travel Iron Team Report for Week 7.5	129
	B.2.30.1 Description of Final Conceptual Design	129

		B.2.30.2	Analyses and Testing	131
		B.2.30.3	Commercialization	131
		B.2.30.4	Team Assessments	133
	B.2.31	Phase 2 Criteria Checklist		133

SECTION B: PHASE 3: PRELIMINARY DESIGN — **137**

MODULE 15
Introduction to Phase 3: Preliminary Design and Planning — 138

B.3	Phase 3: Preliminary Design			139
	B.3.1	Overview		139
	B.3.2	Continue to Build Team Commitment		141
	B.3.3	Phase 3 Planning and Management		141
		B.3.3.1	Preliminary Design Tasks, Schedule, and Labor Budget	141
		B.3.3.2	Phase 3 Task Dictionary	143
		B.3.3.3	Phase 3 Management	145
	B.3.4	Example Travel Iron Team Report for Week 8		145

MODULE 16
Design Drivers and Commercial off the Shelf (COTS) Components: Part 1 — 148

	B.3.4	Design Drivers Research and Analysis		149
		B.3.4.1	Types of Drivers	149
		B.3.4.2	Examples of Design Drivers	149
	B.3.5	Selecting the Commercial off the Shelf (COTS) Components		151
		B.3.5.1	Process for Selecting COTS Items	151
		B.3.5.2	Example of How to Select COTS Items	152

MODULE 17
Design Drivers and COTS Components Selection: Part 2 — 153

	B.3.6	Example Travel Iron Team Report for Week 9		154
		B.3.6.1	Team Planning	154
		B.3.6.2	Sole Plate POC Testing	154
		B.3.6.3	Temperature Settings and Electrical Power Requirement	154
		B.3.6.4	Specifying and Selecting COTS Components	155
		B.3.6.5	Dual-Voltage Capability	157
		B.3.6.6	Configuration Schematic and Wiring Diagram	159

MODULE 18
Product Baseline Preliminary Design: Part 1 — 161

	B.3.7	Size and Specify Fabricated Components	162
	B.3.8	Initial Assembly Plan	163

MODULE 19
Product Baseline Preliminary Design: Part 2 — 164

- B.3.9 Completing the Product Baseline Preliminary Design CAD Solid Model Rev 0 — 165
- B.3.10 Example Travel Iron Team Report for Week 10 — 165
  - B.3.10.1 COTS Specification and Supplier Identification — 165
  - B.3.10.2 Dimensioned Sketches of Fabricated Components — 165
  - B.3.10.3 Assembly Plan — 166
  - B.3.10.4 CAD Model Rev 0 — 166

MODULE 20
Updating the Requirements Matrix: Part 1 — 168

- B.3.11 Updating the Requirements Matrix and CAD Model — 169

MODULE 21
Updating the Requirements Matrix: Part 2 — 170

- B.3.12 Example Travel Iron Team Report for Week 11 — 171

MODULE 22
Systems Engineering and RMS Analyses — 172

- B.3.13 Systems Engineering — 173
- B.3.14 Trade Studies — 175
- B.3.15 Product Design Improvement with Goal Function — 175
  - B.3.15.1 Example of Goal Function Improvement Limited by Project Resources — 175
  - B.3.15.2 Example of Optimizing the Goal Function — 177
- B.3.16 Reliability Engineering — 178
  - B.3.16.1 Basic Concepts — 178
  - B.3.16.2 Describing Failure Rates with the Bathtub Curve — 179
  - B.3.16.3 Reliability for Constant Failure Rate Components — 181
  - B.3.16.4 Example Illustrating How Reliability Can Be Designed into the Product — 184
- B.3.17 Maintainability, Availability, and Warranties — 186
- B.3.18 Logistics Support — 188
- B.3.19 System Safety — 188

MODULE 23
FMEA, DFMA, and DTC: Part 1 — 191

- B.3.20 Failure Mode and Effects Analysis (FMEA) — 192
- B.3.21 Example Travel Iron Team Report for Week 12 — 193

MODULE 24
FMEA, DFMA, and DTC: Part 2 — 195

- B.3.22 Design for Manufacturing and Assembly (DFMA) — 196
  - B.3.22.1 DFM Guidelines — 196
  - B.3.22.2 DFA Guidelines — 198
- B.3.23 Design to Cost (DTC) and Design to Value — 199
  - B.3.23.1 DTC Thinking — 199
  - B.3.23.2 Life Cycle Costs (LCC) — 200
  - B.3.23.3 Example of Life Cycle Costing and Design to Value — 201
- B.3.24 DFMA and DTC Worksheet — 202

MODULE 25
FMEA, DFMA, and DTC: Part 3 — 204

- B.3.25 Worksheet 4: FMEA, DFMA, and DTC — 205
- B.3.26 Example Travel Iron Team Report for Week 13 — 206

MODULE 26
Final Preliminary Design and Commercialization — 207

- B.3.27 Defining the Final Preliminary Design — 208
- B.3.28 Initial Engineering Prototype Planning — 208
- B.3.29 Commercialization — 208

MODULE 27
End-of-Phase 3 Deliverables — 214

- B.3.30 Draft of Preliminary Design Section of the Final Report — 215
- B.3.31 Phase 3 Exit Criteria Checklist — 215
- B.3.32 Team Project Notebook Preparations — 215
- B.3.33 Phase 3 Preliminary Design Review (DR3) — 215
- B.3.34 Example Travel Iron Team Report for Week 14 — 216
  - B.3.34.1 Defining the Final Preliminary Design — 216
  - B.3.34.2 Engineering Prototype Planning — 220
  - B.3.34.3 Worksheet 5: First Draft of Goldsmith Commercialization Stage 3 Report — 220

MODULE 28
Preliminary Design Review and Exiting Phase 3 — 225

- B.3.35 Example Travel Iron Team Report for Week 14.5 — 226
- B.3.36 Phase 3 Exit Criteria Checklist — 226

SECTION B: PHASE 4: DETAILED DESIGN   **231**

MODULE 29
Phase 4 Plan, Prototype Scope, and Requirements   232

B.4	Detailed Design (Phase 4)	233
	B.4.1 Overview	233
	B.4.2 Detailed Phase 4 Planning	234
	B.4.3 Prototype Scope and Requirements	236
	B.4.4 Example Travel Iron Team Report for Week 16	236

MODULE 30
Prototype Analyses, Long Lead Hardware, and Manufacturing Methods   241

B.4.5	Phase 3P: Engineering Prototype Preliminary Design	242
B.4.6	Nonlinearity of the Product Development Process	242
B.4.7	Human Factors Engineering and Industrial Design	242
B.4.8	Long Lead Hardware	243
B.4.9	Manufacturing Methods	243
B.4.10	Preparing for Phase 3P: Engineering Prototype Preliminary Design Review	247
B.4.11	Phase 3P Exit Criteria Checklist	247
B.4.12	Example Travel Iron Team Report for Week 17	249
	B.4.12.1 Integrating New Information into the Design	249
	B.4.12.2 UL Requirements Study	249
	B.4.12.3 Safety Study	251
	B.4.12.4 Connection of Electrical Wires Study	251
	B.4.12.5 Engineering Prototype Requirements Validation Matrix	252
	B.4.12.6 Reliability Analyses	254
	B.4.12.7 Maintainability Analyses	254
	B.4.12.8 Safety Analyses and FMEA	254
	B.4.12.9 Design for Manufacturing and Assembly (DFMA)	254
	B.4.12.10 Design to Cost (DTC)	254
	B.4.12.11 CAD Model	254
	B.4.12.12 Test Rig Design	255
	B.4.12.13 Final Deliverables for Phase 3P	255

MODULE 31
Prototype and Test Rig Drawings, Manufacturing and Test Plans, and DR4 Preparation   256

B.4.13	Phase 4P: Engineering Prototype Detailed Design Overview	257
B.4.14	Detailed Drawing Package Contents	257
B.4.15	Engineering Drawing Format	258
B.4.16	Third Angle Projection	260
B.4.17	Subassembly and Assembly Drawings	262

B.4.18	Detail Views	263
B.4.19	Section Views	264
B.4.20	Questions to Ask When Making a Working Drawing	265
B.4.21	Configuration Management	265
B.4.22	Tolerances	267
	B.4.22.1 Linear Tolerances	267
	B.4.22.2 Tolerances for 3-D Printing	268
	B.4.22.3 Deciding What Tolerance to Use	269
	B.4.22.4 Fits of Mating Parts	269
	B.4.22.5 Surface Finish	270
	B.4.22.6 Tolerance Stacking	271
	B.4.22.7 Questions to Ask When Considering Tolerances	272
B.4.23	Manufacturing Plan	273
B.4.24	Manufacturing Process Planning	274
	B.4.24.1 Analysis of the Detailed Part Design Drawing(s)	274
	B.4.24.2 Preparation of a Routing Sheet for the Part	275
B.4.25	Engineering Prototype Test Plan	279
B.4.26	Final Report Sections	279
B.4.27	Preparation for Phase 3P and Phase 4P Design Review (DR4)	279
B.4.28	Phase 4P Exit Criteria Checklist	279
B.4.29	Example Travel Iron Team Report for Week 18	281
	B.4.29.1 Team Meetings	281
	B.4.29.2 Engineering Prototype Drawing Package	281
	B.4.29.3 Test Rig Drawing Package	281
	B.4.29.4 Phase 5P Manufacturing Plan	281
	B.4.29.5 Phase 5P Testing Plan	281
	B.4.29.6 Preparations for DR4	281
	B.4.29.7 Phase 4P Exit Criteria Checklist	281

MODULE 32
Starting Manufacturing, Prototype Testing Procedures, and Conducting DR4   285

B.4.30	Phase 5P: Engineering Prototype Development and Validation	286
B.4.31	Testing Procedures	286
B.4.32	Example Travel Iron Team Report for Week 19	286

MODULE 33
Engineering Prototype Build Book   289

B.4.33	Engineering Prototype Build Book	290
B.4.34	Example Travel Iron Team Report for Week 20	290

MODULE 34
Incoming Inspection and Manufacturing Troubleshooting   292

B.4.35	Incoming Inspection	293
B.4.36	Manufacturing Issues	295

B.4.37	Example Travel Iron Team Report for Week 21	296

MODULE 35
Deviations and Mid-Course Adjustments — 297

B.4.38	Deviations	298
B.4.39	Mid-Course Adjustments	299
B.4.40	Example Travel Iron Team Report for Week 22	300

MODULE 36
Assembly, First Article Inspection Form, and DR5 Preparations — 301

B.4.41	Prototype Assembly	302
B.4.42	First Article Inspection	302
B.4.43	DR5 Test Readiness Review Preparation	302
B.4.44	Example Travel Iron Team Report for Week 23	303
B.4.44.1	Engineering Prototype Assembly	303
B.4.44.2	Test Rig Assembly	303
B.4.44.3	DR5 Test Readiness Review Preparation	304
B.4.44.4	Change in Temperature Measuring Device	304

MODULE 37
Prototype First Article Inspection, DR5 Presentation, and Final Report Work — 310

B.4.45	Completing Prototype Manufacturing	311
B.4.46	Working on Sections of the Project Final Report	311
B.4.47	Example Travel Iron Team Report for Week 24	312
B.4.47.1	Engineering Prototype Manufacturing	312
B.4.47.2	DR5 Test Readiness Review	312
B.4.47.3	Progress on Final Report	312

MODULE 38
Testing Issues and Repair/Rework — 314

B.4.48	Engineering Prototype Testing Purpose and Issues	315
B.4.49	Example Travel Iron Team Report for Week 25	315
B.4.49.1	Development Testing Setup	315
B.4.49.2	Exploratory Testing	316
B.4.49.3	Exploratory Test 1A: Maximum Sole Plate Temperature and Power Requirement	316
B.4.49.4	Exploratory Test 1B: Maximum Temperature at Other Locations on the Sole Plate Centerline	317
B.4.49.5	Exploratory Test 1C: Temperature Control Settings	320
B.4.49.6	Exploratory Test 1D: Plastic Insulation Plate Temperature	320
B.4.49.7	Development Test 1: Three Temperature Settings	320
B.4.49.8	Development Test 2: Temperature Control Tolerances	321
B.4.49.9	Summary of Engineering Prototype Development Tests	322
B.4.49.10	Progress on Final Report	322

MODULE 39
Test Analyses/Reporting and Updating Engineering Prototype Drawings — 323

B.4.50	Test Analyses	324
B.4.51	Test Reporting	324
B.4.52	Updating Engineering Prototype Drawings	324
B.4.53	Example Travel Iron Team Report for Week 26	324
	B.4.53.1 Development Testing	324
	B.4.53.2 Final Report Writing	324

MODULE 40
Validation Testing and Test Results Design Review 6 (DR6) Preparations — 325

B.4.54	Validation Testing	326
B.4.55	Preparing for Test Results Design Review (DR6)	326
B.4.56	Example Travel Iron Team Report for Week 27	328
	B.4.56.1 Validation Testing	328
	B.4.56.2 Preparations for DR6	328
	B.4.56.3 Production Unit Detailed Design Drawing Package	328
	B.4.56.4 Preparation of the Final Report	328

MODULE 41
Conducting DR6, Updating the Commercialization Plan, and Completing the Final Report — 329

B.4.57	Conducting Test Results Design Review (DR6)	330
B.4.58	Updating the Commercialization Plan	330
B.4.59	Preparing Section 2 of the Project Final Report	330
B.4.60	Final Editing of the Final Report	330
B.4.61	Phase 5P Exit Criteria Checklist	331
B.4.62	Example Travel Iron Team Report for Week 28	333
	B.4.62.1 DR6 Review of Engineering Prototype Test Results	333
	B.4.62.2 Final Report	333
	B.4.62.3 Phase 5P Exit Criteria Checklist	333
	B.4.62.4 Project Notebook	333
	B.4.62.5 Updating the Commercialization Plan	333

MODULE 42
Preparation of Phase 4 Drawing Package and Project Final Presentation — 334

B.4.63	Final Report Presentation Outline and Rubric	335
B.4.64	Phase 4 Production Unit Detailed Design Drawing Package	337
B.4.65	Example Travel Iron Team Report for Week 29	338
	B.4.65.1 Final Presentation	338
	B.4.65.2 Phase 4 Detailed Design Drawing Package	338
	B.4.65.3 Project Notebook	338

MODULE 43
End-of-Project Deliverables — 339

- B.4.66 End-of-Capstone-Project Activities — 340
- B.4.67 Phase 4 Exit Criteria Checklist — 340
- B.4.68 Example Travel Iron Team Report for Week 30 — 340

MODULE 44
End-of-Project Deliverables Due — 344

- B.4.69 Final Report Presentation — 345
- B.4.70 Phase 4 Detailed Design Drawing Package — 345
- B.4.71 Project Closure — 345

SECTION B: PHASE 5: PRODUCTION PROTOTYPE DEVELOPMENT — 347
B.5 Phase 5: Production Prototype Development — 348

SECTION B: PHASE 6: PRODUCTION AND SUPPORT — 349
B.6 Phase 6: Production — 350

SECTION B: PRODUCT DEVELOPMENT FINAL THOUGHTS — 353
B.7 Product Development Final Thoughts — 354

SECTION B: APPENDICES — 357

APPENDIX BA1
360-Degree Teammate Review Form — 358

APPENDIX BA2
ASU Design Project Checklist — 361

APPENDIX BA3
Engineering Analysis for Design — 364

APPENDIX BA4
Product Development Testing — 372

APPENDIX BA5
FMEA Methodology — 378

SECTION C   Final Report Outline — 387

Index   407
About the Authors   413

List of Figures

FIGURE A-1	Module Lesson Sheet Format	3
FIGURE A-2	Overall PPD Process Diagram	6
FIGURE A-3	Engineering Prototype PPD Process Diagram	7
FIGURE A-4	Overall PPD Process Diagram	8
FIGURE A-5	Suggested Master Schedule for Design I Course	11
FIGURE A-6	Suggested Master Schedule for Design II Course	12
FIGURE A-7	Entrepreneurial Mindset (EM) and KEEN Background	15
FIGURE A-8	Inserts to Highlight Key EM Elements in the Handbook Text	16
FIGURE A-9	Symbol to Highlight EM Indicators in the Handbook Text	16
FIGURE A-10	Relationship Between PPD Process and Goldsmith Commercialization Model	20
FIGURE A-11	IPDS Process Applied to DHS SBIR Nonlethal Vehicle-Stopping Program	23
FIGURE A-12	SQUID's Three-Phase Nonlethal Vehicle-Stopping Process	24
FIGURE B-1	Team Charter Form	36
FIGURE B-2	Team Minutes Template	37
FIGURE B-3	Team Drawing: Filtering Dirty Water	42
FIGURE B-4	Generalized Top-Level System Functional Block Diagram	44
FIGURE B-5	Generalized More Detailed System Functional Block Diagram	44
FIGURE B-6	Top-Level Functional Diagram for an Engine Generator Set	44
FIGURE B-7	More Detailed System Functional Block Diagram for an Engine Generator Set	45
FIGURE B-8	Configuration Block Diagram for Engine Generator Set	45
FIGURE B-9	Generic Detailed Capstone Project WBS	52
FIGURE B-10	Example Project Labor Graph for the Proposal	59
FIGURE B-11	Team XYZ Schedule and Labor Chart for Week 9	60
FIGURE B-12	Example Case Phase 1 Schedule by Week	64
FIGURE B-13A	Travel Iron Proposal Slides (Page 1 of 7)	69
FIGURE B-13B	Travel Iron Proposal Slides (Page 2 of 7)	70
FIGURE B-13C	Travel Iron Proposal Slides (Page 3 of 7)	71
FIGURE B-13D	Travel Iron Proposal Slides (Page 4 of 7)	72
FIGURE B-13E	Travel Iron Proposal Slides (Page 5 of 7)	73
FIGURE B-13F	Travel Iron Proposal Slides (Page 6 of 7)	74
FIGURE B-13G	Travel Iron Proposal Slides (Page 7 of 7)	75
FIGURE B-14	Phase 2 Key Objectives	83

FIGURE B-15	Example Phase 2 Schedule 87
FIGURE B-16	Simplified HOQ Graphic 92
FIGURE B-17	Example Goal Function Graph 93
FIGURE B-18	Example Customer Need vs. Engineering Requirements Matrix 98
FIGURE B-19	Customer Need vs. Requirements Matrix with Importance Ratings 99
FIGURE B-20	Travel Iron QFD Matrix with Requirement Target Values 100
FIGURE B-21	QFD for Travel Iron with Requirements Weighting Added (Example is highlighted in gray) 101
FIGURE B-22	QFD Requirements Matrix for Travel Iron Example 102
FIGURE B-23	HOQ for Travel Iron Example 103
FIGURE B-24	Function Block Diagram for the Conceptual Design of Travel Iron 107
FIGURE B-25	The Conceptual Design Exploration and Filtering Process Is Not Linear 108
FIGURE B-26	Generalized Form of Morphological Chart 109
FIGURE B-27	Generalized Process for Selecting the Final Conceptual Design 109
FIGURE B-28	Morphological Chart for the Travel Iron 110
FIGURE B-29	Concept Trade Study for Travel Iron Example Using a Pugh Matrix 115
FIGURE B-30	Travel Iron Final Conceptual Design Components Are Circled 115
FIGURE B-31	Initial UPC Analysis Summary for Travel Iron 118
FIGURE B-32	Updated Travel Iron Requirements Validation Matrix 119
FIGURE B-33	Simple Business Model 124
FIGURE B-34	Example Problem IRR Analysis 126
FIGURE B-35	Example of How to Display Slide Data 128
FIGURE B-36	Travel Iron External View and Wiring Schematic 129
FIGURE B-37	Cross-Section of Travel Iron Final Conceptual Design 130
FIGURE B-38	Sketch of Sole Plate Casting 130
FIGURE B-39	Travel Iron Final Conceptual Design Exploded View 131
FIGURE B-40	Travel Iron Initial IRR Analysis 132
FIGURE B-41	Phase 3 Preliminary Design Flowchart 140
FIGURE B-42	Phase 3 Key Objectives 140
FIGURE B-43	Phase 3 Detailed Schedule 143
FIGURE B-44	Example: Optimization of Valve Weight Reduction for Maximum Incentive Dollars 150
FIGURE B-45	COTS Component Selection Example—Electric Travel Iron Thermostat 157
FIGURE B-46	Dual-Voltage Travel Iron Idea A: Use Same Heating Elements 158
FIGURE B-47	Same Heating Element Requires DPDT Switch 158
FIGURE B-48	Travel Iron Heating Element Options for Using a SPDT Switch 158
FIGURE B-49	Travel Iron Calculations for Resistances of Heating Elements 159
FIGURE B-50	Configuration Schematic of Travel Iron 160
FIGURE B-51	Pictorial Wiring Schematic of Travel Iron 160
FIGURE B-52	Examples of Hand Sketches Suitable for Preliminary Design 162
FIGURE B-53	Fabrication Sketch for Sole Plate 165
FIGURE B-54	Travel Iron CAD Model Rev 0 Exploded View 167

FIGURE B-55	The Vee Model of System Development Applied to the Phased Product Development Process 174	
FIGURE B-56	Flowchart for Conducting Trade Studies 175	
FIGURE B-57	General Format for Reporting Goal Function Progress 176	
FIGURE B-58	Example of Goal Function Optimization 177	
FIGURE B-59	Failure Rate vs. Operating Time for Component A Example 180	
FIGURE B-60	Random Failures Occur When Stress and Strength Probabilities Intersect 181	
FIGURE B-61	Reliability Equations for Components in Series 182	
FIGURE B-62	Reliability Equations for Components in Parallel 182	
FIGURE B-63	Reliability of Two Identical Components in Parallel 184	
FIGURE B-64	Example of the Benefits of Redundancy in Meeting a Reliability Requirement 185	
FIGURE B-65	Implementing the DFMA Process in Product Design 196	
FIGURE B-66	Producer's and Customer's Product Life Cycles 200	
FIGURE B-67	Travel Iron CAD Model Rev 3 External View 206	
FIGURE B-68	Travel Iron CAD Model Rev 3 Internal View 206	
FIGURE B-69	Travel Iron CAD Model Rev 3 Exploded View with Component Identifiers 206	
FIGURE B-70	Outline of Slides for DR3: Preliminary Design Review 216	
FIGURE B-71	Narrative Description of Travel Iron Product 217	
FIGURE B-72	Configuration Block Diagram 218	
FIGURE B-73	Requirements Validation Matrix for Travel Iron 219	
FIGURE B-74	Phase 4 Detailed Design Schedule 235	
FIGURE B-75	Partial Phase 4 Schedule for Week 16 Showing Slips 239	
FIGURE B-76	Copper Soldering Wire Connection to Heating Element Wire 252	
FIGURE B-77	Cross-Section of Travel Iron Engineering Prototype 255	
FIGURE B-78	Sole Plate Test Stand with IR Thermometer 255	
FIGURE B-79	Trolley Wheel CAD Solid Model Created by a Team During Preliminary Design 257	
FIGURE B-80	Parts of an Engineering Part Drawing 259	
FIGURE B-81	Example of an Incomplete Detailed Part Drawing 260	
FIGURE B-82	Possible Viewing Quadrants for a 3-D Object 261	
FIGURE B-83	Projecting a 3-D Object from the Third Angle Quadrant 261	
FIGURE B-84	3-D Object's Front, Top, and Side Views Projected onto an Imaginary Third Angle Projection Box 262	
FIGURE B-85	How the Third Angle Projection of the Part in Figure B-84 Will Display in a Detailed Drawing 262	
FIGURE B-86	Trolley Wheel Assembly Drawing Graphic and Item List 263	
FIGURE B-87	Example of a Detail View 264	
FIGURE B-88	Example of Using Section Views 264	
FIGURE B-89	Example Drawing Numbering System for a Hypothetical Engineering Prototype 266	
FIGURE B-90	Example Drawing Tree for a Hypothetical Engineering Prototype 266	
FIGURE B-91	Example Indented Parts List for a Hypothetical Engineering Prototype 267	
FIGURE B-92	Forms of Linear Tolerances 268	
FIGURE B-93	Example General Tolerance Note on a Drawing 268	

FIGURE B-94	Types of Fits 269
FIGURE B-95	Surface Finishes Possible for Various Manufacturing Processes 270
FIGURE B-96	Applications for Various Surface Finishes 271
FIGURE B-97	Stack-Up Analysis for Plug to Pocket Gap 272
FIGURE B-98	Example Manufacturing Plan Flowchart 273
FIGURE B-99	Stepped AISI 1045 Round Steel Shaft Drawing 276
FIGURE B-100	Routing Sheet for the Stepped Shaft 277
FIGURE B-101	Galvanized Steel Bracket 278
FIGURE B-102	Routing Sheet for Sheet Metal Bracket 278
FIGURE B-103	Manufacturing Plan for Travel Iron Engineering Prototype 282
FIGURE B-104	Travel Iron Prototype Test Matrix for Development and Validation Tests 284
FIGURE B-105	Procedures for Travel Iron Prototype Test 1 287
FIGURE B-106	Engineering Prototype Calculations for Resistance of Heating Elements 288
FIGURE B-107	Inspection Example 293
FIGURE B-108	Electronic Digital Caliper 293
FIGURE B-109	Inspection Sheet for Hole-in-the-Plate Example 294
FIGURE B-110	Component Drawing/Specification Deviation Decision Flowchart 298
FIGURE B-111	Recommended Deviation Form Format 299
FIGURE B-112	Assembled Travel Iron Engineering Prototype 304
FIGURE B-113	Travel Iron Test Rig 304
FIGURE B-114A	Travel Iron DR5 Slides (1 of 10) 305
FIGURE B-114B	Travel Iron DR5 Slides (2 of 10) 305
FIGURE B-114C	Travel Iron DR5 Slides (3 of 10) 306
FIGURE B-114D	Travel Iron DR5 Slides (4 of 10) 306
FIGURE B-114E	Travel Iron DR5 Slides (5 of 10) 307
FIGURE B-114F	Travel Iron DR5 Slides (6 of 10) 307
FIGURE B-114G	Travel Iron DR5 Slides (7 of 10) 308
FIGURE B-114H	Travel Iron DR5 Slides (8 of 10) 308
FIGURE B-114I	Travel Iron DR5 Slides (9 of 10) 309
FIGURE B-114J	Travel Iron DR5 Slides (10 of 10) 309
FIGURE B-115	Travel Iron Engineering Prototype Test Setup 315
FIGURE B-116	Data Analyses and Conclusions for Test 1A 318
FIGURE B-117	Test 1B Results: Temperatures Along Centerline with 95% Confidence Limits 319
FIGURE B-118	Temperature of Plastic Insulation Plate at Maximum Power 320
FIGURE B-119	Engineering Prototype Testing Schedule for End of Week 26 324
FIGURE B-120	Phase 6 Key Objectives 350
FIGURE B-121	Obligation of an Engineer 355
FIGURE BA3-1	Calculus Solution to Electronics Box Volume Optimization Problem 367
FIGURE BA3-2	Numerical Solution to Electronics Box Volume Optimization Problem 368
FIGURE BA3-3	Exact Solution to the Optimization of the Example Function of Two Independent Variables 369

FIGURE BA3-4	Numerical Solution to the Optimization of the Example Function of Two Independent Variables 369	
FIGURE BA3-5	Graph Format for a Single-Value Parametric Study 370	
FIGURE BA4-1	Test Plan Template 374	
FIGURE BA4-2	Experiment Results 375	
FIGURE BA4-3	Calculation of Average X and $(X_i - X)$ 376	
FIGURE BA4-4	Calculation of the Standard Deviation 376	
FIGURE BA4-5	Example Uncertainty Problem and Solution Spreadsheet 377	

List of Tables

TABLE A-1	Team Project Notebook Organization and Content	12
TABLE A-2	ABET Outcomes	13
TABLE A-3	Mastery Level Definitions Based on Demonstrated Behaviors	14
TABLE A-4	Evolution of the Engineering Mindset	14
TABLE A-5	Entrepreneurial Mindset 3C's	16
TABLE A-6	ASU Fulton Engineering EM@FSE 2. *Indicators* 0 Indicators	17
TABLE A-7	Goldsmith Model Key Questions for Each Phase	20
TABLE B-1	Phase 1 Objectives for the Capstone Project	34
TABLE B-2	The Four Stages of Team Development	35
TABLE B-3	Major Sources of Team Conflict	38
TABLE B-4	NAE Grand Challenges for the 21st Century	39
TABLE B-5	Examples of Mechanical Engineering Capstone Projects at ASU	40
TABLE B-6	Outline for Project Proposal	50
TABLE B-7	Generic Capstone Project WBS Dictionary by Phase	53
TABLE B-8	Example Project Plan Hours—Allocation Rev 0	57
TABLE B-9	Example Allocation of Hours Rev 1	58
TABLE B-10	Example Case Task Dictionary for Phase 1	64
TABLE B-11	Case Team Labor Budget for Phase 1 by Task and Week	65
TABLE B-12	List of Recommended Proposal Presentation Slides	68
TABLE B-13	Example Phase 2 Planning Chart in Hours per Week	86
TABLE B-14	Results for Four Designs of the Example Device	93
TABLE B-15	Top-Level Requirements Categories	94
TABLE B-16	Format for Requirements Validation Matrix	95
TABLE B-17	Customer Needs for a Portable Iron	96
TABLE B-18	Design Team's List of Travel Iron Engineering Requirements	97
TABLE B-19	Requirement Weighting Calculation for Operating Reliability Target Value	101
TABLE B-20	Validation Matrix for the Travel Iron Example	104
TABLE B-21	Exploring the Design Space Questions	108
TABLE B-22	Questions to Answer When Initiating a Design Analysis	116
TABLE B-23	Key Points Concerning Design Analyses for the Capstone Project	116
TABLE B-24	Analysis Documentation Format	117
TABLE B-25	Items to Define the Final Conceptual Design	121
TABLE B-26	Slides for Phase 2 Conceptual Design Review	127
TABLE B-27	Example Phase 3 Planning Chart	142

TABLE B-28	Task Dictionary for Phase 3 Tasks	144
TABLE B-29	Phase 3 Task Dictionary for the Travel Iron	147
TABLE B-30	Process for Selecting COTS Items	151
TABLE B-31	Travel Iron Temperature Settings	155
TABLE B-32	List of Travel Iron COTS vs. *Fabricated Components*	156
TABLE B-33	Questions to Be Answered in the Initial Assembly Plan	163
TABLE B-34	Assembly Process for Travel Iron	166
TABLE B-35	NASA's Definition of a System	173
TABLE B-36	System Engineering Topics Covered in This Handbook	174
TABLE B-37	Component A Test Data Failure Rates	179
TABLE B-38	Generic Failure Rates for Common Mechanical Components	183
TABLE B-39	Product Safety Design Tasks	188
TABLE B-40	Steps to Perform a Hazards Analysis	189
TABLE B-41	Hazard Analysis Form	189
TABLE B-42	Hazard Identification Checklist	190
TABLE B-43	FMEA Process	193
TABLE B-44	Items to Define the Final Preliminary Design	208
TABLE B-45	Travel Iron Features and Benefits	217
TABLE B-46	Travel Iron Key Characteristics	218
TABLE B-47	Travel Iron Project Detailed Plan for Phase 4 (Page 1 of 2)	237
TABLE B-47	Travel Iron Project Detailed Plan for Phase 4 (Page 2 of 2)	238
TABLE B-48	Engineering Prototype Requirements Shaded in Grey	240
TABLE B-49	Outline for the Engineering Prototype Phases 3P and 4P Design Review (DR4)	247
TABLE B-50	Summary of UL Requirements	250
TABLE B-51	Safety Decisions and Actions for Production and Prototype Units	251
TABLE B-52	Engineering Prototype Requirements Matrix	253
TABLE B-53	Spare Parts and Tools for Engineering Prototype Unit Maintenance	254
TABLE B-54	Topics for a Minimal Configuration Management Plan	266
TABLE B-55	Typical Tolerances Possible for Various Manufacturing Processes	268
TABLE B-56	Two Key Rules for Selecting a Tolerance	269
TABLE B-57	Suggested Outline for Manufacturing Plan	274
TABLE B-58	Operation Sheet Template	277
TABLE B-59	List of Components and Assigned Liaison Team Members	283
TABLE B-60	Engineering Prototype Build Book Structure	291
TABLE B-61	DR5 Test Readiness Review List of Presentation Slides	303
TABLE B-62	Checklist for Each Section of the Project Final Report	311
TABLE B-63	Final Report Preparation Status as of End of Week 24	313
TABLE B-64	Test 1A Data Sheet	317
TABLE B-65	Test 1B Data Sheet	319
TABLE B-66	Development Test 1 Data Sheet	320
TABLE B-67	Development Test 1 Results	321

TABLE B-68	Development Test 2 Data Sheet	321
TABLE B-69	Development Test 2 Results	321
TABLE B-70	Summary of Engineering Prototype Development Testing Results	322
TABLE B-71	Outline for DR6 Review of Engineering Prototype Test Results	327
TABLE B-72	Summary of Engineering Prototype Validation Testing Results	328
TABLE B-73	Final Report Checklist	333
TABLE B-74	Final Capstone Presentation Slides and Rubrics	335
TABLE B-75	Phase 4 Production Unit Detailed Drawing Package Outline	337
TABLE B-76	Capstone End-of-Course Deliverables	340
TABLE B-77	Key Product Design and Development Actions	355
TABLE BA3-1	Reasons Design Analysis Differs from Textbook Problems	364
TABLE BA3-2	Analysis Documentation Format	366
TABLE BA4-1	Steps in Conducting an Experiment	373
TABLE BA4-2	Steps to Ensure Proper Testing on the Project	374
TABLE BA5-1	Examples of Failure Modes	379
TABLE BA5-2	S Evaluation Criteria	380
TABLE BA5-3	O Evaluation Criteria	381
TABLE BA5-4	D Evaluation Criteria	382

List of Worksheets

WORKSHEET 1   Being Curious   18
WORKSHEET 2   Handbook Section A Quiz   26
WORKSHEET 3   DFMA and DTC   203
WORKSHEET 4   Design Changes due to FMEA, DFMA, and DTC Analyses   205
WORKSHEET 5   Goldsmith Commercialization Stage 3 Report   210

Foreword

Capstone design in engineering, computer science, and technology comprise the culminating experience for undergraduates. While there are variations on the theme and scope, capstone design revolves around student teams who engage the design process to conceptualize, prototype, test, redesign, and build. All of these activities take place against a backdrop of multiple, realistic constraints that necessitate tradeoffs throughout the design process. A team's design must capitalize on the students' prior learning in their academic programs and integrate both technical and nontechnical skills.

Tradeoffs considered throughout the design lead teams to develop their "best" design among several potential "right answers." Capstone design is among the comparatively few courses students take where the process of reaching a solution is as important as getting to a "right answer." Developing expertise by actually engaging in design provides an experience that is foundational to the development of new engineers.

Capstone design is perhaps the most critical curricular experience for students and is their primary exposure to real-world, open-ended design challenges. Successful capstone experiences enable students to learn and apply the design process, leveraging their prior studies to develop solutions against functional requirements, conduct analysis, and identify risks and failure modes as they progress toward prototyping and validation. The capstone experience is critical to student success and, therefore, its instruction is among the most vital to any program.

Teaching student teams to become adept at engineering design requires the instructor to promote systems-level thinking that for many students can feel imprecise. However, the rewards for instructors and their students are compelling, as acquiring new knowledge through the capstone experience results in critical broadening that will serve students well as they move into practice. It is further important to stress that the capstone process is valuable for students who will pursue graduate studies, where research experiences are open-ended, not unlike capstone design. Finally, many students find capstone design the most intensive opportunity for learning and honing the skills that can, at times, receive less attention during traditional coursework but are crucial to success after graduation, such as the ability to work in teams, communicate effectively, and build the soft skills required in practice.

The importance of the capstone to curricula, faculty, and students provides substantial motivation for pedagogical strategies that, on the one hand, supply structure, while on the other, leave ample opportunity for students to iterate and innovate. The supporting course materials are important for students to rely upon for guidance and connection throughout the design process. A superb example is this text, which has proven foundational to the success of mechanical engineering capstone design in the Ira A. Fulton Schools of Engineering at Arizona State University. It has evolved through numerous deliveries of the mechanical engineering capstone and has uniquely benefited from the expertise and dedication of the instructors of the course and authors of this book.

Among the novel attributes of this book is its comprehensive focus and linkage of the design process to other key activities, which include prototyping and manufacturing, the development and validation of prototypes, and the connection of product engineering to the overall commercialization process. These are critical connections and will serve both students and instructors well. This text strikes a terrific balance between innovation and entrepreneurial thinking and a structured approach that will help students focus on their overall goals while also staying on schedule. The way in which this text offers key checkpoints is helpful for students to assess progress and reflect on the team's comprehensive achievements. The book does a wonderful job of imparting detailed guidance on specific areas of the design process, helpful examples, and motivation and perspective concerning the broader aims being achieved both through the course and by the student teams. I have enjoyed working with the authors and am certain this text will continue to serve our students well, far into the future.

Kyle Squires, PhD
Dean, Ira A. Fulton Schools of Engineering and
Executive Vice Provost of Engineering, Computing and Technology
Arizona State University
August 2021

Preface

The purpose of this handbook is to help capstone engineering students and practicing engineers conduct product design and development projects that are successfully completed within given schedule and resource constraints. This handbook originated from the need to increase the number of successful capstone projects in engineering programs. The authors also wrote the handbook to increase the success rate of product design and development projects in industry. The key success factors of the handbook include applying innovative, entrepreneurial, and structured processes to capstone and engineering projects. Many well-written product development books already exist. But none explicitly guides the project team on what to do, when to do it, how to do it, and what the specific outcomes will be. This handbook fills this void.

The inspiration for this handbook was the renowned and successful industry integrated product delivery and support (IPDS) process developed at AlliedSignal Inc. The authors have modified and added more detail to this process. The result is the phased product development (PPD) process, which is the structural backbone of the handbook. The PPD process has two key elements. The first element is the mini-milestones that keep the product design and development team on schedule and within budget. The second element is the use of the phase-exit checklists that ensure that the project team is demonstrating evidence of meeting all Accreditation Board for Engineering and Technology (ABET) outcomes criteria at the stated levels of mastery.

The Kern Family Foundation entrepreneurial mindset (EM) initiative is becoming widely used in both engineering courses and industry. Arizona State University (ASU) is a leader in this effort and has developed specific EM indicators. These indicators are integrated into each handbook subject. The phase-exit checklists also include an assessment of how well the design and development team meets the EM indicators over the course of the project execution.

By following the structured PPD project execution process, about 200 six-member teams at ASU in the past six years have successfully completed their capstone projects on time and within budget; no team failed to complete its project. Moreover, the teams' engineering prototypes were built in time to undergo development and requirements validation testing.

The capstone project course entails much more than the demonstration of known skills and abilities. It is also a time of intense learning. Staying on schedule, understanding the customer's needs, forming a high-performing team, managing sponsor expectations, mitigating risk, and integrating the technical, marketing, and business aspects of an actual product design and development project can be overwhelming. The authors have divided the handbook into learning modules to help teams learn and perform in manageable steps.

The handbook emphasizes to the design and development teams the importance of keeping a project notebook and engineering prototype build book and writing a final report for their project. The capstone design team uses an outline in the handbook to write the final report as the work is being performed.

Product commercialization, which requires an integration of technical product development with marketing and business issues, is also included in the handbook. The Goldsmith Commercialization Model is used to help teams include market and business considerations in the product design and development process.

The authors successfully have used preliminary versions of this handbook for single-discipline, multidiscipline, and universitywide interdisciplinary capstone courses. Student feedback has been helpful in refining the contents of this handbook. Students especially like the travel iron project example that shows students how to apply each concept as the team progresses through the project phases.

The literature indicates that successful team projects, whether they are single-disciplinary, multidisciplinary, or interdisciplinary, share the same key elements: structured process, project management, team dynamics that encourage innovation and manage conflict, documentation, research skills, a commitment to exploring the design space, and effective communications with all stakeholders. This handbook stresses these elements.

Although the handbook was originated for a two-semester mechanical engineering capstone course, it can be easily tailored for use with other engineering disciplines and time frames. Specific guidelines and examples of this tailoring are included in the instructor's manual that is available from the publisher.

This handbook was also written for practicing engineers who are involved in product design and development projects in industry. Both of the authors are professors of practice who have 50 years each of extensive combined academic and industry experience in energy, manufacturing, materials, and design areas. They come from industry and are still active in industry through their respective consulting companies.

Meeting human need through the design and development of new products is a noble effort. We are confident that this handbook will be helpful to teams in learning how to successfully meet customer needs on time and within budget. We wish each team success on their project.

Steven W. Trimble, PhD
Professor of Practice
Ira A. Fulton Schools of Engineering
Arizona State University
July 2021

Abdelrahman N. Shuaib, PhD
Professor of Practice
Ira A. Fulton Schools of Engineering
Arizona State University
July 2021

Acknowledgments

This book has been a team project. The authors are thankful to all for their input, feedback, and support. The inspiration for this book came from Dr. Kyle Squires, who felt that capstone students needed more course structure to be successful. Drs. James Collofello and Gary Lichtenstein championed making the entrepreneurial mindset an integral part of this book. Our school director, Dr. Lenore Dai; Assistant Director for Business Services Mariah Pacey; and Program Chair Dr. Valana Wells provided not only resources, but also encouragement to make the capstone experience valuable to our students. Drs. Luis Bocanegra, James Middleton, and Patrick Phelan and many other professors and staff at Arizona State University (ASU) provided their expertise and guidance.

Preliminary versions of this handbook were successfully used in a university-wide interdisciplinary capstone program called Innovation Space, thanks to its director, Professor Prasad Boradkar. Thanks also go to Dr. Ronald Roedel, director of the multidisciplinary professional science master's solar energy engineering and commercialization program. He encouraged the use of preliminary versions of this handbook for the program's industry projects.

ASU's Student Shop Supervisor Leonard Bucholz provided many of the computer-aided design (CAD) drawings. He and Andre Magdelano have tirelessly worked with our students to turn their project ideas into prototype realities. The successful testing of those prototypes is largely due to the guidance of our laboratory manager, Bruce Steele. Additional CAD drawings were prepared by Brian Ahedor.

Our capstone students and teaching assistants were truly cocreators of this book. They were willing to test many new approaches in preliminary versions of this book and provide important feedback. Industry has also been active in creating this book. Companies such as Northrop Grumman Corporation Space Systems, Raytheon, Honeywell, MTD Southwest, Cummins, and many others provided real-world projects and mentoring. Special thanks go to Mr. Martin Martinez, president of Engineering Sciences and Analysis Corporation, and Mr. Herbert Hayden, president of Southwest Solar Technology, for their time, projects, and monetary support during the initial use of PPD for capstone courses at ASU.

Special thanks go to our wives and families for their support and understanding that writing this handbook was nearly a full-time endeavor.

It has been a pleasure to work with Cognella Academic Publishing. Our editor, Amy Smith, along with our cover designer, Asfa Arshi, and production editors, Abbey Hastings and Alia Bales, have added so much to the organization and quality of this book.

SECTION A

HANDBOOK OVERVIEW

A.1 Purpose of the Handbook

The purpose of this handbook is to help capstone engineering students and practicing engineers conduct projects that successfully complete their project scope within given schedule and resource constraints. This handbook is not just a collection of articles on the aspects of product design and development. It is innovatively structured to lead a student capstone team through a two-semester senior product design and development project. Moreover, it presents product development in the context of having an entrepreneurial mindset. The handbook uses the industry-proven phased product development (PPD) process that features phase-exit checklists.

On the other hand, this handbook is not just for engineering students. It can be used by practicing engineers as a guide to achieving more successful design and development projects. Although industry projects may have different schedule and resource constraints, they still greatly benefit from following the PPD process presented herein. Moreover, the emphasis on the entrepreneurial mindset (EM) brings fresh insight into developing products that creatively and profitably meet customers' needs.

The handbook was originally designed for a two-semester capstone mechanical engineering course at Arizona State University (ASU). However, it presents universal concepts that are applicable to a wide variety of university capstone as well as industry engineering design/development projects. The handbook features lesson modules that break down the process into manageable learning units. Instructors and self-learners can easily tailor these modules to meet their own educational needs.

This handbook focuses on five main project deliverables: project proposal, project notebook, validated engineering prototype, project final report, and project final presentation. The strategies for using this handbook to create these high-quality deliverables are presented.

The handbook is designed to help engineering students demonstrate their ability to meet all the Accreditation Board for Engineering and Technology (ABET) accreditation outcomes as well as the EM indicators. These same outcomes are also desired by industry. Hence, both students and practicing engineers will find this handbook useful.

A.2 Organization of the Handbook

The handbook is divided into the following three major sections:

Section A: Handbook Overview. The overview covers the purpose and organization of the handbook and how to use it during the conduct of a project. The key structural element of the handbook is the PPD process. This is the process that leads the design team through the work required in each sequential phase of the project (i.e., from writing the project proposal to putting the product into production). This section also explains the relationship between the engineering project and (1) the ABET accreditation student outcomes and (2) the Kern Family Foundation's EM initiative. In addition, this section introduces the Goldsmith Commercialization Model and shows how its stages relate to the PPD process.

Section B: Phased Product Development (PPD) Process. This section covers each of the PPD phases. The tasks that must be accomplished in each phase are described in detail. At the end of each phase, there is a phase-exit checklist to make sure that the project team has completed all the necessary design and development work.

Section C: Detailed Final Project Report Outline. This outline will result in a final report that meets all the ABET outcomes and EM indicators. The report includes detailed instructions that will help the team ensure that they have documented the entire product design and development process.

The handbook also features lesson modules. These modules provide an easy way for the capstone student or self-learner to study the material and conduct the project tasks. The lesson module format is shown in Figure A-1. When this handbook is used in an academic setting, there is usually a lecture associated with each module. These lectures reinforce the concepts covered in the handbook; however, the reader can still receive the benefits of the lesson modules without the lectures.

At the end of each module, the activities of a team developing a travel iron are reported. This serves as an example of how to apply the tasks described in that module to an actual project.

Each module begins with a Module Lesson Sheet that appears just before the assigned handbook reading for that module. The Module Lesson Sheet for Module 1 is provided just before Section A.3.

For a two-semester capstone course, it is suggested that modules 1 through 28 be used for the first semester, with two modules assigned each week. For the second semester, modules 29 through 44 are suggested, with one module assigned each week.

MODULE NUMBER

Title

OVERVIEW

Introduces the concepts to be covered and activities that can be assigned to help the reader learn how to apply these concepts to a design project.

LEARNING OBJECTIVES

- This helps the reader identify the key concepts in this module.

PRE-LECTURE ASSIGNMENT

- There is usually a reading assignment that helps the reader proceed through the handbook content.
- There is often an activity listed that will help the reader learn how to apply the concepts covered in the assigned reading.

POST-LECTURE ASSIGNMENT

- The tasks listed here are generally those associated with deliverables for the capstone project team. However, these tasks can also be used by other teams to help solidfy their mastery of the material covered in the reaading assignment and move their project forward.

TEAM DELIVERABLES

- Specific team assignments that are to be submitted for grading.

FIGURE A-1 Module Lesson Sheet Format

MODULE 1

Introduction to the Handbook and Course

OVERVIEW

This module introduces the reader to Section A of the handbook. This section provides an overview of the entire handbook and introduces the key concepts needed for product design and development. For the capstone student, complete the pre-lecture assignment before class and the post-lecture assignment after the lecture. The first assignment is an individual team member assignment while the second one is a team assignment.

LEARNING OBJECTIVES

- Internalize the basic concepts presented in Section A of the handbook:
    - How to use the handbook
    - PPD process
    - EM
    - Goldsmith commercialization process
    - The process of transitioning from student to practicing engineer
    - Student responsibilities
- Become curious about being curious.

PRE-LECTURE ASSIGNMENT

- Skim through the entire handbook and then study Section A.
- Prepare your own outline for Section A of the handbook.
- Complete Worksheet 1 on curiosity.
- Study your course's syllabus.

POST-LECTURE ASSIGNMENT

- Conduct a team meeting to assign each member specific research tasks relative to selecting a societal need and identifying a product that will address that need.
- Conduct assigned societal need research and communicate results to team members.
- As a team, complete the quiz at the end of Section A.

TEAM DELIVERABLES

- Summary of the assigned societal need research conducted by the team to identify a societal need
- Completed quiz at the end of Section A

A.3 How to Use the Handbook

The handbook is both a reference source and a structured guide for conducting an engineering product design/development project. It is suggested that the user do the following:

- Read the preface to understand the authors' motivations in producing this handbook.
- Read the author backgrounds to appreciate the mixture of academic and industry experiences that the authors bring to this handbook.
- Read Section A to:
  - Understand the big picture regarding this handbook's purpose and content
  - Gain insight into the content that is needed for each major project deliverable (i.e., the proposal, project notebook, engineering prototype hardware, final report, and final presentation)

The reader should carefully study the Section A material, which is working knowledge that can be applied at any time while executing the project. A good check of one's grasp of this knowledge is to take the competency quiz provided at the end of Section A.

- Use Section B as a structured guide for conducting each phase of the project.
- Use Section C to help in the writing of each section of the project final report. The project team should write appropriate sections of the final report as the work progresses.

A.4 Phased Product Development (PPD) Process

A.4.1 Overall PPD Process

The heart of this handbook is the PPD process, which is shown in Figure A-2. PPD is based on the industry-proven process that had its origins in the work done by engineers at AlliedSignal Inc., in the 1990s. Similar processes can be found in most aerospace, automotive, and other product development industries. The authors have expanded on the original process in the areas of testing, phase exit criteria, and phase deliverables. Section B of this handbook provides a detailed description of how to accomplish each phase in the PPD process.

As shown in the referenced Figure A-2, the PPD process divides the product development process into six sequential phases. It starts with the project proposal and ends with the product in production. Each phase produces key deliverables, which are shown in the ellipses under each phase box. It is important to note that testing starts with the conceptual design phase. Proof-of-concept (POC) testing can be very simple bench experiments to determine the feasibility of a particular idea. These early tests help identify concepts that will not work if incorporated into the product design. Industry values engineers that are able to create simple POC tests that eliminate unworkable ideas before a lot of effort is spent on them.

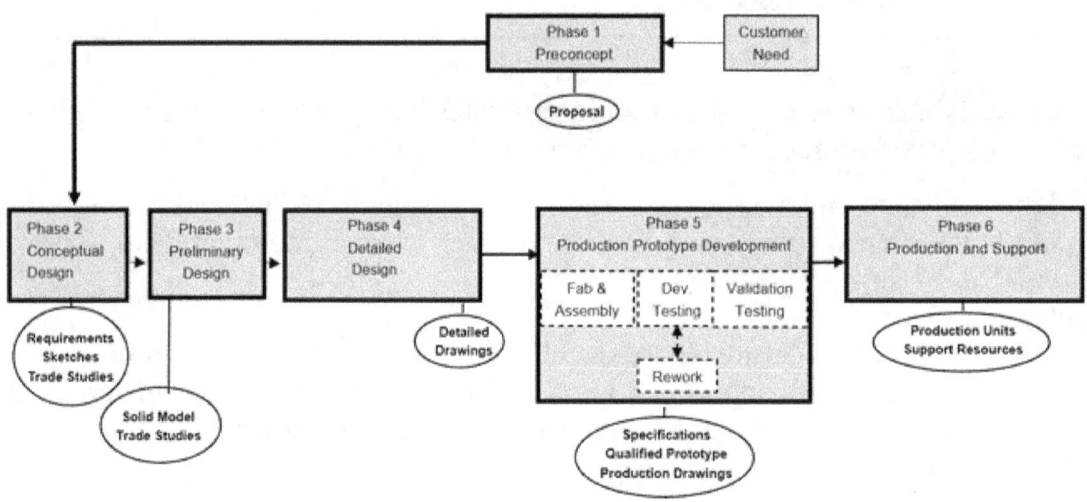

FIGURE A-2 Overall PPD Process Diagram

In this simplified process diagram, it is assumed that only one prototype unit (the preproduction prototype) is built, developed, and validated (see Phase 5 in Figure A-2) prior to the production phase. In most cases, there are one or more engineering prototypes built and tested prior to the preproduction prototype. These engineering prototypes are accomplished during the preliminary design phase or detailed design phase of product development. Information from the engineering prototype testing is used to create a final product detailed design that is then taken into Phase 5 for development of the production unit configuration.

A.4.2 PPD Process Tailored for the Capstone Project

Because one of the main purposes of this handbook is to guide university capstone project teams, it is important to understand that these courses, in general, have schedule and budgetary constraints that preclude the team from taking their product all the way into production. Usually, the team will only take the development process through the fabrication and testing of an engineering prototype. An engineering prototype is different from a preproduction prototype. The preproduction prototype has the production unit configuration. On the other hand, the purpose of the engineering prototype is to gain valuable information that enables the team to finalize their detailed product design with a high probability that when this design is built, developed, and validated as a preproduction prototype, it will meet all engineering requirements. Most often the engineering prototype does not have all the features of the preproduction prototype.

The engineering prototype is a project within the larger product development project. The engineering prototype must go through its own preliminary design, detailed design, and development phases. These phases are labeled *Phase 3P*, *Phase 4P*, and *Phase 5P* to differentiate them from the overall process phases. These engineering prototype development phases are shown in Figure A-3.

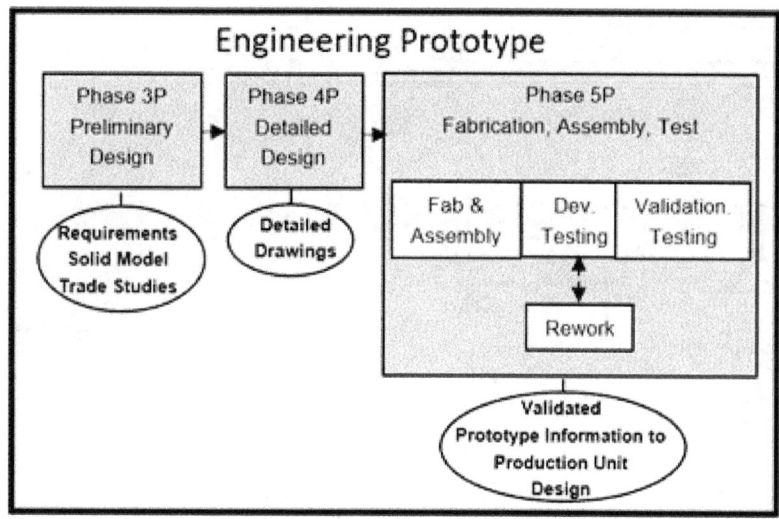

FIGURE A-3 Engineering Prototype PPD Process Diagram

As mentioned previously, the engineering prototype is accomplished during either the preliminary design phase or the detailed design phase of the overall product development project. For the capstone course, the engineering prototype is developed at the beginning of Phase 4. Information from this prototype is incorporated into the detailed design of the production unit. Phase 4 concludes with the preparation of a detailed drawing package for the production unit.

The two-semester capstone course ends at the conclusion of Phase 4. There is not time in the capstone course to address phases 5 and 6. Because the handbook has been prepared for industry engineering teams as well as capstone teams, these topics are also covered in the handbook. Figure A-4 shows the complete product development process. Phases 5 and 6 are in dashed lines to show that they are not part of the capstone project.

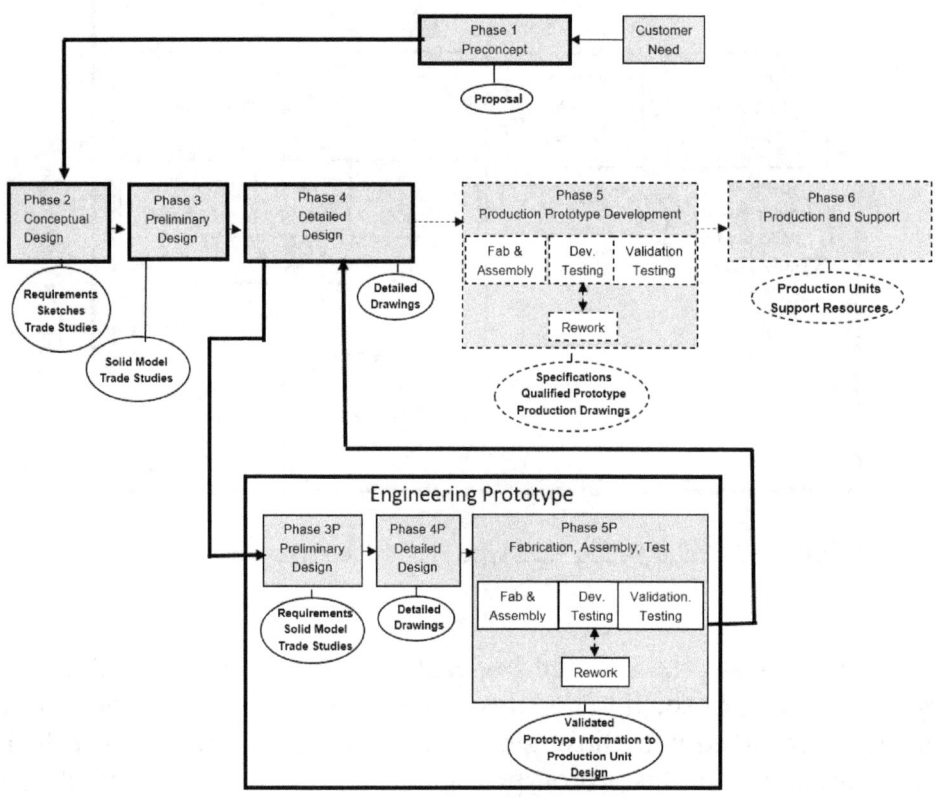

FIGURE A-4 Overall PPD Process Diagram

A.4.3 PPD Process Phase Descriptions

The following paragraphs describe each phase for a two-semester capstone course. However, these descriptions can be tailored for use on any product design and development program.

Phase 1 is Preconcept Design. In this phase, the project team is formed, the team finds a societal need to be addressed, a specific product problem statement is developed, notional ideas for solving the problem are explored, and a project plan including tasks, a schedule, and resource budgets are developed. A proposal to the project sponsor (in this case, the instructor) is the deliverable for this phase.

Phase 2 is Conceptual Design. During this phase, the project team conducts a voice-of-the-customer (VOC) needs analysis and converts these needs into engineering design requirements. The team then selects the key parameter (goal function) that needs to be optimized during the design process. For example, this could be minimizing size, weight, or unit production cost.

Once the engineering requirements are determined, a functional block diagram is constructed to identify the various functions that must be provided by the product to meet the requirements. The design space is then explored through techniques such as research and brainstorming to identify candidate design concepts. A filtering process is used to reduce the options to three or four concepts. These concepts are compared against the design criteria, and the most promising idea is presented in a conceptual design review. As part of the process of characterizing and comparing the different concepts, analyses and POC testing will probably be needed. The final conceptual design is defined with sketches and estimates of weight, size, cost, and performance. A requirements matrix is presented in which an estimate of meeting each design requirement is provided based on engineering judgement, analyses, and testing.

Phase 3 is Preliminary Design. In this phase, the product concept is turned into a robust design that meets the engineering requirements and meets or exceeds the design goal function. Analyses and testing are conducted to show in the requirements matrix that the design will meet the engineering requirements. Analyses such as reliability, maintainability, and safety (RMS) analysis; failure modes and effects analysis (FMEA); design-to-cost (DTC) analysis; design for manufacturing and assembly (DFMA); and material selection are conducted to create a robust design. A key deliverable from this phase is a dimensioned and labeled solid model of the production unit.

Phase 4 is Detailed Design. During this phase, the production unit preliminary design is further analyzed and defined to develop the detailed drawing package for the production unit. As part of Phase 4, the capstone team designs, builds, and tests an engineering prototype. Results from the engineering prototype work are iterated back to the preliminary design phase of the product used in completing the overall preliminary design of the production unit.

The three design subphases of the engineering prototype are described below. The "P" in the phase title stands for *engineering prototype*.

> Phase 3P, Engineering Prototype Preliminary Design, results in a dimensioned and labeled solid model of the engineering prototype. The first task in this subphase is to determine what features of the production unit need to be included in the engineering prototype. Budget and schedule constraints may preclude having an engineering prototype that can meet all the production unit's requirements. Once the engineering prototype requirements are established, proper analyses, similar to those of Phase 3 of the production unit, are conducted to show that the unit will meet the requirements during its testing phase.
>
> Phase 4P, Engineering Prototype Detailed Design, results in an engineering drawing package (with tolerances) and a set of plans on how the engineering prototype will be manufactured, developed, and validated.
>
> Phase 5P, Engineering Prototype Development, results in a unit that meets all its engineering requirements during its validation testing. Results from the engineering prototype subproject are integrated into the Phase 4 detailed design of the production unit.

Phase 5 is Fabrication, Assembly and Testing. This phase involves building a preproduction prototype from the production unit drawings created in Phase 4. The goal of the prior phases is to accomplish enough analyses and testing so that very little development is required for the preproduction prototype. If this prototype fails any of its development tests, it must be reworked into a modified configuration that will be able to pass the validation tests. Any rework changes to the preproduction prototype must also be updated in the production drawings.

Phase 6 is Production and Support. This phase involves the product's production, support of the product once it is in the field, and upgrades of the unit's design based on customer feedback on the early field units.

A.5 Organization of a Typical Capstone Course

A.5.1 Learning Modules

As discussed previously, this handbook can be used for a variety of university capstone courses or industry training courses or for self-learning study. The handbook has been divided into 44 modules. This subsection describes one way the handbook can be used for a two-semester capstone course.

During the first semester, there are two classes each week. A module is used for each class session. The classes are 75 minutes in length. Each class has a 20-to-30-minute lecture followed by team meetings. Teams will complete phases 1, 2, and 3 during the first semester.

During the second semester, there is one 175-minute class per week. Teams will complete Phase 4 during the second semester. Part of Phase 4 is the engineering prototype subproject (i.e., phases 3P, 4P, and 5P).

Another option that has proven to be effective is to flip the class. In this format, the students watch a prerecorded lecture and/or read assigned sections of the handbook before coming to class. The class time is then spent in team meetings, with the instructor available to provide mentoring. The exceptions are the weeks when design reviews take up the entire class time.

A.5.2 Course Schedules

Many engineering schools use the capstone team's project final report and final presentation as an indication of how well the students are demonstrating the ABET outcomes. The handbook guides teams through the process in a way that meets all of the ABET outcomes and EM indicators. Figures A-5 and A-6 provide suggested schedules for the first and second semester, respectively.

A.5.3 Capstone Course Deliverables

The five key course deliverables are described below.

<u>Team project notebook</u>: The team must create and continuously update a team project notebook that contains all the information regarding the project. This is a key element of the overall PPD process that has been proven in industry to be essential for team success. The notebook can be electronic or hardcopy in a three-ring binder. It is important that (1) one of the team members is responsible for keeping it up to date, (2) it is organized and contains the information listed in Table A-1, (3) the team brings the notebook to all team meetings, and (4) the team refers to the notebook often to remember what has been done, what future work must be done, and what the team has agreed upon. If the team uses an electronic format, then the major sections of the notebook should be in separate folders. If a hard-copy notebook is used, there must be tabs between major sections.

Week	1	2		3		4		5		6		7		8		9		10		11		12		13		14		15	
Module	1	2	3	4	5	6	7	8	9	10	11	12	13	14	15	16	17	18	19	20	21	22	23	24	25	26	27	28	

Tasks and Milestones:
- Phase 1 Preconcept Design
- Ph 2 Conceptual Design
- Introduction & Select Teams
- Phase 3 Preliminary Design
- DR1-Proposal Review
- DR2-Conceptual Design Review
- Final Rpt Section 3
- DR3-Preliminary Design Review
- Final Rpt Section 6
- Final Rpt Section 4
- Final Rpt Section 5
- Project Notebook

Module No.	Title
1	Introduction to the Handbook and Course
2	Starting Work on the Proposal (Phase 1)
3	Down-Select to Three Potential Projects
4	Selecting the Project
5	Preparing the Project Proposal
6	Project Schedules and Budgets
7	Finalizing the Proposal
8	Completing Phase 1
9	Introduction to Phase 2 Conceptual Design
10	Capturing the Voice of the Customer
11	Updating Requirements with QFD
12	Exploring the Design Space
13	Selecting, Analyzing and Testing the Conceptual Design
14	Completing Phase 2 Conceptual Design
15	DR2, Introduction to Preliminary Design and Planning
16	Part 1--Design Drivers and COTS Components
17	Part 2--Design Drivers and COTS Components
18	Part 1--Product Baseline Preliminary Design
19	Part 2--Product Baseline Preliminary Design
20	Part 1--Updating the Requirements/Validation Matrix
21	Part 2--Updating the Requirements/Validation Matrix
22	Reliability/Maintainability/Safety (RMS) Analyses
23	Part 1--FMEA, DFMA and DTC
24	Part 2--FMEA, DFMA and DTC
25	Part 3--FMEA, DFMA and DTC
26	Final Preliminary Deign
27	End of Phase 3 Deliverables Preparation
28	DR3 Presentation and Exiting Phase 3

Legend:
- Planned Task
- Completed Task
- 50 % Completed Task
- Slipped Task
- 100% Completed Slipped Task
- Planned Milestone
- Completed Milestone

FIGURE A-5 Suggested Master Schedule for Design I Course

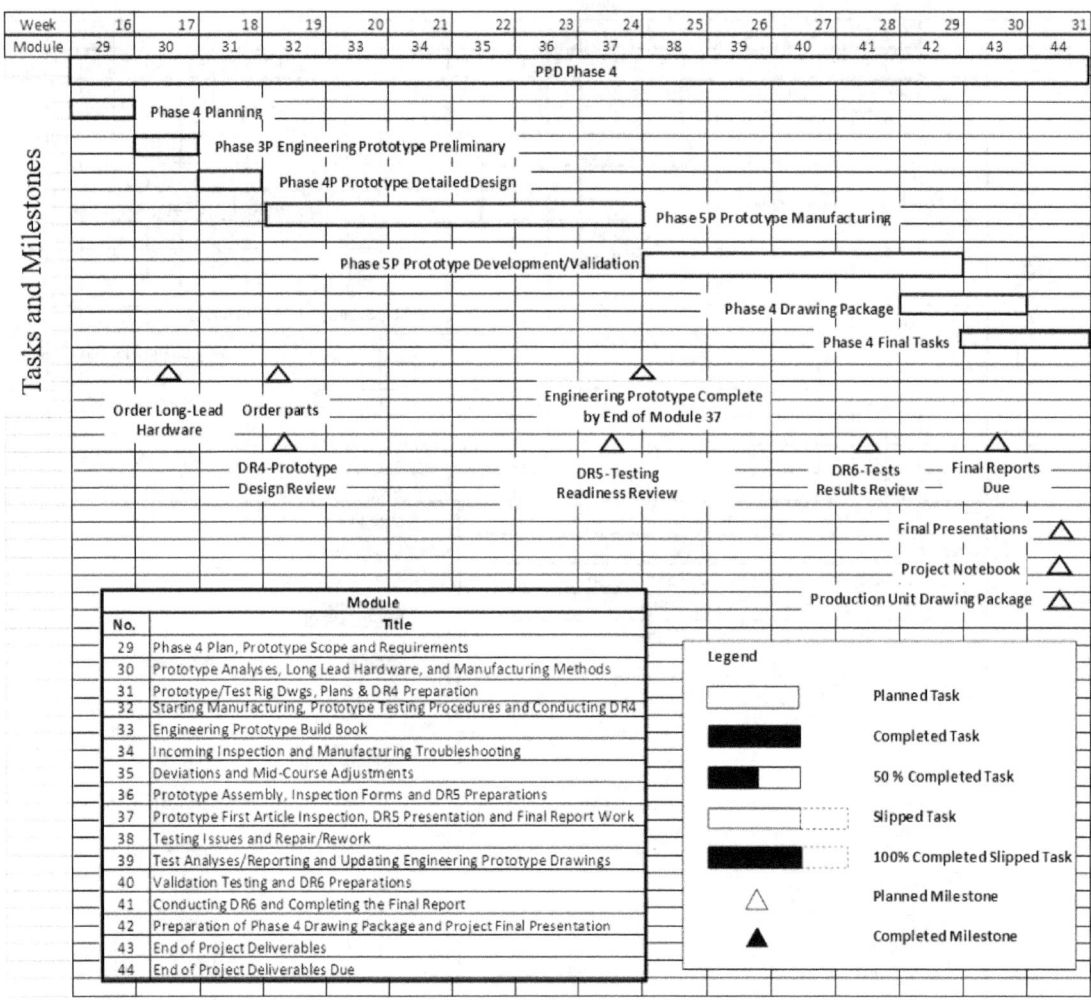

FIGURE A-6 Suggested Master Schedule for Design II Course

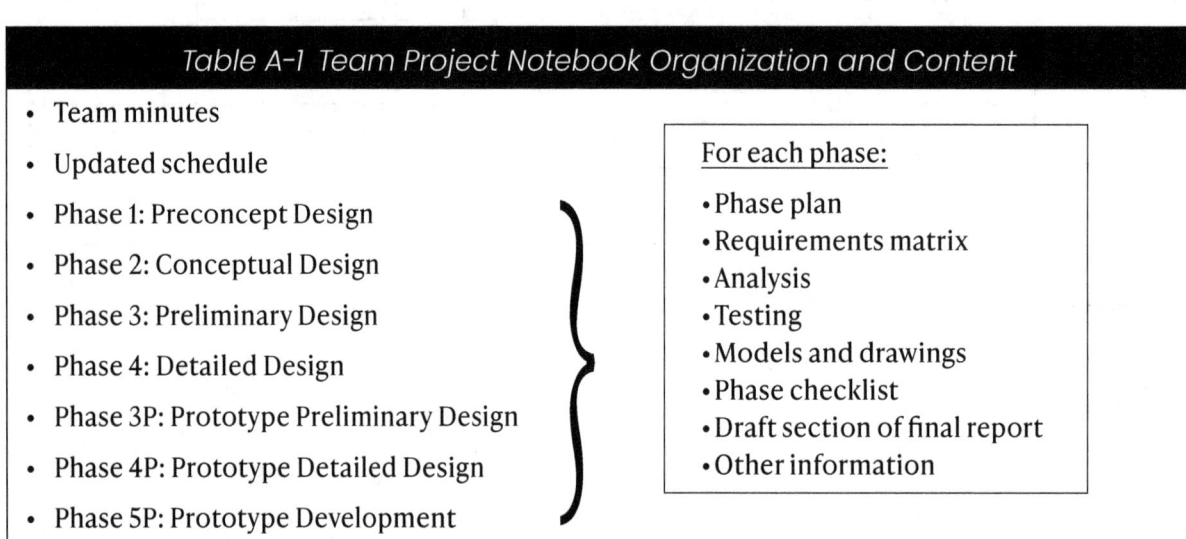

Proposal: This document is written to the potential project sponsor to obtain resources to complete the project. It presents the problem statement and approach for solving it within the resources available. It includes task descriptions, schedules, and budgets.

Final report: This document describes the entire project. It includes final descriptions of both the production unit and the engineering prototype. It also provides a description of each product development phase, including analyses, testing, and trade studies conducted.

Final presentation: This is an oral presentation by the team to the project sponsor that includes slides. It describes not only the final results but also the process followed to achieve these results.

Engineering prototype: Once the production unit is adequately defined, the team will select which product requirements need to be validated with an engineering prototype. The team will then design, build, test, and validate the engineering prototype to meet the engineering requirements.

A.6 ABET Accreditation Outcomes

ABET is the accrediting organization for undergraduate engineering programs. This organization provides a set of outcome criteria for graduating engineering students as listed in Table A-2. At ASU, the student is expected to meet each ABET outcome at a stated level of mastery as defined in Bloom's taxonomy given in Table A-3. ASU has an industrial advisory board, which has reviewed and approved these levels of mastery. It is recommended that these levels be used for other capstone product design and development courses. This handbook is designed to enable capstone teams to demonstrate these levels of ABET outcomes mastery.

Table A-2 ABET Outcomes

ABET Mechanical Engineering Program Outcome	Capstone Team Level of Mastery
1. an ability to identify, formulate, and solve complex engineering problems by applying principles of engineering, science, and mathematics	Analysis
2. an ability to apply engineering design to produce solutions that meet specified needs with consideration of public health, safety, and welfare, as well as global, cultural, social, environmental, and economic factors	Analysis
3. an ability to communicate effectively with a range of audiences	Application
4. an ability to recognize ethical and professional responsibilities in engineering situations and make informed judgments, which must consider the impact of engineering solutions in global, economic, environmental, and societal contexts	Application
5. an ability to function effectively on a team whose members together provide leadership, create a collaborative and inclusive environment, establish goals, plan tasks, and meet objectives	Application
6. an ability to develop and conduct appropriate experimentation, analyze and interpret data, and use engineering judgment to draw conclusions	Analysis
7. an ability to acquire and apply new knowledge as needed, using appropriate learning strategies	Application

Source: ABET, Criteria for Accrediting Engineering Programs, 2019 – 2020. Copyright © by ABET.

Table A-3 Mastery Level Definitions Based on Demonstrated Behaviors

1. At <u>knowledge</u> level of mastery, a student can **define** terms.
2. At <u>comprehension</u> level of mastery, a student can **work assigned** problems and can explain **what** they did.
3. At <u>application</u> level of mastery a student **recognizes** what methods to use and then uses the methods to solve problems.
4. At <u>analysis</u> level of mastery, a student can explain **why** the solution process works.
5. At <u>synthesis</u> level of mastery, a student can **combine** the parts of processes in new and useful ways.
6. At <u>evaluation</u> level of mastery, a student can **create a variety of ways** to solve the problem and then, based on established criteria, **select** the solution method best suited for the problem.

Source: Valana Wells, ABET Assessment Fair Briefing. Copyright © 2013 by ABET.

A.7 Entrepreneurial Mindset

Mindset is the cognitive environment in which one thinks. It is the lens that shapes the information one receives in one's cognitive center. Engineers tend to have a rational mindset. But as the engineering profession has volved, more concepts have been added to the engineer's mindset, as listed in Table A-4. As the complexity of societal life increases, there is a greater need for the engineer to have an entrepreneurial mindset. An entrepreneur is risk tolerant and always looking for opportunities to increase value. Value can take many forms. Businesses often focus on the value of profits, while customers often focus on the lowest price. But value can also include longer-term benefits to both businesses and customers, such as health, clean environment, and peace.

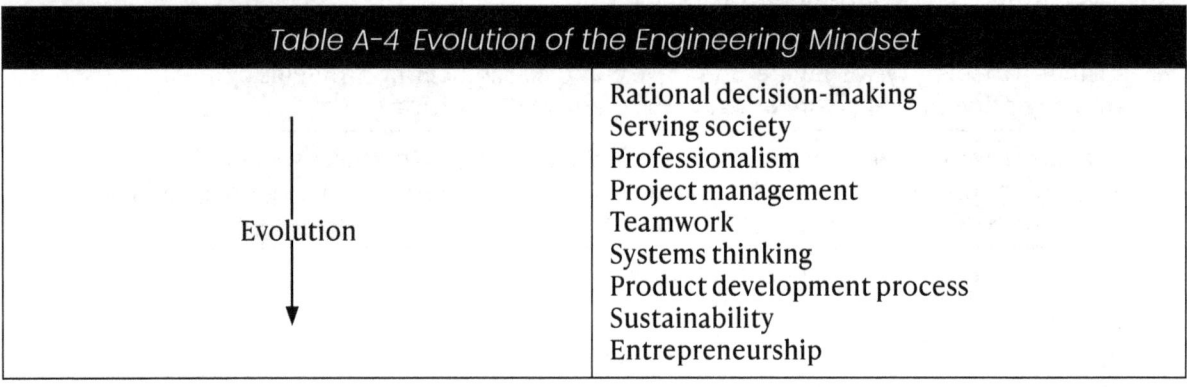

Table A-4 Evolution of the Engineering Mindset

Evolution	Rational decision-making Serving society Professionalism Project management Teamwork Systems thinking Product development process Sustainability Entrepreneurship

Like a growing number of American engineering programs, the Ira A. Fulton Schools of Engineering (FSE) at ASU consider gaining an EM to be a vital part of becoming a practicing engineer. ASU has partnered with the Kern Family Foundation to instill engineers and engineering students with an EM. Details on the Kern Family Foundation's focus on EM is provided in Figure A-7.

Entrepreneurial Mindset Background

This summary is based on the following references:
1. AC 2008-771: building an entrepreneurial engineering ecosystemfor future generations: the kern entrepreneurship education network, J. Blessing, K. Mekemson, D. Pistrui, American Society for Engineering Education, 2008.
2. Engineeringunleashed.com Website accessed July 27, 2020
3. Entrepreneurship.engineering.asu.edu/asu-keen/ accessed July 27, 2020

- "…the Kern Family Foundation established the Kern Entrepreneurship Education Network (KEEN) with the mission to create an action-oriented, entrepreneurial mindset among engineering, science and technical undergraduates. The KEEN initiative, launched in 2005, represents a new and unique entrepreneurial approach to improving undergraduate education in the U.S. (Ref. 1)

- KEEN is a collaborative partnership of colleges and universities dedicated to graduating "engineers with an entrepreneurial mindset so they can create personal, economic, and societal value through a lifetime of meaningful work." (Ref. 2)

- "An entrepreneurial mindset (EM) is a habit of mind geared toward action. It is a learned behavior; a way of thinking about the world and acting upon what you see. EM empowers people to identify opportunities and create value in any context. Engineers equipped with the entrepreneurial mindset understand the bigger picture, recognize opportunities, evaluate markets, and learn from mistakes to create value for themselves and others. " (Ref. 2)

- "The ASU Kern Grant is engineering leadership, faculty and staff partnering with the Kern Entrepreneurial Engineering Network (KEEN) to integrate the entrepreneurial mindset (EM) in engineering education on a national scale. The project aims to: 1) develop resources that support mentorship of engineering faculty in building a community of practice that values quality teaching, impactful research, and instilling a collaborative and entrepreneurial mindset; and (2) impact ASU engineering students through curricular and co-curricular approaches by weaving the entrepreneurial mindset into the Fulton Schools of Engineering." (Ref. 3)

FIGURE A-7 Entrepreneurial Mindset (EM) and KEEN Background

As discussed on the Kern Family Foundation website, the EM has three key elements, which are listed in Table A-5.

Table A-5 Entrepreneurial Mindset 3C's
- **Curiosity**—Have a constant curiosity about our changing world and employ a contrarian view of accepted solutions. - **Connect**—Habitually connect information from many sources to gain insight and manage risk. - **Create value**—Create value for others from unexpected opportunities as well as persist through and learn from failure.

A shorthand for remembering that there are three key elements to the EM is EM = C^3. Throughout this handbook, there are box insets like the ones in Figure A-8 to highlight a key EM element in the text.

Entrepreneurial Mindset: *Curiosity*

Entrepreneurial Mindset: *Connection*

Entrepreneurial Mindset: *Creating Value*

FIGURE A-8 Inserts to Highlight Key EM Elements in the Handbook Text

The ASU Fulton Knowledge and Innovation Center is coordinating EM efforts and has assembled a list (see Table A-6) of EM indicators for assessing how well students are demonstrating EM principles. This list (called the EM@FSE 2.0 indicators) has universal application to product development projects.

At the end of each major PPD process phase is a phase-exit checklist. Within the checklist are the EM@FSE 2.0 indicators. To help the reader correlate the checklist with the handbook's EM material, the symbol shown in Figure A-9 appears next to the information in the text.

EM-(a)

Letter associated with one of the EM@FSE 2.0 Indicators. In this example, it is (a) Critically observes surroundings to recognize opportunity.

FIGURE A-9 Symbol to Highlight EM Indicators in the Handbook Text

Table A-6 *ASU Fulton Engineering EM@FSE 2. Indicators*

a) Critically observes surroundings to recognize opportunity
b) Explores multiple solution paths
c) Gathers data to support and refute ideas
d) Suspends initial judgement on new ideas
e) Observes trends about the changing world with a future-focused orientation/perspective
f) Collects feedback and data from many customers and customer segments
g) Applies technical skills/knowledge to the development of a technology/product
h) Modifies an idea/product based on feedback
i) Focuses on understanding the value proposition of a discovery
j) Describes how a discovery could be scaled and/or sustained, using elements such as revenue streams, key partners, costs, and key resources
k) Defines a market and market opportunities
l) Engages in actions with the understanding that they have the potential to lead to both gains and losses
m) Articulates the idea to diverse audiences.
n) Persuades why a discovery adds value from multiple perspectives (technological, societal, financial, environmental, etc.)
o) Understands how elements of an ecosystem are connected
p) Identifies and works with individuals with complementary skill sets, expertise, etc.
q) Integrates and synthesizes different kinds of knowledge

Source: ASU EM@FSE, "ASU Fulton Engineering EM@FSE 2.0 A-Q Indicators." Copyright © 2016 by ASU EM@FSE. Reprinted with permission.

A.8 Being Curious Worksheet

Successful practicing engineers are curious about the society in which they live. They make the time to stay aware of what people are doing and what their needs are. They have a strategy for gaining this information through personal observation and the news media. They are not satisfied with sound bites; they want to know the real story. Worksheet 1 is for team members to expand their curiosity about societal needs and the state of the art of relevant engineering capabilities.

Entrepreneurial Mindset: *Curiosity*

Worksheet 1 Being Curious

Name: _____

Curiosity is one of the key elements of the EM. Design engineers must not only have analytical and testing skills; they must also be creative, and to be creative, they need to be curious. That curiosity must go beyond course assignments or job tasks. The design engineer must be curious about society at large, including professional, technological, social, economic, political, and environmental issues. This worksheet helps the design engineer to address this important topic.

1. How do you practice curiosity outside the classroom? Do you have a standard set of habits that make time for being curious? For example, how do you learn about current events happening in the world?

2. What magazines, journals, websites, and other forms of information do you regularly use to address the following:

    a. World events?

    b. National events?

    c. Local events?

    d. Developments in engineering?

    e. Other areas about which you are regularly curious?

EM-(a)

3. What do you do when you become curious about something new? How do you make the time to learn more about the curiosities you identify?

4. What new habits can you develop that gives you opportunities and time to be curious?

A.9 Commercialization Process

This handbook primarily looks at product development from the technical perspective of designing the physical product to meet customers' needs and the associated product engineering requirements. Actually, getting the product from an idea to something available to the customer involves more than technical effort. Most important, an entity—that is, a business—has to sponsor the development effort, provide the means of production, distribute the product to customers, and support the product in the field. That business has to make a profit. An explanation of how the business makes that profit is called the *business case*. To properly meet customers' needs and satisfy the profit motive of the business, the market in which customers and competing producers reside must be addressed.

> **Entrepreneurial Mindset:**
> *Creating value*

The overall process of taking an idea on how to meet a customer need to the delivery of that product to the customer is called *commercialization*. It requires work in the three areas described above: technical, market, and business. Commercialization is an interdisciplinary process. The technical activities must take into account the requirements of the marketing and business activities.

The authors of this handbook like the Goldsmith Commercialization Model[1] for describing this process. As shown in Figure A-10, this model breaks down commercialization into six sequential stages. In each stage, there are technical, marketing, and business activities that must be accomplished before the overall effort can move into the next stage. At the end of each stage, the sponsor has the opportunity to review the phase results and determine whether the project still has an acceptable probability of success.

> **Entrepreneurial Mindset:**
> *Connections*

---

1. Adapted from Goldsmith, H.R. 1995. *A Model for Technology Commercialization*. Mid-Continent Regional Technology Transfer Centre Affilliate's Conference. NASA Johnson Space Centre, Houston.

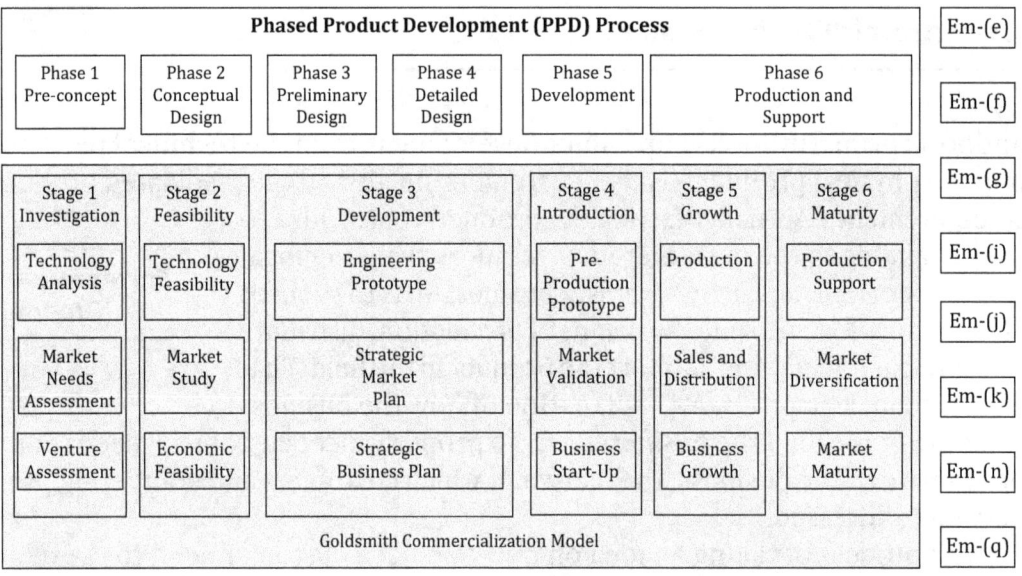

FIGURE A-10 Relationship Between PPD Process and Goldsmith Commercialization Model[2]

Note that the column of EM indicator boxes at the right of Figure A-10 show which indicators correlate with this part of the handbook.

Table A-7 lists the key questions that are addressed in each commercialization stage. The capstone project addresses the first two stages and results in a working engineering prototype. In Stage 1, a potential need is identified, the concept of a technical solution is selected, and a notional business case is prepared. In Stage 2, the market in broad terms is quantified, a demonstration model is created that meets customers' needs, and the business case's feasibility is determined.

Table A-7 Goldsmith Model Key Questions for Each Phase	
Phase	Key Questions
1. Investigate	Is there a potential need, product, and business case?
2. Feasibility	Does a market exist, is the product doable and affordable, and is the business case consistent with the company's strategic plan?
3. Development	Does the product meet the customers' and business' requirements? Are the market and distribution channels ready? Has a business organization been developed for the project?
4. Introduction	Do the initial product units meet the customers' requirements and the business' objectives?
5. Growth	Is the product in full production, meeting customer expectations, growing in market share, and yielding acceptable profitability?
6. Maturity	Is the project expanding into a profitable product line serving increasing market segments?

---

2. University of Nebraska Omaha. n.d. *Goldsmith Technology Commercialization Model*. https://www.unomaha.edu/nebraska-business-development-center/technology-commercialization/goldsmith-technology/index.php.

The PPD process and the Goldsmith model can be correlated. Phases 1 and 2 of PPD process are part of the investigation stage of the Goldsmith model. Phase 3 and the engineering prototype of the PPD process correlates with the feasibility stage of the Goldsmith model. The other Goldsmith model phases are beyond the scope of the capstone project. During each PPD process phase, the project team considers how their efforts fit into the overall commercialization of the product being developed.

A.10 Transitioning from Student to Practicing Engineer

Students tend to be overcommitted. There are always more assignments than there is time to properly complete them. Many successful students are good at cutting corners and streamlining their problem-solving strategies to save time. They put a premium on saving time. This strategy often works in school because the problems tend to be simple and well-defined.

Real-world problems are different, as they tend to originate as issues. A major task is finding the right problem within the issue. These problems are usually complex with incomplete information. The pathway to a solution is usually unclear. There are often roadblocks and changing conditions. Assumptions made early on are often proven incorrect as the project progresses. Industry problems take time—a lot of time. The process by its very nature is not efficient (there are many false starts and mid-course corrections), yet the process is effective—that is, it results in a solution that solves the complex problem.

Entrepreneurial Mindset: *Curiosity*

EM-(n)

EM-(o)

Many good students have trouble solving industry problems. They want to cling to their old student approaches rather than embrace the approaches used by successful professionals. The PPD process has been designed to transition the student into following effective approaches to solving complex industry problems associated with the development of a new product. This process enables the student to be a critical thinker who challenges every assumption, documents their work, and looks for multiple candidate solutions before settling on a final approach.

A.11 Sustainability

Sustainability is an important mindset for today's engineers. Ecological economists such as Herman Daly[3] have made it clear that the world's ecosystem has limits. The products that engineers design now and, in the future, to meet human needs must be more ecologically friendly. This calls for more efficiency in the use of energy and raw materials as well as a commitment to recycling. It also calls for the realization that economic growth without bounds is not thermodynamically possible. The entrepreneurial engineer values sustainability and gives it due weight in design tradeoffs.

A.12 Responsibilities and Expectations of the Capstone Course Student

It should be noted that in most university undergraduate engineering programs, not all of the skills and abilities needed for capstone project work have been mastered prior to entering the course. Therefore, students must gain new engineering skills and then quickly demonstrate their use to the level of mastery required by the ABET outcomes. Many students are weak in their retention of needed engineering skills from prior courses. Many (likely most) of the students will require additional outside-of-class time to bring their engineering skill sets to a level of mastery needed for completing their capstone project. Because most students work to pay for school in addition to carrying a full load of coursework in their senior year, the student must make the capstone project their highest priority and budget their time accordingly. It is recommended that capstone students seek the help of subject-area experts during the execution of their project in areas where they lack expertise. This is similar to hiring these kinds of experts as consultants in industry by product development engineers. A good and available source of subject-area experts for capstone teams are their faculty professors who teach relevant subjects. Faculty are usually available and willing to help during their office hours.

Teamwork is vital in completing projects successfully. Capstone course students must attend all class meetings and participate fully in all project team meetings. Each student has individual coursework deliverables in addition to team deliverables.

Part of each student's responsibilities is to hold team members accountable for finishing their project deliverables on time and at a high level of engineering quality. Failure of a student to hold another team member accountable is an unacceptable level of performance for both students.

---

3. Daly, Herman E. 1996. *Beyond Growth: The Economics of Sustainable Development.* Boston: Beacon Press.

A.13 Case Study of the PPD Process in Industry

This handbook uses the engineering program capstone project as the primary example of how to successfully design and develop a customer-focused product. However, this process is equally applicable to industry projects. To illustrate this point, the case study of developing a nonlethal vehicle-stopping device for the Department of Homeland Security (DHS) is presented.

Engineering Sciences Analysis Corporation (ESA) is an engineering firm that provides detailed computer analyses for the aerospace industry and develops new product ideas for a wide variety of industry applications, ranging from baseball bats to laser weapons. Martin Martinez is the company's founder and president. He was one of the engineers, along with author Trimble, who developed the original integrated product development and support (IPDS) process at AlliedSignal. Martinez left AlliedSignal to start his own engineering firm, ESA. IPDS is a fundamental process in his company.

In 2006, Martinez read a Small Business Innovation Research (SBIR) request for proposal by DHS for a nonlethal vehicle-stopping device. He came up with the unique idea of using a remotely activated strap to ground the spinning vehicle drive shaft to the car frame, thus stalling the engine and putting the vehicle in a controlled skid that brings the vehicle to a safe stop.

Martinez asked author Trimble to be the program manager. Together, they won all three phases of this SBIR program and achieved the first DHS SBIR program in Arizona to go all the way from innovative idea to commercialized product. Figure A-11 shows how the SBIR phases were correlated with the IPDS product development process that has now evolved into the PPD process presented in this handbook.

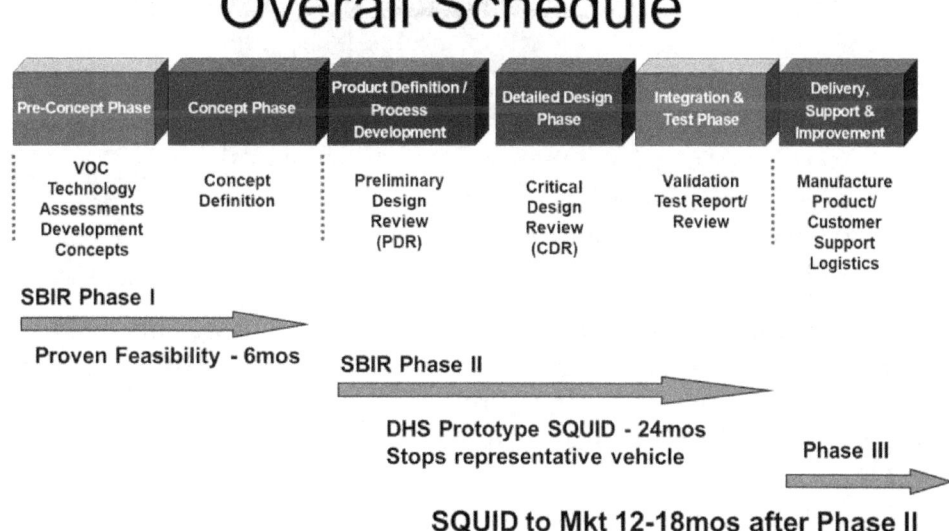

FIGURE A-11 IPDS Process Applied to DHS SBIR Nonlethal Vehicle-Stopping Program[4]

---

[4] IPDS Process Applied to DHS SBIR Non-Lethal Vehicle Stopping Program," Engineering Science Analysis Corporation.

As shown in Figure A-12, the device is deployed in three phases. In Phase 1, the law enforcement officer places the device on the road and arms it. In Phase 2A, the officer remotely deploys the barbed straps that attach to the vehicle's front tires and are wrapped around the wheels' A frames. In Phase 2B, the car's front bumper passing over the device causes tentacles with paint balls filled with glue at their ends to be ejected into the vehicle's drive shaft. These tentacles stick to the drive shaft and are wrapped around it. Each tentacle is tied to the other end of a strap. Once this end of the strap is wound around the drive shaft, it becomes taut, as the other end is tied to the vehicle's A frame. This results in Phase 3 with the drive shaft rotation stopping, the engine stopping, the vehicle skidding to a safe stop, and the apprehension of the people inside the vehicle.

This device was featured in the April 2009 issue of *Popular Science* magazine. A description of the device is also on the DHS' website.[5]

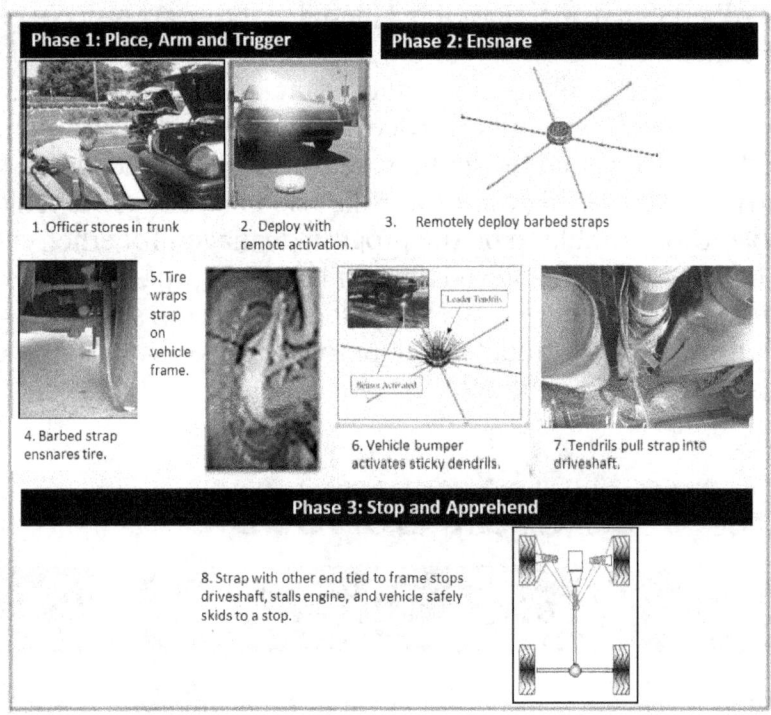

FIGURE A-12 SQUID's Three-Phase Nonlethal Vehicle-Stopping Process[6]

Martinez and his engineers have an EM that was able to see opportunity in the midst of the issues associated with this product's development. A key success factor for this program was following the Goldsmith Commercialization Process that emphasized finding the VOC early in the program and growing the business plan as the product is being developed. This included

---

5. Department of Homeland Security. 2009, July 6. SQUID: The Long (and Sticky) Arms of the Law. https://www.dhs.gov/science-and-technology/squid-long-and-sticky-arms-law.

6 Adapted from "SQUID's Three-Phase Non-Lethal Vehicle Stopping Process," *https://www.dhs.gov/science-and-technology/squid-long-and-sticky-arms-law,*

interviews with a large number of law enforcement agencies, including Border Patrol, local and state police, and the military.

DHS program managers were impressed with the process ESA followed in developing their device and attributed the process to ESA's overall success. This is the same process covered in this handbook. As this case study shows, this handbook is indeed valuable to practicing industry engineers as well as university capstone project students.

A.14 Review Quiz for Section A

Worksheet 2 provides a review quiz for Section A. The answers are in Appendix AA1.

Credits

Fig. A-10: University of Nebraska Omaha, College of Business Administration, Nebraska Business Development Center, "The Goldsmith Commercialization Model," https://www.unomaha.edu/nebraska-business-developmentcenter/technology-commercialization/goldsmith-technology/index.php. Copyright © 2020 by University of Nebraska Omaha.

Fig. A-11: Engineering Sciences Analysis Corporation, "IPDS Process Applied to DHS SBIR Non-Lethal Vehicle Stopping Program." Copyright © by Engineering Science Analysis Corporation.

Fig. A-12: Adapted from Source: https://www.dhs.gov/science-and-technology/squid-long-and-sticky-arms-law.

Worksheet 2 Handbook Section A Quiz

Name: _____

A. List the titles of the three sections of the handbook.

    Section A _____ Section B _____
    Section C _____

B. There are six levels of mastery. Complete the list in order.

    1 _____ 2 _____ 3 _____
    4 _____ 5 _____ 6 _____

C. The PPD process has six phases. Name the phases in order.

    1 _____ 2 _____ 3 _____
    4 _____ 5 _____ 6 _____

D. Engineering design requirements are based on these: _____

E. List the five main deliverables for the example capstone course in any order.

    1 _____ 2 _____ 3 _____
    4 _____ 5 _____

F. List the three key EM elements.

    1 _____ 2 _____ 3 _____

True-or-False Questions Regarding the Example Capstone Course

G. Each project team must meet each ABET criterion at least at the analysis level.	T	F
H. Teams must manage their projects to stay on schedule.	T	F
I. Team minutes must be taken and stored in the project notebook.	T	F
J. The PPD phase-exit checklist must be completed and approved before exiting the phase.	T	F

Section A
Appendices

APPENDIX AA1

Worksheet 2 Answers

Worksheet 2 Handbook Section A Quiz

Name: _____

A. List the titles of the three sections of the handbook.

　　Section A　　Handbook Overview
　　Section B　　Phased Product Development (PPD) Process
　　Section C　　Final Report Outline

B. There are six levels of mastery. Complete the list in order.

　　1　Knowledge　　　2　Comprehension　　3　Application
　　4　Analysis　　　　5　Synthesis　　　　6　Evaluation

C. The PPD process has six phases. Name the phases in order.

　　1　Preconcept design　　　2　Conceptual design
　　3　Preliminary design　　　4　Detailed design
　　5　Production prototype　　6　Production

D. Engineering design requirements are based on this:　voice of the customer

E. List the five main deliverables for the example capstone course in any order.

　　1　Proposal　　　　　　2　Project notebook　　3　Final report
　　4　Final presentation　　5　Engineering prototype hardware

F. List the three key entrepreneurial mindset elements.

　　1　Curiosity　　2　Connections　　3　Creating value

True-or-False Questions Regarding the Example Capstone Course

	T	F
G. Each project team must meet each Accreditation Board for Engineering and Technology (ABET) criterion at at least the analysis level.	T	**(F)**
H. Teams must manage their projects to stay on schedule.	**(T)**	F
I. Team minutes must be taken and stored in the project notebook.	**(T)**	F
J. The PPD phase-exit checklist must be completed and approved before exiting the phase.	**(T)**	F

SECTION B

PHASED PRODUCT DEVELOPMENT (PPD) PROCESS

Section B
Phase 1: Preconcept Design

MODULE 2

Starting Work on the Proposal (Phase 1)

OVERVIEW

To start a project, the team needs to plan what is to be done. Then a proposal is made to a potential sponsor to obtain funding to proceed. This module starts the proposal process.

LEARNING OBJECTIVES

- Gain an overall understanding of Phase 1: Preconcept Design.
- Understand how to become a high-performing team.
- Comprehend the process of selecting a societal need and understanding what the customers' needs are.
- Comprehend the process of going from customer needs to engineering design requirements.

PRE-LECTURE ASSIGNMENT

- Read Section B.1 to gain an overall understanding of Phase 1.
- Study the material covered in sections B.1.1–B.1.6.

POST-LECTURE ASSIGNMENT

- Conduct a team meeting to 1) complete the team charter, 2) review what the team has learned about potential societal needs, 3) make a list of these societal needs, 4) brainstorm design projects for each listed societal need, 5) decide what further research needs to be done and make team member assignments, and 6) document all team activity in the team notebook.

TEAM DELIVERABLES

- Team charter
- List of the team's candidate societal needs, with a brief description of each project
- Minutes of the team meetings
- Team activity as documented in the team notebook

B.1 Phase 1: Preconcept Design

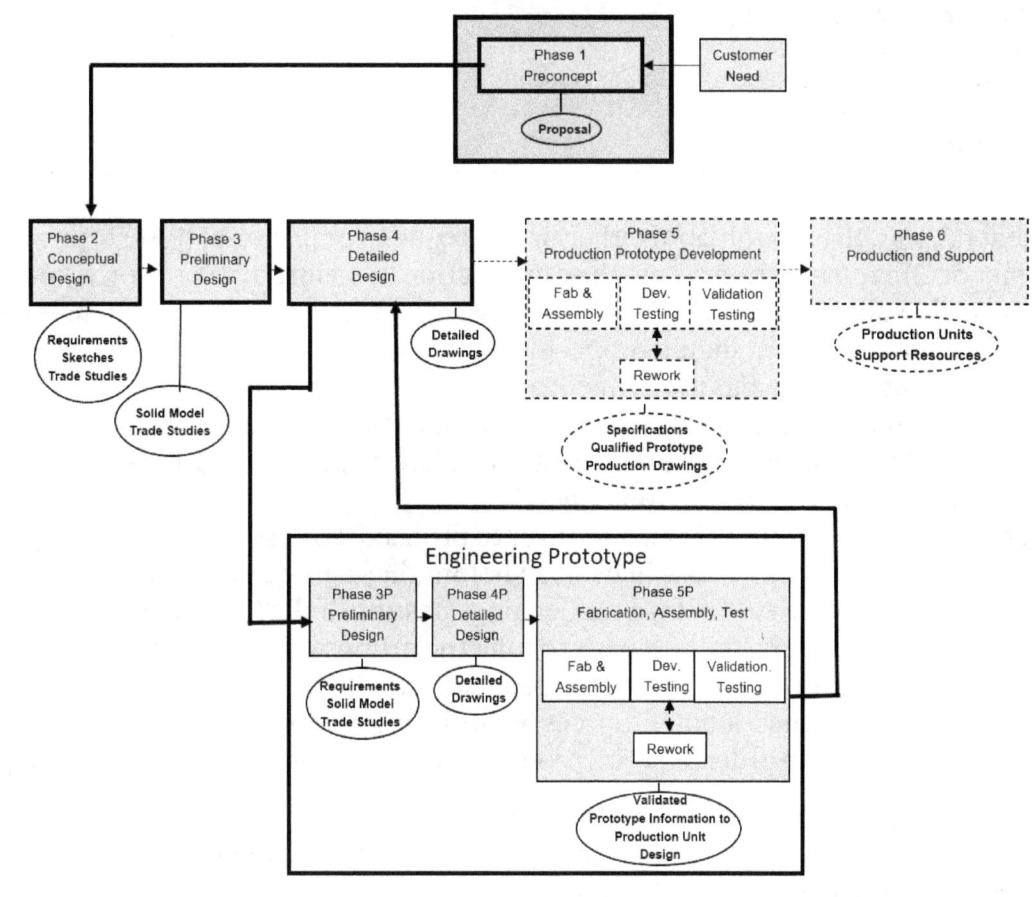

B.1.1 Overview

The first activity for Phase 1 is the formation of the team as discussed in Subsection B.1.2. The first task of the team is to have each team member read and study Section A of this handbook. This should take only a few hours prior to the first team meeting. During the first team meeting, a plan for Phase 1 needs to be formulated. <u>It is imperative that each team member has a clear understanding of all the concepts in Section A of this handbook before the team starts the Phase 1 planning process. Each team member should take the quiz at the end of Section A prior to the first team meeting. Each team member should bring their completed quiz to the team meeting to have objective proof that they are ready to start Phase 1.</u>

The objectives for Phase 1 of the capstone project are listed in Table B-1.

Table B-1 Phase 1 Objectives for the Capstone Project

1. Each team member has read and studied Section A of this handbook.
2. Form the project team.
3. Prepare a Phase 1 project plan.
4. Select a customer need.
5. Establish a solution-neutral problem statement.
6. Prepare a preconcept design.
7. Prepare a project plan proposal.

In general, this phase is initiated either by (1) a request for proposal (RFP) from an outside customer or (2) a request from within the organization by an internal sponsor. In many cases, the sponsor has already done a voice-of-the-customer (VOC) assessment and therefore has a clear idea of what the team should propose.

In capstone courses, teams are formed, and then the team usually conducts research to find an important customer need. Selecting a meaningful customer need requires curiosity. Once the need is identified, the team must formulate a specific solution-neutral problem statement that addresses this customer need.

Most projects need funding from a sponsor to proceed. Sponsors want a plan (in the form of a proposal) on how the funds will be used before they agree to fund the project. So who funds the proposal activity? That is a challenging question. If the RFP comes from an outside customer, it is usually up to the company to fund the proposal. If it is an inside RFP, the manager issuing the RFP also needs to fund the proposal activities. Whoever is funding this initial work can't afford to spend a lot of resources on the proposal. So the proposal activity must be efficient and effective. In order to have a proposal, the following tasks must be accomplished:

- Select a customer need and characterize it.
- Formulate a product problem statement that will address the need.
- Identify an initial list of product requirements.
- Create a functional block diagram for the product based on the initial requirements.
- Identify a product preconcept that will perform the functions in the block diagram and therefore meet the initial requirements.
- Use this preconcept as a starting point for identifying the necessary tasks to transform this idea into a production unit.
- Write the proposal based on the product's preconcept and the tasks needed to achieve a production unit.

There is not enough time during the proposal process to identify the best product concept, so the team selects a reasonable preconcept to use during the proposal process. Once the sponsor accepts the proposal and funds the project, the team will need to go back and explore the design space in more detail to identify the best concept. Likewise, the customer need will have to be studied in more detail after the project is funded.

The project proposal addresses the RFP sponsor's desire to have the following questions answered before they feel motivated to fund the project:

- Is this the right team to develop the product?
- Has the team identified a good customer need?
- Does the team have a good technical approach for meeting this need?
- Is there a viable market for this product?

- Is there a value proposition that justifies the funding?
- Is the requested amount of funding adequate to develop the product?
  - Does the team understand what tasks need to be done?
  - Will the team be able to complete the project within the time frame allowed and available resources?
  - Is there a good plan for mitigating project risks?
  - Will the team be able to manage the project and accommodate unexpected developments?

An excellent way for the team to see if they have a winning proposal is to list the five key reasons why the sponsor should fund the team to do the project. These key success factors should be highlighted in the proposal.

Section C provides a detailed outline of the project final report. The final report includes a copy of the proposal. The team should use the outline presented in Section C for preparing the proposal.

B.1.2 Teamwork

Most product development in industry is a team effort. The quality of the final product in meeting the customers' need is dependent on having skilled team members who are committed to the project. In capstone courses, the teams are assigned by the instructor, or the students self-organize. Either way, once the team members are identified, it is up to each member to help the team evolve into a high-performing one.

EM-(p)

B.1.2.1 Stages of Team Development

As shown in Table B-2, there are four stages of team development. Initially, people come together to form a team, and their intention is to work together. However, differences of opinion and conflict on how to proceed begin to ripple through the team. This storming stage is completely normal in the team's early work. The important point is that the team needs to manage the conflict by jointly agreeing on how decisions will be made and conflict will be resolved. This process is called *norming*. It involves establishing ground rules concerning how team members will respect each other and take advantage of other team members' insights, opinions, and expertise. When the team follows these norming ground rules, it begins to become a high-performing team that works effectively together to accomplish the team's objectives.

Table B-2 The Four Stages of Team Development

Forming–Forming members come together and start to work together.
Storming–Differences of opinion and conflict about how to proceed occur at this stage.
Norming–The team learns to manage conflict and move toward productivity.
Performing–The team effectively works together to accomplish the team's objectives.

Source: Tuckman, Bruce W. and Mary Ann C. Jensen. 1977. Stages of Small-Group Development Revisited. Group & Organization Studies 2, no. 4 (December): 419–27. https://doi.org/10.1177%2F105960117700200404.

B.1.2.2   Team Roles

Once teammates have been chosen, it is essential that team roles be taken. Students often become confused about this subject. Some students think that the team leader is the "boss" and that the team's success is the team leader's responsibility. This can be a fatal flaw for the team. Each team member must take on equal responsibility for the team's success. However, the team does need someone to coordinate the team's activities and make sure that the phased product development (PPD) process is being followed. Some teams select a team leader for the duration of the project. Other teams rotate this responsibility. High-performing teams split the team leader task into various roles, such as meetings coordinator, proposal manager, systems engineering manager, test manager, and final report editor. The important point is that each team member needs to take responsibility for certain aspects of the team's workload. That means not only being accountable for the work getting done but also enrolling other team members to help when the work requires more than one member. An important role is for someone to be the quality control monitor. This person's task is to review the work of the team and determine if it is of proper quality to fit the expectations of the sponsor. The job of the quality control monitor is to identify work that needs to be improved. It is the role of the team to improve the work.

B.1.2.3   Team Charter

The team charter is a contract among the team members and the sponsor of the team that describes what the team will do and how it will do it. A sample team charter form is provided in Figure B-1. Students should fill out a draft charter form and meet with the sponsor (in this case, the instructor). The charter is negotiated between the sponsor and the team. The final document is signed by each team member and the sponsor. Team members should only sign the charter if they are willing and capable of living up to the agreements stated in this document.

Team Charter

1. Team Name:

2. Team Motto:

3. Team Objectives:

4. Team Vision (how will team look as it works to meet its objectives)

5. Success Factors (What will the team do to be successful?)

6. How will the team interface with the sponsor?

7. Describe when and where the team will meet for team meetings.

8. Team Roster and Agreement Signature

Name	Phone	Email	Hours/Week Pledged	Signature

FIGURE B-1   Team Charter Form

B.1.2.4 Team Minutes

It is vital that the team take minutes of each team meeting and that these minutes be placed in the team project notebook. The team must include in the minutes any decisions relative to what is to be done, by whom, and when. They should also state in the minutes all discussions and their resulting conclusions. At the beginning of each team meeting, the team should refer to the prior meeting's minutes to refresh their memories relative to assignments given and conclusions reached. Figure B-2 provides a sample team minutes template.

Note that there is a metrics box in the upper-right corner of the template. The team should consult their phase plan and determine what percentage of the phase labor budget should have been spent by the meeting date. Then they should enter the actual percent of the labor spent and determine what their mood is regarding their progress in completing the phase. The range is 0 to 10, with 0 being very worried and 10 being completely happy with the team's progress.

```
                    Team Minutes Template
Team Name: _____  Date _____
Attendees: _____ _____
_____ _____           Metrics as of this date
Project Description:                   Phase % Labor Budgeted     _____
Topics Discussed and Decisions Made:   Phase % Actual Labor Spent _____
1.                                     Mood of Team (0 to 10)     _____
2.
3.
4.
5.
6.
7.
8.
9.
10
11.
12.
Other Information
```

FIGURE B-2 Team Minutes Template

B.1.2.5 Sources of Team Conflict

Most major team conflicts, some of which are listed in Table B-3, arise from a small set of sources. During the norming stage, the team should discuss each of these sources of conflict and decide as a team how to deal with these issues if they should arise.

Table B-3 Major Sources of Team Conflict

- One team member becomes dominant and wants to make all the decisions.
- The team leader has a different vision from that of the rest of the team.
- A team member does not regularly attend team meetings.
- A team member attends team meetings but does not participate.
- A team member does not complete their team-assigned tasks on time and/or the quality of their work is not acceptable.
- A team member is not able to complete a team-assigned task but does not inform the team of this situation.

B.1.2.6 Holding Teammates Accountable

Team members should offer to help any members who are struggling to meet their assignments. An important task of each team member is to hold the other members accountable for producing their assigned tasks on time and of high quality. Holding a fellow teammate accountable is a difficult and uncomfortable task, but it must be done. This is part of being on a team and is actually a service to the entire team, including the nonperforming team member. Holding a team member accountable is best done during a team meeting with all of the team members participating and using established team accountability tools. These tools include the team charter, revolving action item list (RAIL), and individual 360-degree team member peer reviews. The 360-degree teammate evaluation form is provided in Appendix BA13.

To properly confront a teammate about poor performance, it is important to focus on the behavior and not the person. Examples of inappropriate behavior include failing to attend team meetings, not participating in team activities, not completing team-assigned tasks on time, and not properly documenting the assigned team task. The team should enter into a dialogue with the nonperforming teammate to find out the root cause of the problem. Often the team can help this teammate resolve the root cause once it is known.

Efforts by the team to hold a nonperforming teammate accountable should be documented in the team minutes. If the problem persists, the team should meet with the sponsor. The team must have the proper documentation to show that the teammate is not performing. The sponsor, in general, will try to facilitate a resolution of the conflict with the team. The goal is to arrive at a list of measurable tasks and events that the nonperforming teammate must complete to show that they have changed their behavior. In both capstone courses and industry projects, most sponsors will remove members who continue to be non-performing from the team to maintain the team's overall high performance.

B.1.3 Selecting the Capstone Project

The capstone project should fulfill a societal need with a prototype product that can be designed, manufactured, developed, and validated within the course's schedule and resource constraints. The engineering required by the project should be within the team's skill set. If it is not within its current skill set, the team must have a concrete plan for obtaining these missing skills during the course of the project.

Students should take adequate time to select a project that is meaningful and has a scope of work that fits the course resource constraints. A good starting point is to look at the Grand Challenges for the Twenty-First Century created by the National Academy of Engineering (NAE) as listed in Table B-4. In addition to the specific challenges listed in Table B-4, the NAE

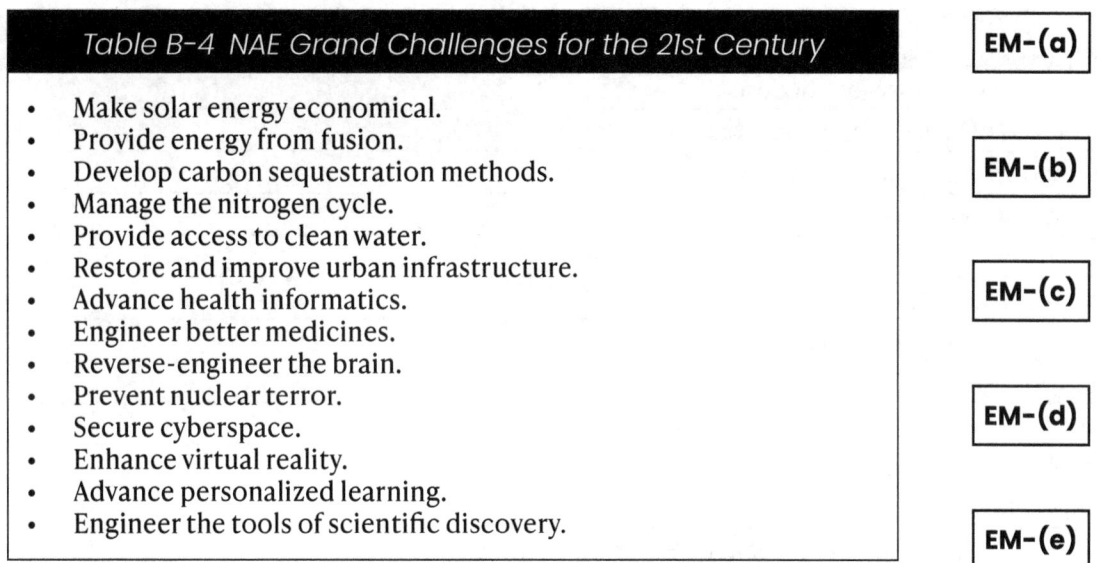

has broaden the challenges to the following four areas: health, security, sustainability, and joy of living.

There is limited time in the capstone course available for selecting the project. This means that the team must efficiently brainstorm candidate customer needs, investigate them, and select a need that the team believes will result in a product that can be developed within the course's time frame. The team must select the project by the end of Module 3.

The selection of a customer need and subsequent decision on the type of product to meet that need requires the team to (1) be curious about what societal needs exist and (2) identify what societal needs resonate with the team's interests and capabilities. It is important to explore this issue prior to making a decision.

Once the societal need is identified, the team needs to identify several candidate product concepts that will meet this need. There is limited time during the proposal to conduct an exhaustive search for product concepts, so a representative candidate preconcept is chosen to make the proposal. This preconcept will be reevaluated during Phase 2, when the team has more resources available.

The mechanical engineering capstone project at Arizona State University (ASU) is a two-semester course typical of capstone projects at many universities. The goal of this course is to design and develop an innovative product that meets a societal need. Table B-5 lists some of the capstone projects accomplished at ASU.

Table B-5 Examples of Mechanical Engineering Capstone Projects at ASU

Accident investigation quadcopter
Auto-focus off-road vehicle light bar
Automatic field irrigation system
Autonomous backyard lawnmower
Balloon rescue device
Camera suspension system
Cummins engine ISX-15 oil consumption measurement system*
Deployable neck brace
Drone-based shark attack victim response and repellent system
Fire shut-off valve system
Fluidized bed demonstrator
Freeway wrong-way driver accident deterrent system
Go pro camera stabilization gimbal
Homing unmanned underwater vehicle contest
Hospital 360° virtual reality TBI rehabilitation platform*
Hospital pressure redistribution bed
Hot compost bin
Human-powered vehicle ASME competition
Improved car jack design
KitchenAid jar opener for people with arthritis
Lunar void space astronaut ascent-descent device*
Modification of the Craftsman Bolt-on drill*
Portable rain room*
Portable home gym
Pull start mechanism for lawn trimmer*
Recon robot
Robot for corrosion detection inside pipes: 2019 NACE competition*
Formula SAE pneumatic shifter
Formula SAE race car suspension system
SAE super mileage engine test rig
Sarvonious wind turbine
Semi-autonomous patient transfer and transport bed
Small, low-head hydropower plant
Smart-wire actuator
Stent dislodgement force tester*
Subterranean exploration device
Trailer sway prevention system
Underwater device for monitoring water quality in freshwater irrigation canals

Projects for specific companies

At ASU, the project checklist given in Appendix BA2 is used to approve a project selection. The team should use this checklist.

To illustrate many of the elements of the PPD process, the authors have selected *the design and development of a travel iron* as the case study. This case study covers many of the design issues associated with electromechanical devices.

B.1.4 Understanding the Basic Physics Involved

Once the customer need is identified, it is vital to understand the basic physics involved. The best way to start the process is to make a simple box and label it "the product." Label the inputs and outputs of the box and draw arrows. Then environmental conditions are listed along with any interfaces with other products, operators, etc. Once this is done, the physics are to be further defined with graphics such as free-body diagrams, control volumes, and control masses. The defining equations, such as Newton's second law of motion—i.e., force = mass × acceleration (F = ma)—should be listed with each of the variables defined.

EM-(g)

For example, a team wanted to address the need for clean water in developing countries. They started by making a sketch of dirty water, as shown in Figure B-3. They drew water molecules along with microbes and inorganic matter on the left side of a piece of paper. On the right side, they drew only water molecules to represent the clean water. Between these two water states, they drew the cross-section of a filter that had passages large enough to allow the water molecules to navigate but too small for the microbes or inorganic matter to pass through. Then one of the team members said that the pressure on the dirty-water side would be a function of the passages' diameter. From this simple drawing, the team went on to create a hydraulic model and investigate the size of microbes relative to the size of water molecules.

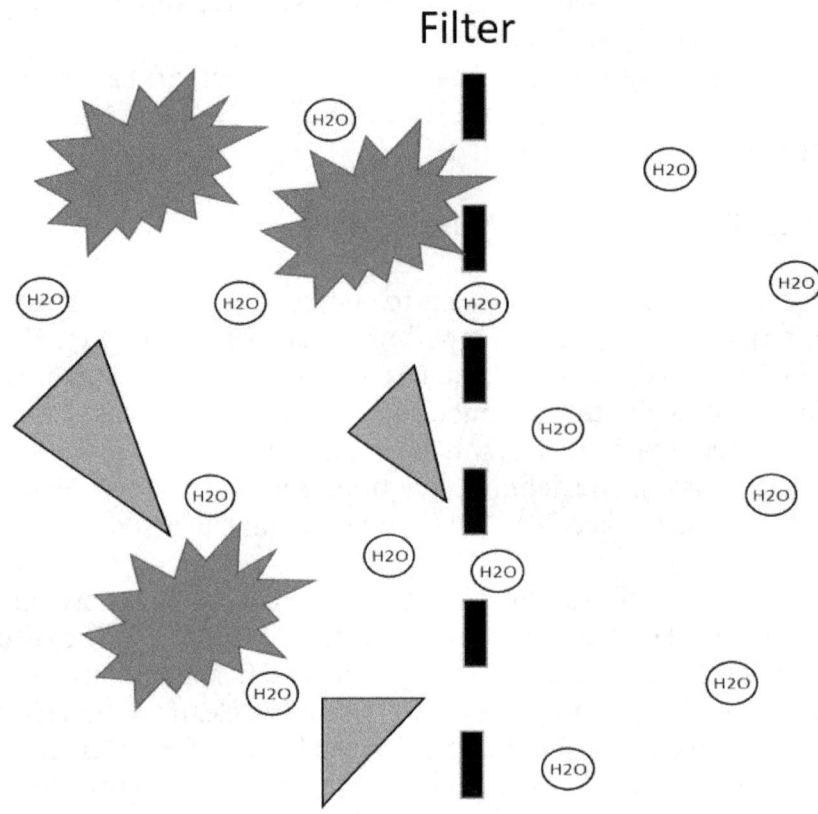

FIGURE B-3 Team Drawing: Filtering Dirty Water

B.1.5 Creating the Problem Statement

Once the customer need is understood, the team must identify a specific problem that addresses this need. The problem must lead to a product that can be developed within the time available in the capstone course. The problem statement should be one sentence that clearly identifies what is needed and why.

B.1.6 Initial Product Requirements

The customer needs must be converted into an initial set of measurable engineering design requirements. For the proposal effort, the team should focus on the key requirements. Once the project gets the approval from the sponsor to move into Phase 2, the team will take more time and effort to arrive at a detailed set of requirements.

MODULE 3
Project Selection

OVERVIEW
The team has started the process of researching societal needs and thinking about potential products to address these needs. During this module, the team will select their project and create function block diagram and physical decomposition diagram.

LEARNING OBJECTIVES
- Apply the process of selecting a concept for the proposal effort.
- Comprehend the process of creating a functional block diagram.
- Comprehend the process of physical decomposition to arrive at a configuration block diagram.

PRE-LECTURE ASSIGNMENTS
- Study the material covered in sections B.1.7 and B.1.8.
- Bring a list of at least three potential capstone projects to the class session.
- For each project, describe the societal issue and specific customer needs to be addressed by the product to be developed.

POST-LECTURE ASSIGNMENT
- Conduct a team meeting to 1) discuss the potential projects identified by each team member, 2) select the final project, 3) list key requirements, 4) create functional and configuration block diagrams, and 5) document all team activity in the team notebook.

TEAM DELIVERABLES
- Description of the elected project.
- List of key requirements for selected project.
- Functional block diagrams and configuration block diagram.
- Minutes of the team meetings
- Team activity as documented in the team notebook

B.1.7 Selecting the Team Project

After considering the team's list of potential customer needs, the team selects the customer need they want to address in their project. The team will then create a problem statement and a list of key requirements. This is then followed by creating function and configuration block diagrams for the project.

A good way to start the analysis of any product is to model it as a simple system by making a block representing the system and then showing the inputs and outputs as shown in Figure B-4. The system can then be divided into its functions, which can be related to each other as shown in Figure B-5. A top-level functional block diagram and a more detailed functional block diagram for an engine generator set are shown in figures B-6 and B-7, respectively.

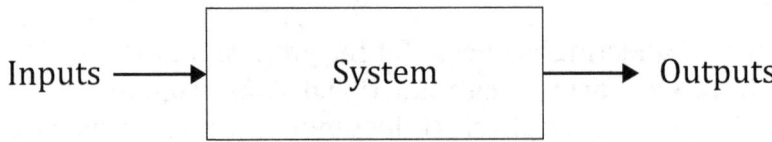

FIGURE B-4 Generalized Top-Level System Functional Block Diagram

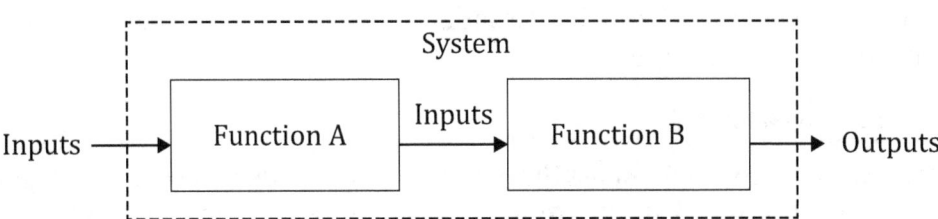

FIGURE B-5 Generalized More Detailed System Functional Block Diagram

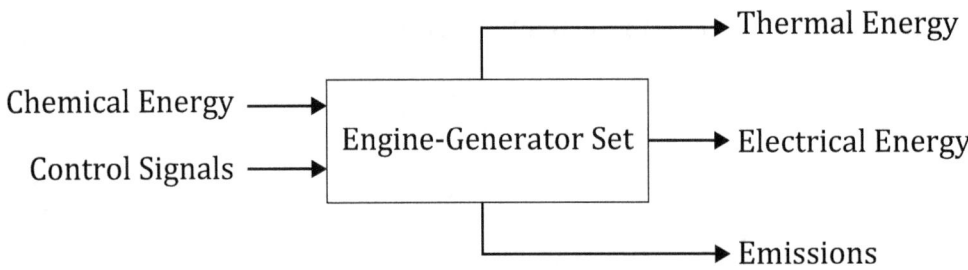

FIGURE B-6 Top-Level Functional Diagram for an Engine Generator Set

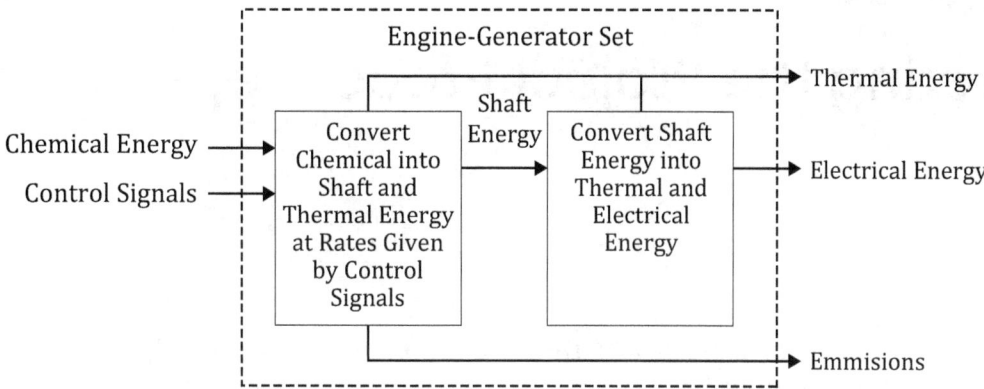

FIGURE B-7 More Detailed System Functional Block Diagram for an Engine Generator Set

Configuration (physical decomposition) block diagrams have blocks for the components rather than the functions. A configuration block diagram for the example engine generator set is shown in Figure B-8. This figure could be expanded to include the components within the engine and generator.

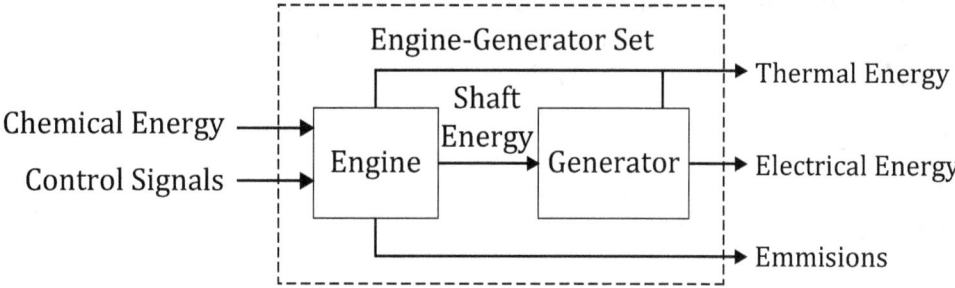

FIGURE B-8 Configuration Block Diagram for Engine Generator Set

B.1.8 Example Travel Iron Team Report for Week 2

As discussed in Section B.1.3, an example project team will be used to illustrate the design/development concepts that are covered each week. Ideally, the example project should feature an innovative approach to solving a current societal need. For simplicity, the handbook authors selected the development of an affordable travel iron. Although this type of device is already in production by a number of companies, the authors are assuming that there is still room in the market for an additional producer of affordable travel irons.

During Week 2, the travel iron team formed, started a team notebook, and prepared its team charter. The team explored societal needs and identified three potential projects.

MODULE 4

Selecting the Project Preconcept

OVERVIEW

This module covers the process of selecting a concept for the proposal. It is often called the *preconcept* because its function is to provide a basis for establishing the proposal schedule and resource budgets. Once the project is funded, the team will start Phase 2: Conceptual Design. During this phase, a more detailed effort will be employed to find the final product concept that will eventually go into production. This module also introduces the concept of value proposition.

LEARNING OBJECTIVES

- Comprehend the process of selecting the final concept for the proposal.
- Comprehend what must be done to define the proposal concept.
- Comprehend the task of identifying the value proposition.
- Apply the above topics to the preparation of the project proposal.

PRE-LECTURE ASSIGNMENT

- Study the material covered in sections B.1.9 and B.1.10.
- Prepare a comparison chart for the three candidate project concepts of your team from Module 3.
- Select the product concept for the proposal and prepare a value proposition for this product.

POST-LECTURE ASSIGNMENT

- Conduct a team meeting to 1) select the proposal product concept, 2) prepare a value proposition for that product concept, and 3) document all team activity in the team notebook.

TEAM DELIVERABLES

- Preconcept design, which is clearly defined by a narrative description, including functional and configurational block diagrams, a sketch or solid model, and a table of features and their associated benefits (see Section B.1.9).
- Statements of the value propositions for the customer and the sponsor of the preconcept design (see Section B.1.10)
- Minutes of the team meetings
- Team activity as documented in the team notebook

B.1.9 Preconcept Design

In order to arrive at a preconceptual design, an initial set of engineering requirements must first be developed based on the VOC research done by the team. The next step is to make a function block diagram for the product that will address each design requirement. At this point, the team needs to brainstorm concept ideas that will provide the needed functions. An effective method is to have each member sketch several concepts on pieces of paper. The team should have at least 10 concept ideas.

Next, the team evaluates each concept idea and compares it to the other ideas. A more formal process will be presented in the section on conceptual design. At this point, the team can take a less structured evaluation approach because the product concept will be studied in detail once the proposed project has been funded.

Once the team selects a preconcept for the proposal, the functional block diagram can be developed into a configurational block diagram and a sketch of how the product might look.

This preconcept product configuration will be used to envision a representative design and development program that can then be broken down into tasks and task descriptions. The preconcept design should be defined by a narrative description that includes a sketch or solid model with components labeled; a functional block diagram; a list of key parameters such as size, weight, input requirements, and output values; and a table of features and their associated benefits.

It should be noted that the team will discard the preconcept when they start the conceptual design process. In other words, a new and more complete set of requirements will be developed, and a structured conceptualization process will be followed to make sure the entire design space is explored.

B.1.10 Value Proposition

A value proposition is a statement of why customers should buy a product or service. They should buy it because it adds value to their lives better than any competing product or service. A good start at a value proposition is to fill in the following sentence:

> Value proposition: For (customer) who is dissatisfied with (alternative service or product), this product or service is a (product description) that provides (a solution to what problem).

Entrepreneurial Mindset: *Creating value*

The experience of Swiss mechanical watch makers provides a good illustration of how to create a value proposition. Before the advent of electronic watches, Swiss mechanical watches were valued due to their superior ability to keep time. Electronic watches are even better at keeping time than Swiss mechanical watches and have lower production costs, which leads to lower selling prices. The Swiss watch manufacturers needed a new value proposition. Swiss watches are a marvel of craftsmanship and a form of art. In fact, watches are one of the few types of jewelry that most men wear. So the watch can be a piece of jewelry as well as a device to keep time. The new value proposition for the Swiss watch makers became the following:

| Swiss watch makers' value proposition: | Discerning men who want more than a timepiece wear a Swiss mechanical watch as a statement of their success and appreciation for the finer things in life. |

So far, the discussion has been about creating a value proposition for the customer. There is also a need for a value proposition for the project sponsor. The project needs to provide more value to the sponsor than do other investment opportunities.

It is important to think about the larger definition of value. More customers and, hence, producers are interested in products that provide utility and also serve society in other ways, such as protecting the environment.

As the team moves through the different stages of product development, they need to keep returning to their stated value propositions for the customer and the sponsor as they receive feedback from both constituencies. An important rule is to remember that it's not what value you propose, it's the value the customer perceives.

The sponsor wants to be confident that the sale of the production unit will yield a reasonable return on the resources invested in developing the product. There must be a compelling value proposition. This requires an initial connection with the potential market to estimate customer demand as a function of price point. The team must also estimate costs associated with developing, producing, marketing, distributing, and supporting the product. They must do some research to gain enough facts to put together a meaningful value proposition. Once these facts are shared with the team, the team needs to arrive at an initial value proposition to feature in the proposal to the sponsor. This proposition must be compelling in order to gain funding from the sponsor. The value proposition should not be limited to the classic return-on-investment approach.

EM-(i)

EM-(n)

EM-(o)

EM-(h)

MODULE 5

Preparing the Project Proposal

OVERVIEW
The purpose of the project proposal is to gain resources from the sponsor to conduct the project. This module covers the process of writing the proposal.

LEARNING OBJECTIVES
- Apply the proposal preparation process to the team's project.
- Comprehend the skill of writing a project development scenario and then apply this knowledge to writing a development scenario for the project.

PRE-LECTURE ASSIGNMENT
- Study the material covered in sections B.1.11 and B.1.12.
- Prepare a more detailed proposal outline than the one presented in Table B-6 and bring it to class.
- Write a one-page scenario of how the project's product should be developed. Include brief descriptions of how the major tasks will be accomplished for each development phase. Bring the scenario to class.

POST-LECTURE ASSIGNMENT
- Conduct a team meeting to complete the following tasks: 1) Outline the project proposal. For each major section of the proposal, add bullets for the topics and concepts that need to be covered in that section. 2) For each team member, assign a draft section of the proposal to include the bulleted items listed by the team. 3) Each team member shares their product development scenario with the team. 4) The team develops the final development scenario for use in the next module. 5) The team documents all team activity in the team notebook.
- Each team member prepares a draft of their sections of the proposal.

TEAM DELIVERABLES
- Final development scenario for use in the next module
- Draft of the project proposal
- Minutes of the team meetings
- Team activity as documented in the team notebook

B.1.11 Preparing the Project Proposal—Part 1

This section presents the procedure for assembling the project proposal. This proposal covers the entire project. Its purpose is to (1) define the customer need and the preconcept for meeting it, (2) provide a top-level presentation of the value proposition, (3) discuss the technical, management, and risk mitigation approaches the team is taking, and (4) present the basic project plan in terms of tasks, schedules, and budgets.

The project proposal is a top-level plan and estimate, one that will be refined at the start of each new phase of the project. The top-level outline for the project proposal is given in Table B-6. A detailed outline for the project proposal is provided in Section C of this handbook. The capstone team should study this outline and follow it to complete the proposal.

Table B-6 Outline for Project Proposal

1. Introduction
2. Customer need and engineering requirements
3. Preconcept design
4. Value proposition
5. Strategies to address key issues
6. Technical approach
7. Project management approach
8. Risk management plan
9. Work breakdown structure (WBS) and WBS dictionary
10. Project schedules
11. Labor loading and labor budget
12. Monetary budget
13. Project success factors

Brief descriptions of how to prepare the project plan's sections follow.

B.1.11.1 Introduction
This section tells the reader the proposal's purpose and defines the audience and team's makeup.

B.1.11.2 Customer Need and Engineering Requirements
This section briefly describes the customer need and provides a list of customer wants. The customer wants are then transformed into an initial set of engineering requirements.

B.1.11.3 Preconcept Design
This section describes the preconcept design in terms of a product sketch and description of how the preconcept meets the initial engineering requirements.

B.1.11.4 Value Proposition
In this section, the team presents the value propositions for the customer and the project sponsor.

B.1.11.5 Strategies for Addressing Key Issues

Every project has some key issues that must be addressed. It could be, for example, that the time to market must be short; the design activity requires complicated, finite element analyses; or the validation testing will require expensive test equipment. This section of the proposal identifies the main issues and discusses how they will be addressed during the project's conduct. For example, ASU provides only $100 per team member for the engineering prototype's manufacturing and testing.

B.1.11.6 Technical Approach

This section tells the reader how the team will develop the product. For example, the team may feel that several prototype iterations will be necessary to address all the engineering requirements. Due to the limited resources available to the capstone team, their engineering prototype will only address certain features of the production unit. Another example is that validation testing of some engineering requirements may be too costly, so the validation will be done on a similarity-to-other-products basis. The technical approach should be brief but provide the reader with an overall understanding of how the product will be developed.

B.1.11.7 Management Approach

A key success factor for any project is effective project management. This section of the proposal tells the reader how the team will plan, execute, monitor, and adjust the project's conduct so that it will complete on schedule and within budget. This section should be brief but effective in convincing the reader that the team will be able to manage any unplanned issues that present themselves during the project's conduct.

B.1.11.8 Risk Management

Once the team has completed the project's baseline plan, they must review each task to determine what risks are involved. For example, the team may have assumed that they will be able to obtain an inexpensive controller from an overseas company in China. What is the risk that there may be an import issue? What is the backup plan if this controller is not available? Often the team will add one or more tasks regarding a backup design of some portion of the selected design that might prove undesirable once the team moves further along the development process.

EM-(1)

B.1.11.9 Work Breakdown Structure (WBS) and Task Dictionary

It is important to remember that the project proposal's purpose is to define the project well enough in terms of schedule, budget, and scope to obtain funding from the sponsor to proceed. The project proposal provides the sponsor with a reasonable expectation of how the project will proceed, but it is recognized that there is not much visibility of how the project will actually unfold. Hence, the proposal has to be in the ballpark, but it certainly will not be, in baseball game terminology, at home plate or even within the infield.

A good method of planning is to write a scenario of how the project could be accomplished. Because the team has imperfect information, the scenario will be imperfect as well, but it will be close enough to yield an effective project plan.

Figure B-9 presents a generic detailed WBS that identifies the tasks that will need to be accomplished during the project. Studying this WBS will help the team write their scenario.

The scenario should describe in broad terms the action items completed during each phase of the project.

The WBS for the proposal is less detailed than the one shown in the referenced Figure B-9; however, it should provide the sponsor with enough visibility to understand what will be done during each project phase. Once the project is approved, the team, at the beginning of each phase, will prepare a more detailed WBS for that phase and the associated phase project plan.

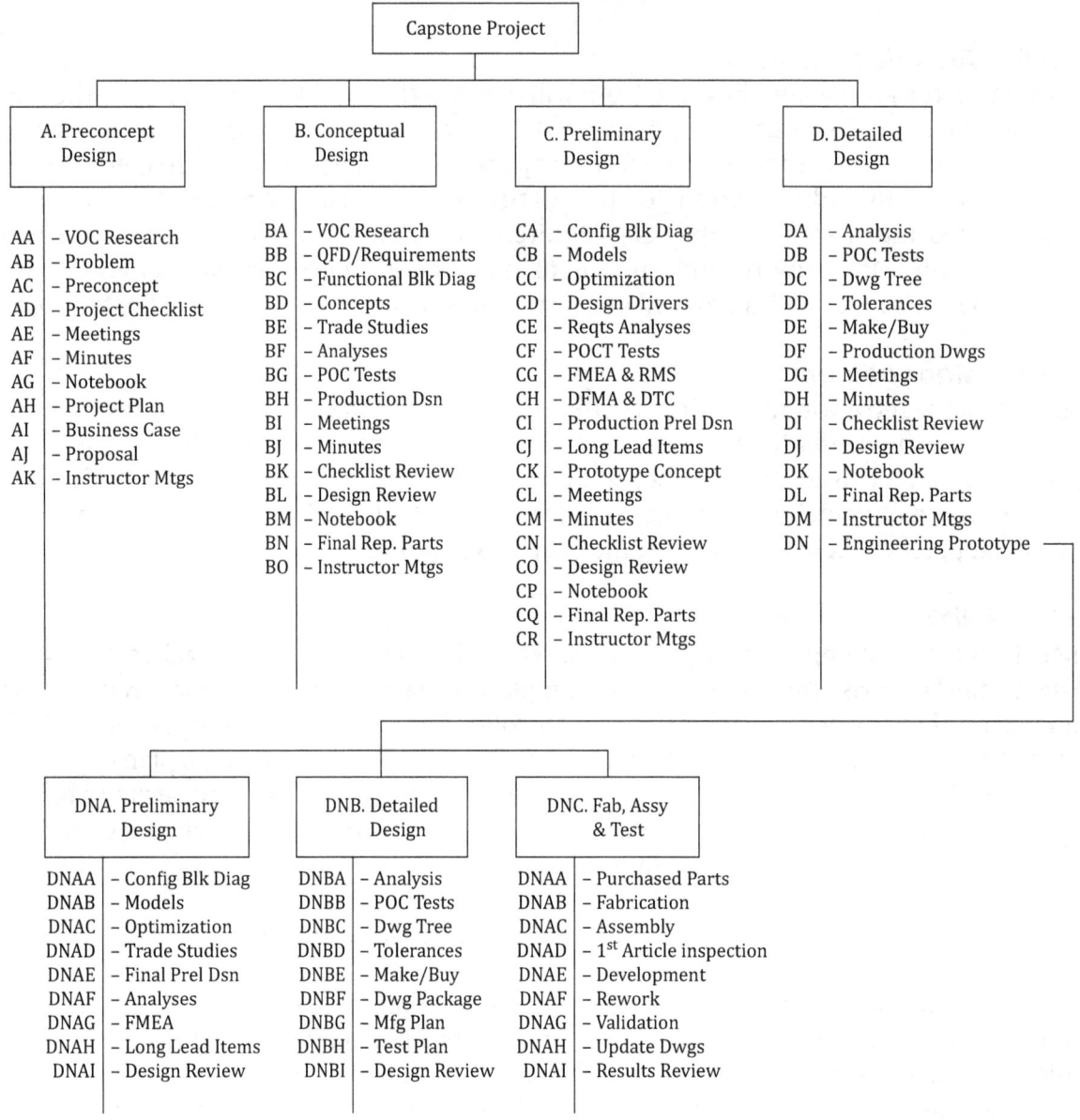

FIGURE B-9 Generic Detailed Capstone Project WBS

For the project proposal, the task descriptions can be summarized by phase. Table B-7 provides a generic WBS Dictionary by phase.

Table B-7 Generic Capstone Project WBS Dictionary by Phase

Phase	Tasks to Be Performed
1. Preconcept Design	Select customer need, define preconcept design, and prepare project proposal.
2. Conceptual Design	Explore design space, identify options, and select and define final concept.
3. Preliminary Design	Perform trade studies and analyses to convert the design concept into a dimensioned computer-aided design (CAD) model with the components identified and the performance defined. This includes analyses that will lead to a robust design.
4a. Detailed Design Planning	Plan the activities for detailed design, including the tasks to design, build, and test an engineering prototype.
3P. Engineering Prototype Preliminary Design	Establish engineering requirements for the engineering prototype and perform analyses and proof-of-concept testing to arrive at a dimensioned solid-model CAD that will meet the engineering requirements.
4P. Engineering prototype detailed design	Complete any detailed analyses and prepare a drawing package for the engineering prototype and any test equipment required. In addition, prepare manufacturing, development, and validation plans for the engineering prototype.
5P. Engineering prototype development	For the engineering prototype, fabricate parts, procure hardware, and assemble the prototype and any test equipment. Conduct development tests and necessary rework to prepare the prototype for validation testing. Conduct validation testing. Document the activities of this phase in a build book that is part of the project notebook.
4b. Detailed Design Drawing Package	Integrate the engineering prototype subprogram results into the production unit detailed design. Conduct other analyses and tests as needed. Prepare the production design drawing package.
4c. Detailed Design Final Tasks	Complete the final report, provide a final presentation, and submit the project notebook to the sponsor.

B.1.12 Example Travel Iron Team Report for Week 3

As stated previously, the authors have included the design and development of a travel iron as an example capstone project. A summary paragraph of the project for Week 3 is given below.

The team selected the development of an affordable travel iron as the capstone project. They created a proposal outline and assigned sections to the team members. In addition, they prepared the WBS and its dictionary. The team noted that the engineering prototype would be a subproject that started at the beginning of Phase 4: Detailed Design.

MODULE 6

Project Schedules and Budgets

OVERVIEW
This module covers the process of converting the product development scenario into a set of specific tasks. These tasks are addressed at the phase level for the proposal. More detailed schedules and budgets will be developed at the beginning of each project phase, starting with Phase 2.

LEARNING OBJECTIVES
- Comprehend how to convert a project development scenario into a specific project schedule and labor budget.
- Apply this knowledge to the development of a top-level (at the phase level) project schedule and labor budget for the team's project.

PRE-LECTURE ASSIGNMENT
- Study the material covered in sections B.1.13–B.1.14.

POST-LECTURE ASSIGNMENT
- Conduct a team meeting to complete the following tasks: 1) Add more detail to the phase descriptions given in Table B-7. Be sure to include risk mitigation tasks. 2) Review and, if needed, modify the project schedules given in figures A-5 and A-6. 3) Prepare a team labor budget chart. Be sure to account for potential scheduling issues, such as fall and spring break, mid-terms, and finals.

TEAM DELIVERABLES
- Updated Table B-7 of the team project
- Updated team project schedules in figures A-5 and A-6
- Team labor budget chart
- Minutes of the team meetings
- Team activity as documented in the team notebook

B.1.13 Preparing the Project Proposal—Part 2

B.1.13.1 Project Schedules
The project schedules should follow the format given in figures A-5 and A-6.

B.1.13.2 Project Labor Budget
The capstone project has calendar time and team member labor constraints. The phases must occur according to the project schedules presented in figures A-5 and A-6. Each team member should spend at least 10 hours outside of class each week on the project. The capstone project is the culminating activity for the undergraduate engineering degree, so at least 10 hours per week is a reasonable constraint. With the limitations of calendar time and labor hours per week, the team can determine how many team labor hours are available for each phase of the project. Table B-8 provides an example based on a team of six students.

The team needs to compare their allocation of hours (see the example in Table B-8) with the tasks listed in the generic WBS (see Table B-7). Does the allocation for each phase match with the team's project scenario? If not, then the allocations of hours need to change. For example, it is assumed that the team recognizes that they will need a detailed computational fluid dynamics (CFD) analysis of the engineering prototype during detailed design, yet the team is not experienced with this type of analysis. So the team decides to allocate some of their hours during phases 2 and 3 for the team to learn how to do CFD analysis before the analysis is done in phases 3P and 4P. This revised allocation of hours is shown in Table B-9.

Once the allocation-of-hours task is completed, the results can be shown on a labor chart. The cumulative hours per week for the team are plotted versus semester weeks. This is the budget line as shown in Figure B-10.

As the project progresses, the team will monitor the project by creating an actual labor line. They must also add one more line: the estimate-to-complete (ETC) line. This line shows how the team expects to spend labor for the rest of the project. This process is discussed in more detail in Section B.1.14.

Table B-8 Example Project Plan Hours—Allocation Rev 0

First Semester

Semester Week	1	2	3	4	5	6	7	8	9	10	11	12	13	14	15	Hours/Phase
Phase 1: Project Proposal		60	60	60	30											210
Phase 2: Conceptual Design					30	60	60	30								180
Phase 3: Preliminary Design								30	60	60	60	60	60	60	60	450
Team Hours/Week	0	60	60	60	60	60	60	60	60	60	60	60	60	60	60	840
	0	60	120	180	240	300	360	420	480	540	600	660	720	780	840	

Second Semester

Semester Week	16	17	18	19	20	21	22	23	24	25	26	27	28	29	30	31	Hours/Phase
Phase 4: Planning	60																60
Phase 3P: Engr Prototype		60															60
Phase 4P: Engr Prototype			60	60	60	60	60	60									360
Phase 5P: Engr Prototype									60	60	60	60	30				270
Phase 4: Dwg Package													30	60			90
Phase 4: Final Tasks															30	30	60
Team Hours/Week	60	60	60	60	60	60	60	60	60	60	60	60	60	60	30	30	900
	900	960	1020	1080	1140	1200	1260	1320	1380	1440	1500	1560	1620	1680	1710	1740	
Total Project Hours																	1740

Table B-9 Example Allocation of Hours Rev 1

Hours Increased to Accommodate Gaining CFD Knowledge

First Semester

Semester Week	1	2	3	4	5	6	7	8	9	10	11	12	13	14	15	Hours/Phase
Phase 1: Project Proposal		60	60	60	30											210
Phase 2: Conceptual Design					30	60	60	30								180
Phase 3: Preliminary Design								30	60	60	60	60	60	60	60	450
CFD Model Development					10	10	10	10	10	10	10	10	10	10	10	110
Team Hours/Week	0	60	60	60	70	70	70	70	70	70	70	70	70	70	70	950
	0	60	120	180	250	320	390	460	530	600	670	740	810	880	950	0

Second Semester

Semester Week	16	17	18	19	20	21	22	23	24	25	26	27	28	29	30	31	Hours/Phase
Phase 4: Planning	60																60
Phase 3P: Engr Prototype		60															60
Phase 4P: Engr Prototype			60	60	60	60	60	60									360
Phase 5P: Engr Prototype									60	60	60	60	30				270
Phase 4: Dwg Package													30	60			90
Phase 4: Final Tasks															30	30	60
Team Hours/Week	60	60	60	60	60	60	60	60	60	60	60	60	60	60	30	30	900
	1010	1070	1130	1190	1250	1310	1370	1430	1490	1550	1610	1670	1730	1790	1820	1850	
Total Project Hours																	1850

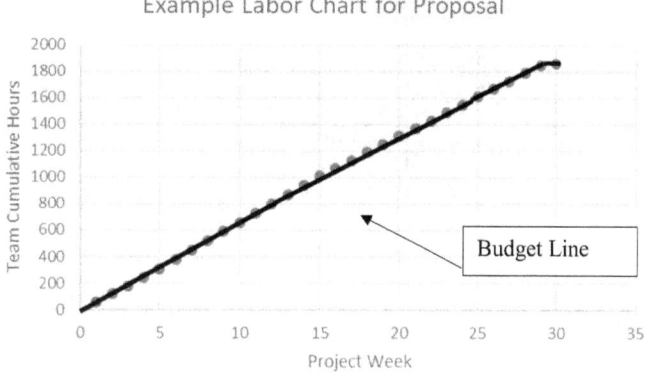

FIGURE B-10 Example Project Labor Graph for the Proposal

B.1.13.3 Project Monetary Budget

In addition to labor hours, the team must also establish the monetary budget. For example, at ASU, the capstone budget is only $100 per team member. This is a major constraint that must be addressed in the proposal.

B.1.13.4 Five Key Success Factors

The purpose of the proposal is to convince the sponsor to provide resources for the project. A powerful communication tool involves ending the proposal by telling the sponsor the five key factors that will make this proposed project successful. The team needs to spend some quality time identifying these factors. Sometimes teams realize that they are missing a key factor and revise the proposal to incorporate it.

B.1.14 Using a Labor Chart and Schedule to Manage the Project

Once the proposal is accepted by the sponsor, the team should update their schedules and labor chart on a weekly basis. Updating involves the following steps:

1. Enter the reporting date on the chart.

2. Enter the cumulative hours actually spent as of the reporting date and extend the line for the number of actual labor hours spent to that point.

3. Draw an estimate-to-complete (ETC) line showing the team's best judgment of how they will spend their labor hours to the end of the program.

To illustrate how these weekly update charts can be used to help manage the project, the following example is provided:

B.1.14.1 Example: Managing Team XYZ

As shown in Figure B-11, Team XYZ submitted their project's schedule and labor chart for the reporting period through the end of Week 9 to their sponsor.

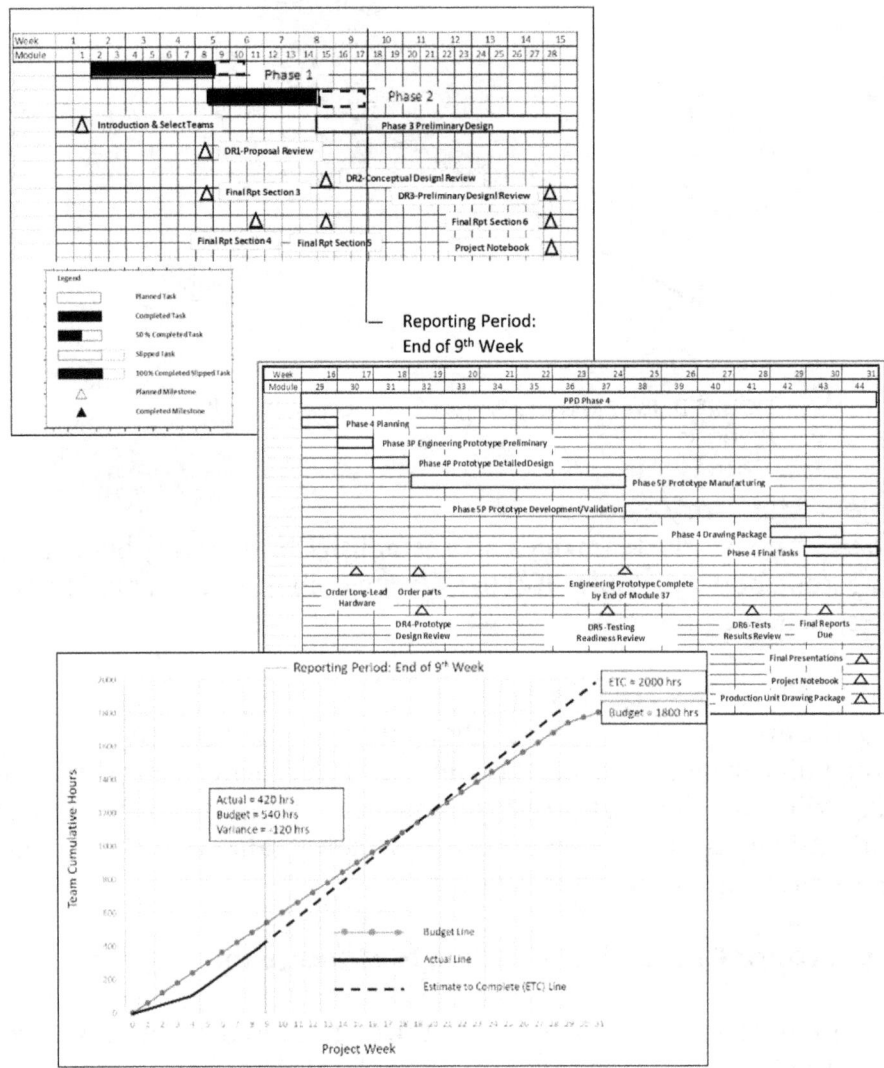

FIGURE B-11 Team XYZ Schedule and Labor Chart for Week 9

The sponsor looked at these documents and made the following observations that then led to the following questions.

Sponsor Observations:

1. The team completed Phase 1 but slipped the schedule by one week.

2. The team completed Phase 2 but slipped the schedule by 1.5 weeks.

3. As of the reporting date, the team has just completed Phase 2 and started Phase 3.

4. The team plans to be back on schedule by the end of Phase 3 and complete the remainder of the project on schedule.

5. The team has spent only 420 hours of their budgeted 540 hours. This means they have only spent 78 percent of their budgeted hours to date.

6. For the first four weeks, the team's labor slope was less than the budget slope. However, the team increased their slope at that point and has remained on that slope through Week 9.

7. The team has included an estimate to complete (ETC) line. This line's slope is the same as the actual line slope starting in Week 5.

8. The team plans on completing the project on time but will spend 200 hours more than budgeted.

Sponsor Questions:
Question 1: Why did the team slip schedule?
Question 2: Is the team's recovery plan reasonable?
Question 3: How can the team reduce their ETC from 2,000 hours to the original budget of 1,800 hours?
Team XYZ provided the following narrative to the sponsor along with the schedule and labor chart.

Team XYZ Narrative:
For the first four weeks of the project, some of the team members were not investing their pledged time on their assigned tasks. As a result, the team began to slip schedule. At the end of Week 4, the team held a meeting during which all team members agreed to get back on schedule by increasing their ETC line to a slope slightly greater than the proposed budget slope. This results in the project completing 200 hours over the original project budget of 1,800 hours. The actual line follows this agreed-to slope for weeks 5-9. This shows that the team should be able to continue on this slope for the rest of the project. The team also will try to find ways of being more efficient so that it can complete the project on budget, but it is not confident enough at this point to change the ETC line.

Sponsor's Conclusions:
The team has resolved the issue of not all members spending adequate time on the project. The team has demonstrated that they are able to stay on their new labor slope. The team currently plans to complete their project on time; however, they will be 200 hours over budget. The team will try to find ways to reduce their ETC to 1,800 hours. The team's efforts to reduce their ETC of 2,000 hours to 1,800 hours should be monitored and encouraged.

Key Points from This Example:

1. The team updates and manages their schedule and labor chart on a weekly basis during the project.

2. The team uses the schedule and labor chart as a means of communicating with the sponsor.

3. The team uses the ETC labor line and the schedule bars to communicate how they plan on completing the project.

4. The team is careful to provide the reporting date on both the schedule and labor chart.

MODULE 7

Finalizing the Proposal

OVERVIEW

The proposal should conclude with five key reasons why the project will be successful. The team should include a detailed plan for Phase 1 (Preconcept Design) in the proposal to show how subsequent phases will be planned at the beginning of the phase. The proposal's commercialization plan should be based on the Goldsmith Commercialization Model. Once a full draft of the proposal is created, each team member is responsible for editing the entire document and accepting accountability for all the contents.

LEARNING OBJECTIVES

- Apply the five key success factors idea to the project proposal.
- Apply project planning knowledge to the development of a detailed project schedule and labor budget for Phase 1.
- Comprehend the Goldsmith Commercialization Model.
- Apply the competencies of proposal writing and teamwork to arrive at a quality project proposal.

PRE-LECTURE ASSIGNMENT

- Study the material covered in sections B.1.15–B.1.19.

POST-LECTURE ASSIGNMENT

- Conduct a team meeting to complete the following tasks: 1) Agree on the five key factors; 2) prepare a detailed Phase 1 plan that includes a list of tasks, schedule, and labor budget; and 3) create a plan for completing the proposal and having it edited and approved by all team members.

TEAM DELIVERABLES

- Detailed Phase 1 plan, including a list of tasks, schedule, and labor budget
- Plan for completing the proposal and having it edited and approved by all team members
- Minutes of the team meetings
- Team activity as documented in the team notebook

B.1.15 Five Key Success Factors

The proposal should conclude with five key reasons why the project will be successful. This is a powerful communications approach that instills confidence in the sponsor that the project is worthy of funding.

To identify the success factors, start with a list of what the team thinks are the key questions that the sponsor needs answered before they are comfortable funding the project. Some of these potential questions are listed as follows:

- Does the scope of this project match with the available resources and schedule?
- Does the team have the necessary competencies to execute the project?
- Is the team committed to completing the project on time and within budget?
- Is the proposed product design consistent with basic physics?
- Is there a market for this product?
- Is there a business case for this product?
- Has the team included a risk mitigation plan?

B.1.16 Phase 1 Project Plan

It should be noted that the project proposal addresses the project on a phase basis. Specific tasks within that phase are planned at the beginning of it with the constraints of schedule and labor hours established in the project plan. This planning process starts with Phase 2 when the project proposal has been approved.

In preparation for planning at the beginning of Phase 2, the team should practice this skill by documenting what was done in Phase 1 in a planning format. To illustrate this planning process, an example case project with the following information is assumed:

- Table B-10: Example case team lists the tasks and task descriptions for Phase 1.
- Figure B-12: Example case team prepares Phase 1 schedule based on the Table B-11 tasks.
- Table B-11: Example case team prepares a hypothetical labor loading for a six-person team with each team member spending 10 hours per week outside of class on the project.

Table B-10 Example Case Task Dictionary for Phase 1

Task	Description
Research	Research societal needs, select a need, and use primarily secondary sources.
Problem	From the customer needs, determine the initial engineering requirements.
Preconcept	Identify preconcept options and down-select to final preconcept and define.
Business case	Determine top-level value propositions for the customer and sponsor.
Project checklist	At the end of the phase, fill out the checklist and have it approved.
Project plan	Prepare a project plan for the entire project and a detailed plan for Phase 1.
Proposal	Prepare the proposal and obtain sponsor approval.
Meetings	Assume most meetings will be held during class hours for general planning.
Minutes	A team member will be designated to take minutes of meetings and put them in the notebook.
Notebook	Maintain the project notebook with all material developed on the project.
Instructor meetings	Allocate time for the team to meet with the instructor during office hours.

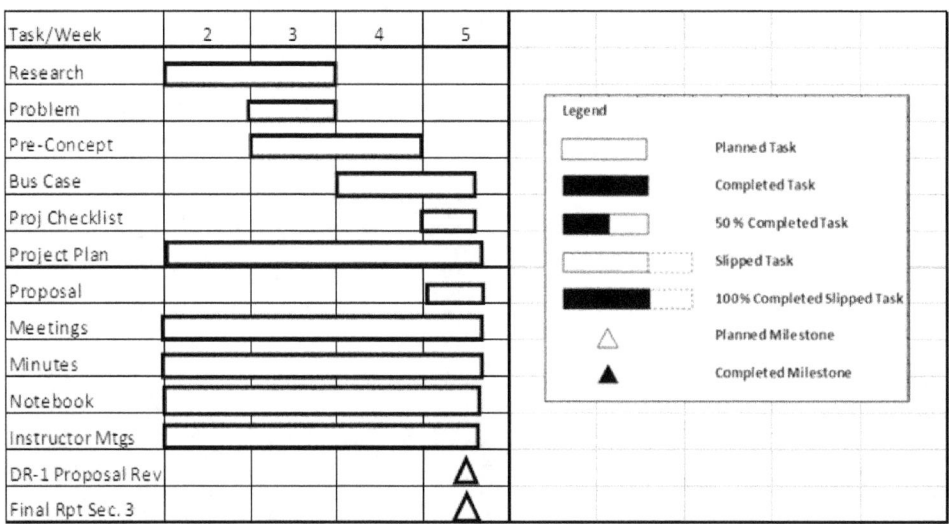

FIGURE B-12 Example Case Phase 1 Schedule by Week

Table B-11 Case Team Labor Budget for Phase 1 by Task and Week[1]

Task/Week	2	3	4	5
Research	39	13		
Problem		6		
Preconcept		20	20	
Bus Case			5	3
Proj Checklist				6
Proj Plan	6	6	20	3
Proposal				6
Meetings	6	6	6	6
Minutes	1	1	1	1
Notebook	2	2	2	2
Instructor Mtgs	6	6	6	3
	60	60	60	30

After studying the example case given above, the team should gather together the actual work, schedule, and labor used for Phase 1 of their project. Based on these data, the team should then put these data into the form of a plan by creating documents similar to Table B-10, Figure B-12, and Table B-11.

B.1.17 Commercialization Aspects of Phase 1

The PPD process Phase 1 (Preconcept Design) is part of the investigation stage of the Goldsmith Commercialization Model. During Phase 1, the team identifies a societal need and a potential customer with that need. It also completes a top-level assessment of the market for products to meet this societal need. Then, the project team identifies a preconcept for the product that has the potential to meet that need. Next, the team prepares a notional business case that indicates there may be a profitable business associated with bringing this product to market. PPD process Phase 2 continues to address the investigative phase of the Goldsmith Commercialization Model.

B.1.18 Final Steps in Completing the Proposal

Once a full draft of the proposal is created, each team member is responsible for editing the entire document and accepting accountability for all its contents. The proposal should follow the outline given in Section C of the Handbook. The team should be especially careful regarding grammar and spelling. The sponsor sees the proposal's quality as a key factor in deciding whether to fund the project.

B.1.19 Example Travel Iron Team Report for Week 4

This week, the team completed the project proposal. Major tasks included finalizing the commercialization plan, preparing a detailed plan for Phase 1, and identifying the five key success factors for this project. Each of these items are discussed in this section.

---

1 Note: Week 5 is only 30 hours per the schedule in Figure B-12.

B.1.19.1 Commercialization Plan—Market Analysis

The team explored the travel iron market by doing secondary research through the internet and found the following information:

- There is a worldwide market for clothes irons.
- The market is segmented into irons for home and travel and commercial and industrial uses.
- Most irons feature steam, but dry irons are still popular and usually less expensive.
- In the United States, there are home-use and travel-use markets.
  - Numerous brands and offerings
  - Most if not all are imported
  - Emphasis is on ease of use
  - Retail price ranges from $20 to $90
  - Preliminary survey did not find any offerings emphasizing United States-made and/or environmentally friendly travel irons.

The travel iron example assumes that the team is located in the United States. Currently, there is a segment of US customers who want more locally produced (i.e., United States-made) products. There is also another segment of US customers who want more environmentally friendly products. Based on this brief research effort, the team concluded that there is a niche in the US customer market for locally made and environmentally friendly travel irons.

It should be noted that desire for locally made items is not limited to customers in the United States. Other countries are also interested in having more manufacturing in country. For example, one of the authors was involved in marketing a United States-based company's proprietary solar power systems in South Africa. The local power company made local manufacture of selected items in the solar power system a requirement.

The travel iron project team realized that the major challenge regarding the use of locally made components is finding suppliers of items such as thermostats that are currently produced off shore. The team also realized that using environmentally friendly materials and manufacturing processes is a large scope of activity. Both of these requirements had to be refined to fit within the limited resources of the capstone project. The team concluded that the goals would be (1) to have more than 80% (by cost) of the travel iron locally made and (2) that at least one major component of the iron would be environmentally friendly.

B.1.19.2 Commercialization Plan—Business Case

The team based their project on the assumption that the locally made travel iron would have a unit production cost of about $4 and the sales volume would be 5,000 units per year. The product would be sold to companies such as Sharper Image that appeal to customers who are looking for traditional items with a value-added twist.

EM-(a)

The team agreed that this business concept would be good for an organization seeking to employ relatively low-skilled assemblers who historically have had a difficult time finding employment in the United States. This would include, for example, individuals who have a significant disability.

Even if the team's travel iron has a somewhat higher price point than that of current market alternatives, the fact that it is locally made, environmentally friendly, and potentially manufactured by workers with major disabilities provides a reasonable market entry strategy.

MODULE 8

Completing Phase 1

OVERVIEW

Phase 1 (Preconcept Design) is completed with the team preparing and presenting a proposal review with the sponsor and having the sponsor sign off on the completed Phase 1 exit checklist. The team members will conduct the first 360-degree teammate evaluation using the form given in Appendix BA1.

LEARNING OBJECTIVES

- Apply communications skills to produce a high-quality project proposal presentation.
- Evaluate the work done in Phase 1 while completing the Phase 1 exit checklist.

PRE-LECTURE ASSIGNMENT

- Read sections B.1.20–B.1.22.
- Each team member edits and approves the project proposal draft.
- Each team member reviews the Phase 1 exit checklist and makes a list of documents that are in the project notebook to support completion of each checklist item.

POST-LECTURE ASSIGNMENT

- Conduct a team meeting to complete the following items: 1) edit and approve the project proposal, 2) prepare the Phase 1 review presentation, and 3) present the Phase 1 exit checklist and supporting documentation to the sponsor for their review and approval.
- Conduct a team meeting to practice the Phase 1 review presentation.
- Complete the 360-degree team member peer evaluation using the forms in Appendix BA1.

TEAM DELIVERABLES

- Edited and approved project proposal
- Phase 1 design review (DR1) presentation slides
- Phase 1 exit checklist and supporting documents
- Minutes of the team meetings
- Team activity as documented in the team notebook

B.1.20 Phase 1 Preconcept Design Review (DR1)

In addition to a written proposal, most sponsors want the team seeking funding to give a stand-up slide presentation. The sponsor not only wants to hear what is already in the written report, but they also want to determine how well each team member understands what is being proposed. The sponsor also wants to evaluate how well the team works together to make an effective presentation. The presentation must be crisp with effective slides and a team who can professionally answer the sponsor's questions.

EM-(m)

The length of this presentation varies according to the sponsor's availability. ASU capstone teams are usually limited to about a 10-minute presentation and five minutes for questions. Table B-12 provides a list of slides that are used for ASU capstone proposal presentations.

Table B-12 List of Recommended Proposal Presentation Slides

1. Title slide
2. Issue selected and problem statement
3. Physics involved
4. Customer needs
5. Engineering requirements
6. Preconcept design
7. Value propositions
8. Market analysis
9. Approach for conducting project
10. Risk mitigation plan
11. Schedule
12. Labor and monetary budgets
13. Five key success factors

The team should practice their presentation until it is crisp and accomplished within the time constraints given by the sponsor. Each team member should understand the proposal well enough to answer any questions the sponsor may pose after the formal presentation.

B.1.21 Example Travel Iron Team Report for Week 4.5

During the first half of Week 4, the team prepared and presented their proposal presentation as preconcept design review (DR1). Figure B-13 shows the presentation slides on the following seven pages. The shaded boxes include general instructions for what should be in each slide.

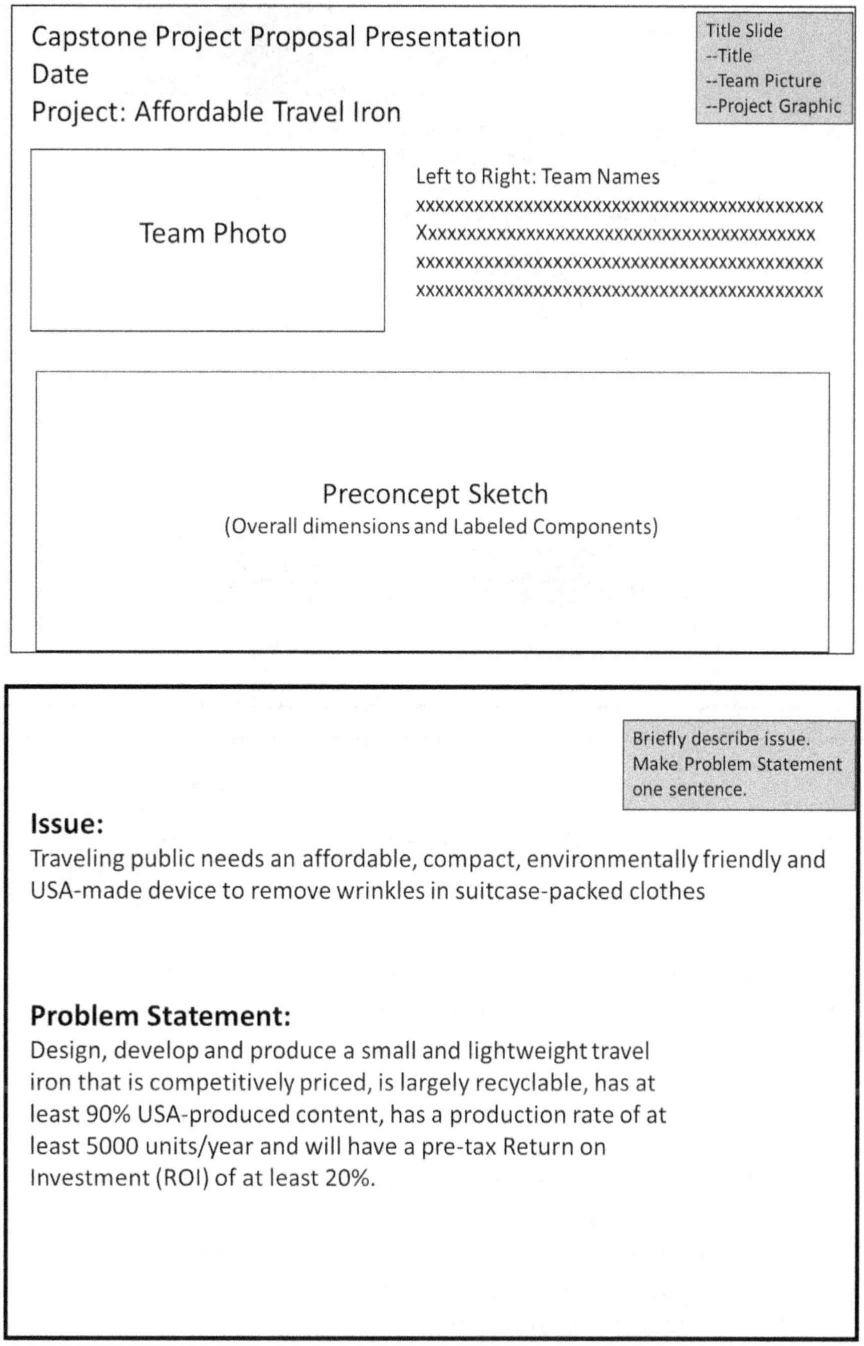

Slide 1

Capstone Project Proposal Presentation
Date
Project: Affordable Travel Iron

[Team Photo]

Left to Right: Team Names
xxxxxxxxxxxxxxxxxxxxxxxxxxxxxxxxxxxxxxxx
Xxxxxxxxxxxxxxxxxxxxxxxxxxxxxxxxxxxxxxxx
xxxxxxxxxxxxxxxxxxxxxxxxxxxxxxxxxxxxxxxx
xxxxxxxxxxxxxxxxxxxxxxxxxxxxxxxxxxxxxxxx

Preconcept Sketch
(Overall dimensions and Labeled Components)

Title Slide
--Title
--Team Picture
--Project Graphic

Slide 2

Issue:
Traveling public needs an affordable, compact, environmentally friendly and USA-made device to remove wrinkles in suitcase-packed clothes

Problem Statement:
Design, develop and produce a small and lightweight travel iron that is competitively priced, is largely recyclable, has at least 90% USA-produced content, has a production rate of at least 5000 units/year and will have a pre-tax Return on Investment (ROI) of at least 20%.

Briefly describe issue. Make Problem Statement one sentence.

FIGURE B-13A Travel Iron Proposal Slides (Page 1 of 7)

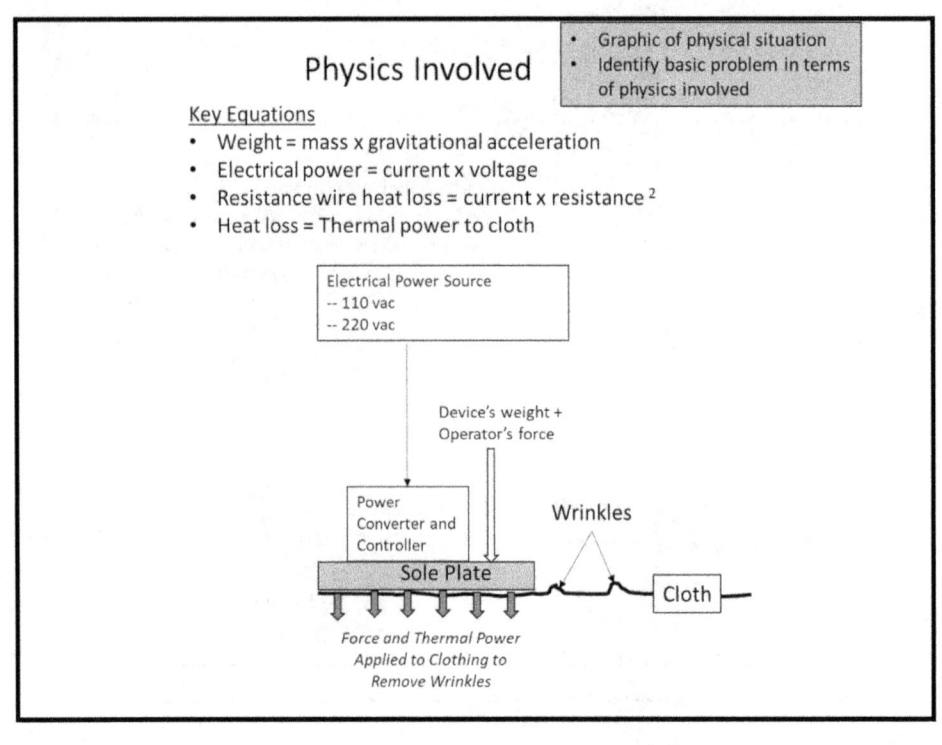

Physics Involved

- Graphic of physical situation
- Identify basic problem in terms of physics involved

Key Equations
- Weight = mass x gravitational acceleration
- Electrical power = current x voltage
- Resistance wire heat loss = current x resistance ^{2}
- Heat loss = Thermal power to cloth

Customer Needs

- Who is customer
- Research basis
- Simple list

- Customer—people traveling with clothes in suitcases, etc.
- Research Basis—secondary sources--internet sites
- Needs
    - Affordable
    - Lightweight
    - Compact
    - Dual voltage (110 vac and 220 vac)
    - Variable temperature
    - Meets Underwriters Laboratory (UL) Requirements
    - Able to remove wrinkles in various types of material
    - Does not need to use steam

FIGURE B-13B Travel Iron Proposal Slides (Page 2 of 7)

Initial Engineering Requirements

- Simple list
- Must be measurable

Requirement	Description
Retail Price	Less than $20
Weight	Less than 2 lbs
Size	Approximately 5 x 3 x 4 in inches
Voltage	110 vac and 220 vac
Temperature	275 to 400 deg F
Max. Temp.	600 deg F
Stability	Rest upright on a 20 deg incline
Ironing	Remove wrinkles in cotton cloth

Preconcept Design

- Hand Sketches
- Labeled
- Overall Dimensions

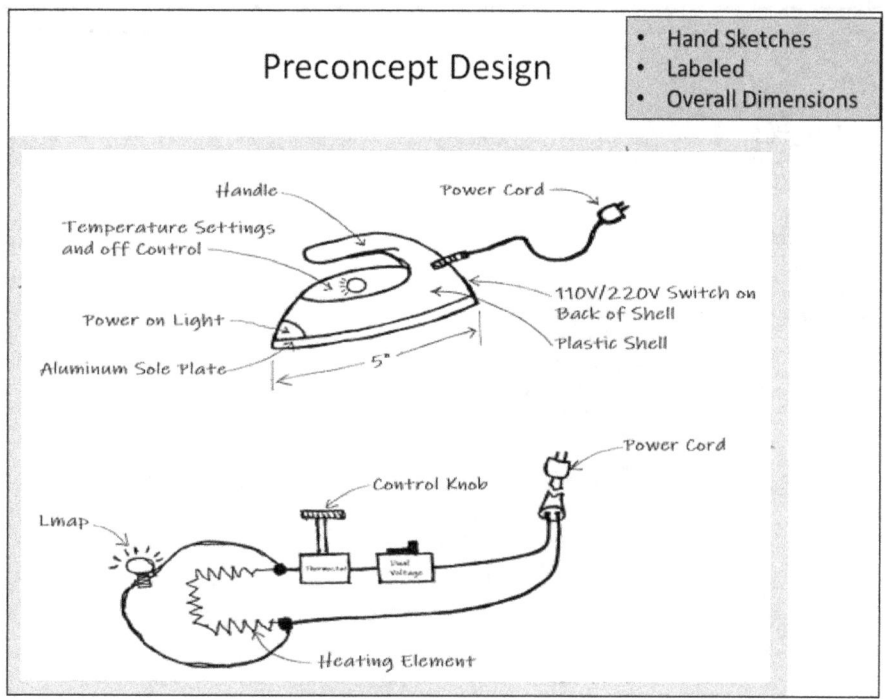

FIGURE B-13C Travel Iron Proposal Slides (Page 3 of 7)

Value Propositions

- Customer

  Affordable, compact, lightweight, dual voltage, environmentally friendly, US-made product for the traveler to easily remove wrinkles from packed clothing.

- Producer

  Design that meets the market need for an environmentally friendly and US-made travel iron that is easy to produce, provides a profitable revenue stream and yields a ROI of greater than 20%.

Market Analysis

- There is a world-wide market for clothes irons.
- Market segmented into irons for home, travel, commercial and industrial uses.
- Most irons feature steam, but dry irons are still popular and usually less expensive.
- In the US home-use and travel-use market
  - Numerous brands and offerings
  - Most if not all are imported
  - Emphasis is on ease of use
  - Retail price ranges from $20 to $90
  - Preliminary survey did not find any offerings emphasizing US-made and/or environmentally friendly
- Travel iron market entry strategy
  - Appeal to customers who want to buy-American and want to protect the environment
  - Keep retail price in the $20 to $30 range
  - Select a producer with assembly workers that customers want to support
    - Disabled
    - Native American
    - Job training
    - Etc.
- Target a factory with a production rate of 5000 units/year and a unit production cost (UPC) of less than $10.

FIGURE B-13D Travel Iron Proposal Slides (Page 4 of 7)

Approach for Conducting Project

- Develop an entrepreneurial mindset and apply it to the project
- Maximize use of COTS hardware*
- Use recycled plastic for outer shell
- Select US made components and assemble in the US
- Use proof of concept testing to retire risk early
- Have notebook editor and quality control engineer
- Hold regular team meetings with full participation
- Limit scope to committed labor hours
- Make a risk reduction plan part of Phase 2 planning
- Utilize faculty as subject experts
- Select engineering prototype requirements based on the most difficult to achieve product requirements and limit scope to fit $600 budget including instrumentation and test rigs.
- Write final report sections as the work is being done.

*Commercial Off-the-Shelf

Risk Mitigation Approach

- Brainstorm potential risks and identify ways of risk mitigation
- Include a risk mitigation approach table in the proposal

Risk	Mitigation
Not completed on-time	Reduce scope if needed
Over labor budget	Reduce scope if needed
Team member not committed	Hold each other accountable each team meeting and use RAIL
Unexpected performance during Phase 5P	Analyses and POC tests early in project to retire prototype development risks
Lack of US suppliers	Early supplier involvement
Not meeting sponsor's expectations	Frequent meetings with sponsor
Lack of team communications	Effective meetings with agendas and minutes posted in Project Notebook

FIGURE B-13E Travel Iron Proposal Slides (Page 5 of 7)

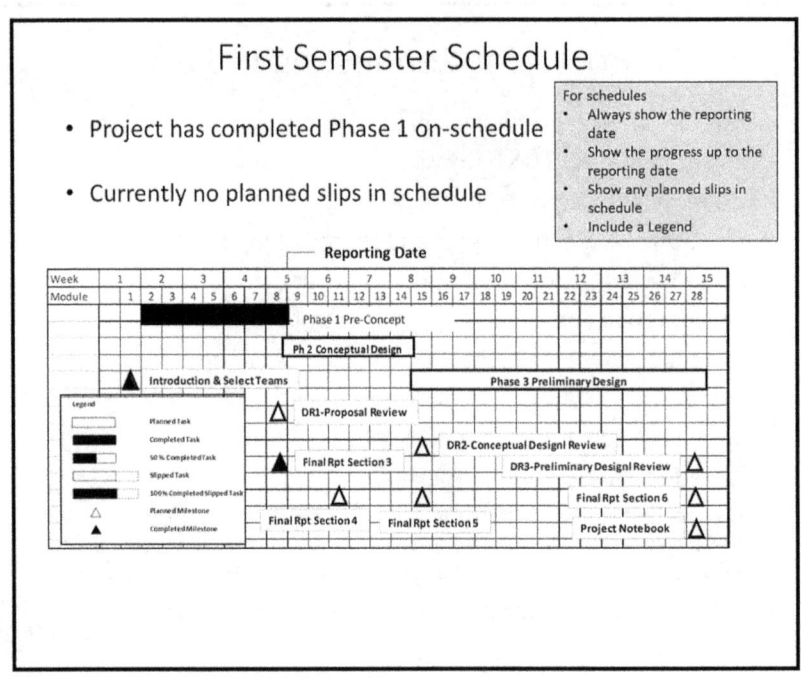
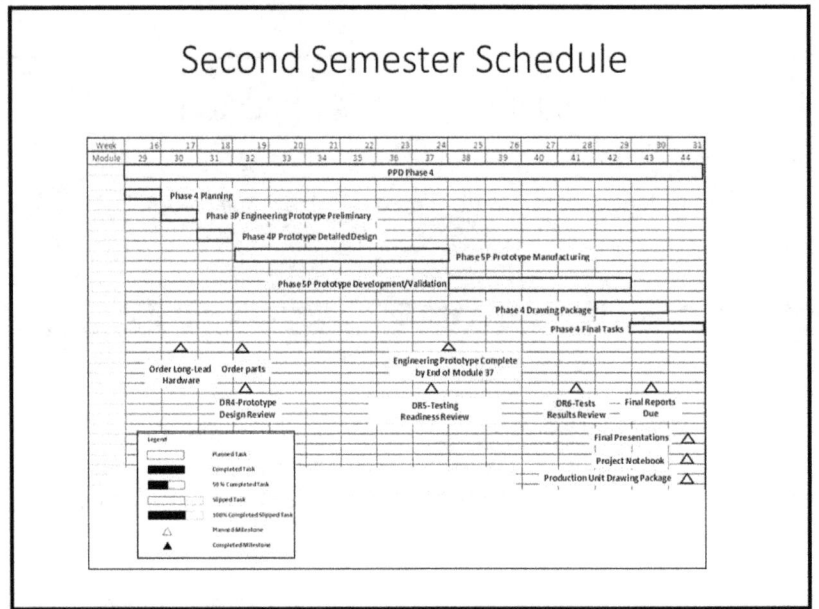

FIGURE B-13F Travel Iron Proposal Slides (Page 6 of 7)

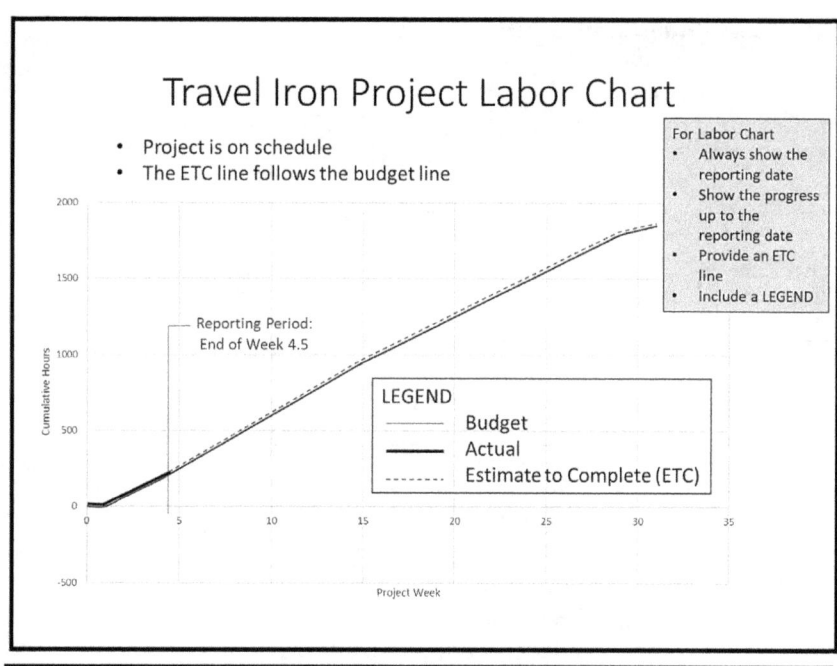

FIGURE B-13G Travel Iron Proposal Slides (Page 7 of 7)

B.1.22 Phase 1 Exit Criteria Checklist

Below is the exit checklist for this phase. All items must be completed before the sponsor will verify that the team has exited this phase of the project. The checklist is divided into two parts. The first part covers the general items related to Phase 1. The second part is a checklist of the 17 EM@FSE 2.0 entrepreneurial mindset (EM) indicators.

For the EM indicators portion of the checklist, the team must briefly summarize how the indicator was addressed. There must also be references to pages in the team project notebook where this information is provided in detail.

Exit Criteria Checklist: Phase 1: Preconcept Design

Team Name: _____   Team Leader: _____

Members: _____   _____

_____   _____

_____   _____

Date Started: _____   Date Completed: _____

Done	Exit Criteria	Comments
	1. Team formed, team roles selected, and team charter prepared	
	2. Project notebook initiated	
	3. Problem statement worksheet submitted and approved	
	4. Team project checklist submitted and approved	
	5. Process of arriving at the problem statement using problem solving techniques is documented as an appendix for the final report	
	6. Preliminary requirements list prepared	
	7. Preconcept design review completed and approved	
	8. Project plan formatted as Section 3.2 of the final report submitted and approved	
	9. Notebook checked and approved by sponsor	

Yes or No	EM Indicators Exit Criteria	Questions Related to Criteria
	a. Critically observes surroundings to recognize opportunity	Was the team properly curious about societal needs? Did the team address the grand challenges in their search for a societal need? Did the team list the multiple ways they explored the environment to arrive at an initial list of important societal issues and their associated problem statements?
	b. Explores multiple solution paths	Did the team list a number of potential societal needs?
	c. Gathers data to support and refute ideas	After listing potential customer needs, how did the team learn about each need so that the final selection could be based on facts?
	d. Suspends initial judgement on new ideas	How did the team suspend judgment during brainstorming?
	e. Observes trends about the changing world with a future-focused orientation/perspective	How did the team explore evolving customer needs on a global basis?
	f. Collects feedback and data from many customers and customer segments	Once a customer need was selected, how did the team learn about the customer base and its segmentation?
	g. Applies technical skills and knowledge to the development of a technology or product	Did the team understand the basic physics of the selected customer need, and did the team adequately define the preconcept?
	h. Modifies an idea or product based on feedback	How did the team interface with the instructor to obtain feedback on their ideas?

(continued)

	i. Focuses on understanding the value proposition of a discovery	State the value proposition for both the customer and the producer. Reference where this is stated in the notebook.
	j. Describes how a discovery could be scaled and/or sustained, using elements such as revenue streams, key partners, costs, and key resources	Why does the team think this product idea is sustainable in terms of adequate profit for the producer and affordability for the consumer? Where is a business analysis located in the notebook?
	k. Defines a market and market opportunities	Did the team characterize the market in terms of size, segmentation, competitors, and market entry opportunities?
	l. Engages in actions with the understanding that they have the potential to lead to both gains and losses	Did the team address project risks and include a risk mitigation plan in their proposal?
	m. Articulates the idea to diverse audiences	Did the team write the proposal for the reader and not for the team? Did the proposal follow the outline in Section C of the handbook? Did the team take the time to properly review and edit the proposal?
	n. Persuades regarding why a discovery adds value from multiple perspectives (technological, societal, financial, environmental, etc.)	Did the team provide multiple perspectives in the value propositions for both the customer and the producer?
	o. Understands how elements of an ecosystem are connected	How did the team include sustainability in the value proposition?

(continued)

	p. Identifies and works with individuals with complementary skill sets, expertise, etc.	Did the team go through the four phases of team development and demonstrate that it is now a high-performing team?
	q. Integrates and synthesizes different kinds of knowledge	How did the team improve their marketing and business acumen to properly integrate these considerations into the technical design of the preconcept?
Approved by	Name and Title	Completion Date

Section B
Phase 2: Conceptual Design

MODULE 9

Introduction to Phase 2: Conceptual Design

OVERVIEW

In this phase, the team will determine the engineering requirements, explore the design space, down-select from several candidates to the final conceptual design, and, through analysis, determine how well the final conceptual design meets the engineering requirements. This module covers the initial work on Phase 2 that includes building team commitment and defining a more detailed Phase 2 plan.

LEARNING OBJECTIVES

- Understand what tasks are needed to complete Phase 2.
- Apply project management principles to create a detailed Phase 2 plan.

PRE-LECTURE ASSIGNMENT

- Read and study handbook sections B.2.1–B.2.3.
- Each team member makes a list of the Phase 2 tasks and indicates on which tasks they would like to take the lead.

POST-LECTURE ASSIGNMENT

- Conduct a team meeting to 1) create a team charter and 2) prepare the Phase 2 detailed plan, including task assignments for each team member.

TEAM DELIVERABLES

- Team charter
- Phase 2 detailed plan including task assignments for each team member
- Minutes of the team meetings
- Team activity as documented in the team notebook

B.2 Phase 2: Conceptual Design

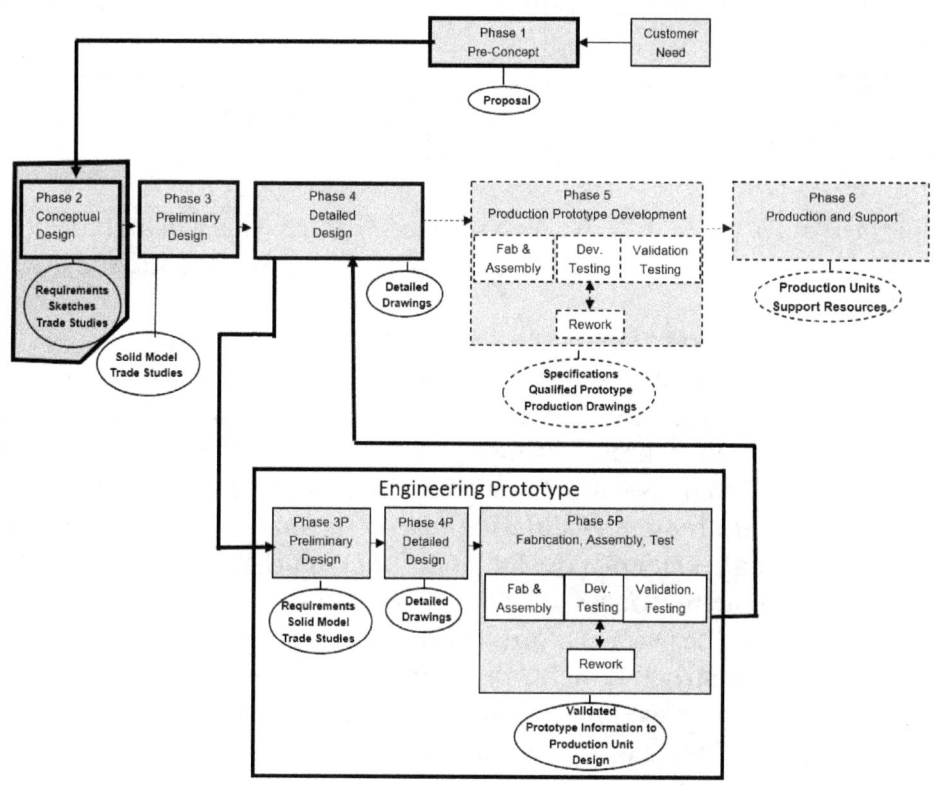

B.2.1 Overview

Figure B-14 summarizes the key objectives for Phase 2.

1. Reconvene the team and hold a successful kickoff meeting.
2. Refine the technical approach and project plan.
3. Refine the requirements based on expanded VOC research.
4. Explore the design space and identify potential conceptual design candidates.
5. Use a Pugh matrix to evaluate the candidate designs and select the final conceptual design.
6. Conduct analyses and proof-of-concept testing as needed to support the above activities.
7. Document the conceptual design phase.
8. Conduct a conceptual design review.
9. Complete Phase 2 Exit Criteria Checklist and obtain instructor/sponsor approval.

FIGURE B-14 Phase 2 Key Objectives

In general, Phase 2 is initiated when notification is received that the sponsor has selected the proposed project. At this time, the team reconvenes. The sponsor may want to make some changes to the proposed statement of work and/or budgets. The team negotiates with the sponsor to make sure that the new scope of work can be

EM-(h)

accomplished for the desired budgets. At the end of the negotiations, the final agreements are stated in a contract. The contract can simply be the updated proposal signed by each member of the team and the sponsor. A contract should be created whether the sponsor is an outside entity or a sponsor inside the organization. The contract should be placed in the team project notebook.

The purpose of the conceptual design phase is to establish design requirements, explore the design space, identify candidate concepts, select a final conceptual design, perform initial analyses, conduct proof-of-concept (POC) testing (if applicable), and hold a conceptual design review.

The first task for the team is to create a detailed plan for Phase 2. This will include how the team will do research, establish requirements, analyze/test selected concepts, and down-select a final conceptual design.

Entrepreneurial Mindset: *Connections*

The next task for the team is to refine their understanding of what the customer wants through market research. These qualitative customer needs are then translated into product requirements that are measurable using a quality function deployment (QFD) process. Progress toward meeting these requirements is recorded in the requirements validation matrix. From the list of requirements, the team selects a design goal function that will drive the design refinement process.

Entrepreneurial Mindset: *Curiosity*

Next, the team explores the design space to identify concepts that will potentially satisfy the product functional requirements established in Phase 2. The team uses a morphological chart to identify optional ways of providing each product function. The team then creates candidate overall designs by selecting a specific option for each function. The team generates concepts with the potential of fulfilling the design requirements. A function is simply a task that is performed by a device, system, or process. From the set of candidate design concepts, the team selects three or four of the most promising concepts for further evaluations. Simple feasibility analyses and POC testing are used to inform the down-select process.

At this point, each of the candidate concepts needs to be refined. The first step is to convert the product function block diagram into a configuration block diagram for each candidate concept. This is followed by making sketches of each concept. The team then performs top-level analyses and testing to determine how well each candidate concept meets the design requirements and design goal function.

Once the team has a good understanding of each candidate concept, the final conceptual design is determined using a Pugh comparison matrix. Based on the results of the comparison matrix, a final conceptual design is defined in terms of a labeled sketch that shows key relevant technical specifications such as dimensions, a performance characteristics table, initial analysis results entered in the requirements validation matrix, and a starting value for the design goal function.

Entrepreneurial Mindset: *Create value*

The Goldsmith Commercialization Model is then applied to the final conceptual design. The business case is revisited with more refined estimates of the sales volume and selling price.

The development and production costs are also updated. These data are combined to arrive at a new estimate of the profitability of taking the product to market.

A conceptual design review is then held among the team, sponsor, customer (maybe), and project consultants to review the team's work and make recommendations for improving the final conceptual design.

As Phase 2 work is accomplished, it is entered into the project notebook. Near the end of Phase 2, the work is documented in the Phase 2 section of the final report draft. Conceptual design review 2 (DR2) is then held with the sponsor. When all Phase 2 actions are complete, the team presents a completed Phase 2 exit criteria checklist to the sponsor for their review and signature.

B.2.2 Building Team Commitment

It is vital that all team members understand their roles and responsibilities on the team and make a commitment to be active team members. Many teams have each member show their commitment by signing a roles-and-responsibilities agreement before it is entered into the project notebook.

For the capstone project, the team membership generally stays the same after the project proposal is accepted. However, in industry there is often a change in the team membership. When this occurs, the team needs to participate in a new round of forming, storming, norming, and performing.

B.2.3 Refining the Phase 2 Project Plan

Before the team starts work on Phase 2, they need to revisit the schedule and labor/dollar budgets for Phase 2. This activity starts with each team member carefully studying this chapter. Next, the team should make a list of Phase 2 tasks, estimate the labor hours needed to complete each task, and determine who on the team will work on what tasks. Finally, the team needs to create a detailed Phase 2 Gantt chart. For capstone projects, these tasks need to be distributed so that the labor commitments for each team member are approximately level and comprise at least 10 hours per week.

Table B-13 presents an example team labor estimate for a team of six members each working outside of class for 10 hours per week. Figure B-15 provides the associated Gantt chart for Phase 2.

Table B-13 Example Phase 2 Planning Chart in Hours per Week

Task/Week	5	6	7	8
Research	12	6		
Requirements		12		
Functional Block Diagram		6		
Concepts		15		
Trade Studies			20	1
Analyses	3	6	10	1
POC Tests	2	2	2	
Prototype Design			3	1
Production Design			6	1
Meetings	6	6	6	4
Minutes	2	2	2	1
Checklist Review				3
Design Review				6
Notebook	2	2	2	2
Final Report Parts			6	10
Instructor Meetings	3	3	3	
Total Hours Per Week	30	60	60	30

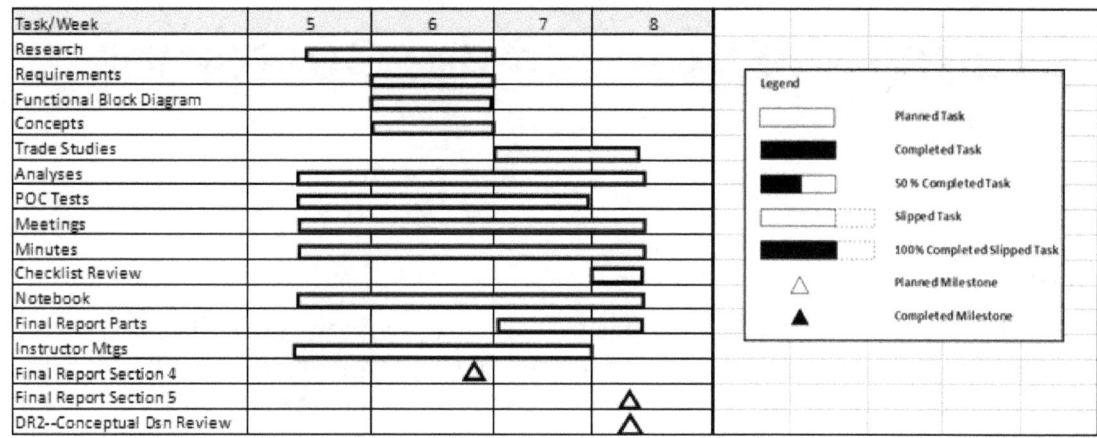

FIGURE B-15 Example Phase 2 Schedule

B.2.4 Example Travel Iron Team Report for Week 5

The team prepared the detailed plan for Phase 2 and started the voice-of-the-customer (VOC) research.

MODULE 10

Capturing the Voice of the Customer

OVERVIEW

This module discusses the process of learning about customers' need. The first task is to conduct secondary research into what others have learned from customers. This is followed up by members of the design team actually talking to customers to learn what they need and want to solve their problem.

LEARNING OBJECTIVES

- Become knowledgeable about and apply secondary research to learn what the customer needs.
- Become knowledgeable about and apply primary research to learn what the customer needs.

PRE-LECTURE ASSIGNMENT

- Read Section B.2.5.
- Each team member writes on a sheet of paper their understanding of what the customer needs and then uses a library search engine to identify at least three articles that describe that need from the customer's perspective. The sheet of paper should include a brief summary of each article. This paper should also provide at least three ideas of how primary research can be done to further understand customers' needs.

POST-LECTURE ASSIGNMENT

- Conduct a team meeting to 1) review and discuss each team member's pre-lecture secondary research findings and ideas regarding primary research into the customer's needs, 2) decide how the team will do primary customer research, and 3) do that research. Note: The capstone schedule is aggressive, so the team needs to quickly do this research.
- Conduct a follow-up team meeting to review the secondary and primary customer research and arrive at a list of customer needs, a description of the market for this customer need, a segmentation of this market, and an estimate of the market size.

TEAM DELIVERABLES

- List of customer needs
- Description of the market for this customer need, segmentation of this market, and estimate of the market size
- Minutes of the team meetings
- Team activity as documented in the team notebook

B.2.5 Capturing the Voice of the Customer (VOC)

The entire capstone project depends on the team understanding what the customer wants (i.e., hearing the VOC). In a two-semester capstone course, there is usually little over a week for the team to do research on the VOC. The team must be extremely efficient during this time to adequately understand what the customer wants. Due to calendar time constraints, a majority of the VOC research must come from secondary sources.

Entrepreneurial Mindset: *Connections*

(EM)-(a)

The product development process starts with the identification of an unmet societal need. Entities that will buy a product to meet that need are called *potential customers*. The sum of all potential customers (buyers) and sellers is called a *potential market*. The major customer need was identified during the proposal activity. Now that the project has been funded, the team must go back and explore the potential market for the product.

EM-(c)

EM-(d)

This exploration task starts with research into what has already been learned about what the customer needs. This is called *secondary research*. Books, magazines, professional journals, US patent office information, and the internet should be explored to uncover more about who the customers are, what they need, and how those needs should be prioritized.

EM-(e)

EM-(f)

Secondary research is followed by primary research, in which the team actually talks to potential customers and determines what their needs are. There are a number of methods for conducting primary research. There is limited time in the capstone course to conduct primary research, so it must be planned carefully to yield representative information about the potential customer population. An effective method is to divide (segment) the customers into separate groups. The significant groups in terms of buying power need to be identified and their list of prioritized needs must be determined. Two effective methods for doing this are individual customer interviews and customer focus groups. In both cases, the team needs to develop a set of questions that will uncover the true needs of the customer.

Care must be taken to select specific customers for interviewing. Questions for these interviews need to be formulated such that the responses can be easily interpreted into specific customer needs and their priorities. In addition to surveying individual potential customers, the team can form a group of representative customers (called a *focus group*) to discuss the product need with the team and their ideas about what characteristics the solution to the need should have. The team should consult a good textbook on market research to help plan the VOC activities. Interviews have to be well planned to ensure that the data gathered will be representative of the potential customer market. Survey instruments can be highly structured, such as in the following statement:

> On a scale of 1 to 10, with 1 being strongly disagree and 10 being strongly agree, what is your opinion of this statement: I would purchase a portable travel iron if it did a reasonable job of removing creases in my clothes and it cost less than $20, in 2018 dollars.

On the other hand, a group of potential customers (focus group) might be given these open-ended questions:

> How many of you would consider buying a compact travel iron, and what would influence your decision?

A two-semester capstone course does not provide time to adequately conduct VOC research. The team must realize that they have only a short amount of time to gain an understanding of what customers need. The team must be especially efficient and focused with regards to this all-important aspect of the design process.

MODULE 11

Updating Requirements with QFD

OVERVIEW

The quality function deployment (QFD) process is presented as an effective way of going from customer needs to measurable engineering requirements. The design of a travel iron is used as an example. Students apply QFD to their team project. The team also selects a goal function.

LEARNING OBJECTIVES

- Understand the QFD process and the concept of a goal function.
- Apply QFD to the example of the design of a travel iron.
- Apply QFD and the goal function to the team project.
- Document the team's work in a draft of Section 4 of the project final report.

PRE-LECTURE ASSIGNMENT

- Read sections B.2.6–B.2.10.
- Each team member prepares their own (1) customer needs vs. engineering requirements matrix and (2) a statement of the design's goal function. A hard copy of this document is due at the beginning of the lecture.
- Consult Section C of the Handbook to understand how Section 4 of the project final report should be written.

POST-LECTURE ASSIGNMENT

- Conduct a team meeting to complete the following tasks: 1) Review each team member's idea about how the QFD analysis for the team project should look. 2) Synthesize the individual team members' inputs into an overall team QFD analysis for the team project. Make sure that the team arrives at a measurable value for each engineering requirement and create an overall QFD spreadsheet. (3) Decide on the goal function for the project. (4) Use the detailed Phase 2 project plan to create assignments. (5) Have a frank discussion about how well each team member is meeting their team responsibilities. This discussion should focus on behaviors and not personalities.
- Conduct additional work and team meetings if the QFD and goal function to the project needs further exploration.
- Document the team's work in a draft of Section 4 of the project final report.

TEAM DELIVERABLES

- Design goal function
- Product requirements validation matrix
- Section 4: Engineering Requirements of the final report
- Minutes of the team meetings
- Team activity as documented in the team notebook

B.2.6 Establishing Design Requirements

B.2.6.1 Quality Function Deployment (QFD)

EM-(g)

The product's design is defined by a set of measurable engineering requirements. These requirements are derived from a set of customer needs called *voice of the customer* (VOC). VOC items tend to be qualitative, while engineering requirements are quantitative. To go from customer needs to engineering requirements, a process called *quality function deployment* (QFD) is used.

An important tool in the QFD process is the house of quality (HOQ). This is a graphical way of relating customer needs to engineering requirements. There is a wealth of knowledge on this subject. Due to time constraints, a simplified HOQ as discussed in this section is recommended for capstone projects.

This simplified HOQ is shown conceptually in Figure B-16. The customer needs and engineering requirement categories are listed. A measurable target value for each engineering requirement is then determined. For each customer need, appropriate engineering requirements are identified with a level of importance in the needs/requirements cross-matrix. The roof of the graphical house is a cross-matrix of engineering requirements that indicates the relationships among the engineering requirements. A level of importance is assigned to each customer need and engineering requirement. A scale of 1 to 10 is usually used. An example of how to use QFD is provided later in this section when the design of a travel iron is used as an example.

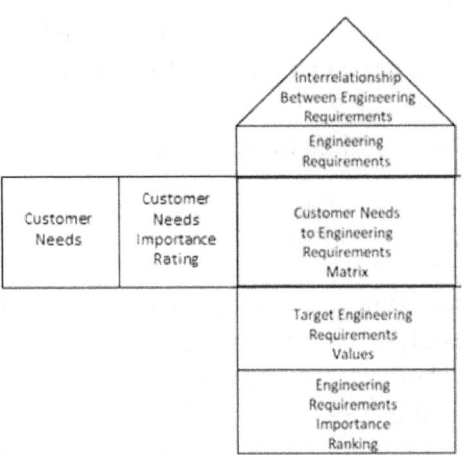

FIGURE B-16 Simplified HOQ Graphic

B.2.6.2 Goal Function

The team selects one of the requirements as the optimization variable. This means that the product must be designed to meet all the other requirements and minimize or maximize the optimization variable. A minimum or maximum goal is established by the team.

As an example, consider a project team that is designing a product to meet certain engineering requirements at the least unit production cost (UPC) when manufactured in lots of 100 items. The team has set a UPC of equal to or less than $5 per unit. Figure B-17 shows the team's progress in meeting the goal function. During preliminary design, redesigning the product's housing decreased the UPC, but it still exceeded the goal. Later in preliminary design, the product's shaft was switched from one manufactured in house to a component off-the-shelf item. This reduced the UPC below the minimum goal. Further design work during detailed design was able to reduce the UPC to about $3.50, which is substantially below the minimum goal.

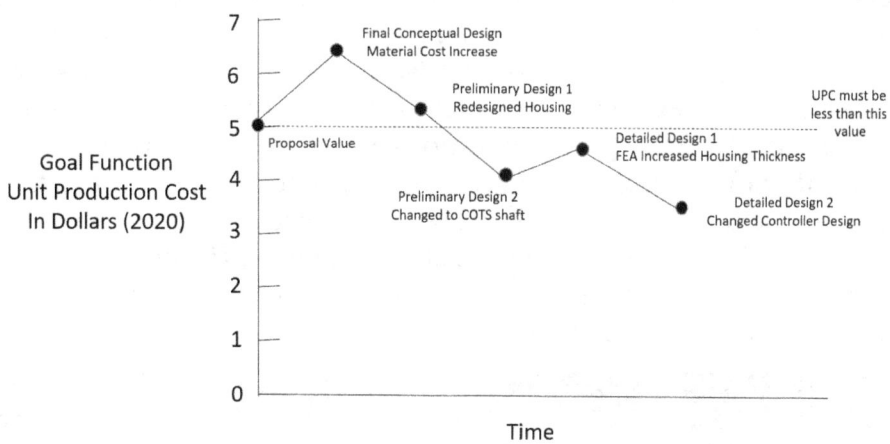

FIGURE B-17 Example Goal Function Graph

It should be noted that a goal function can have more than one design variable. However, the function then needs to have percentage weighting factors. For example, let the goal function be of the form:

Goal function = 60% (weight in pounds [lbs]) + 40% (UPC in dollars)

The above goal function is for a device where its weight can be reduced by using more costly components. Table B-14 shows the goal function for four different designs. In this case, the objective is to minimize the goal function. Design C is the preferred design because its goal function is the lowest of the four designs.

Table B-14 Results for Four Designs of the Example Device			
Design	Weight, lbs	UPC, dollars	Goal Function Value
A	5.00	10.00	7.00
B	6.00	8.00	6.80
C	7.00	5.00	6.20
D	8.00	4.00	6.40

B.2.6.3 Engineering Requirements

The definition of requirements and specifications vary from one company to another. In this book, requirements are the characteristics the customer wants the product to have. The specifications are what characteristics the product actually has. The goal is to have the specifications meet or exceed the requirements.

The project team must then convert the VOC into a set of measurable engineering requirements. Capstone design teams tend to not have enough requirements. They list only the key constraints, such as weight or cost. Although customers may say that they are mostly interested in low weight or affordable cost, it usually turns out that they also want the product to perform under a number of required conditions. For example, the United States Air Force usually wants its aircraft systems to operate from -40 ºF to 125 ºF. They often require electronic boxes to be able to sustain a bench top drop of 3 feet.

Table B-15 provides a list of top-level requirement categories. The project team should refer to this list as they prepare the project requirements. Each requirement must be stated on a measurable (quantitative) basis so that it can be validated.

Table B-15 Top-Level Requirements Categories

- Functional requirements—product's required performance
- Environmental conditions—the product must perform over these ranges of conditions
- Operating life—length of time the product is designed to be operational
- Reliability—probability that the product fails to function during its useful life
- Maintainability—level of ease in maintaining the product
- Structural performance—the product must function with these loadings and safety factors
- Electrical requirements—the product must function with these characteristics
- Size and dimensions—applicable size limits
- Interfaces—how the product interfaces with other components, the environment (for example, mounting), and the user
- Design and construction—the product must conform to the following:
  - Materials, processes, and parts—limits due to corrosion, safety, etc.
  - Component parts—use of a common parts list, etc.
  - Name plate and product marking—identification of the product
  - Design control—method of configuration management
  - Drawings—format and approval procedures
  - Design and construction changes—waivers, material review board, etc.
  - Interchangeability—whether matched bores are allowed, etc.
  - Safety—warnings, analyses, special issues such as radiation, etc.
- Packaging and labeling

A requirements document is incomplete without an explanation of how each requirement will be validated. The four methods of validation are as follows: inspection, demonstration, analysis, and test.

- <u>Inspection</u> is used to validate certain requirements (e.g., the product must have a warning sign that conforms to Occupational Safety and Health Administration (OSHA) requirement 29 *Code of Federal Regulations (CFR)* 1910.145). In this case, inspecting the product for the appropriate sign is all that is needed.
- <u>Demonstration</u> is used to validate certain requirements (e.g., the product must have a mean time to repair (MTTR) of less than 1.5 hours). In this case, a person must actually perform the required repair operation on a timed basis.

- Analysis is used to validate the requirement by calculation. For example, a detailed FEA analysis must be performed to show that the product's housing can sustain a proof pressure of 100 pounds per square inch absolute with a safety factor of 2.
- Testing requires the product to perform under specified conditions. For example, the product must convert 24.0 +/- 1.0 volts dc to 60.0 +/- 0.1 hertz ac at standard day conditions of temperature and pressure using calibrated measurement techniques.

The U.S. military has developed a good set of requirements for most of its equipment. These requirements are provided in the military standards. These are often referred to as *MIL-SPECS*. Military specification documents are divided into sections. Normally, the requirements are listed in Section 3: Requirements. The procedures for validating that it does indeed meet the requirements are provided in Section 4: Quality Assurance Provisions.

The following specification is used as an example: MIL-S-19557D (AS) dated 29 January 1988, General Specification for Aircraft Air Turbine Starters.[1]

B.2.6.4 Requirements Validation Matrix

A concise way of communicating the requirements and their method of validation is provided by the requirements validation matrix. This matrix also provides a way of documenting the progress in validating the design. Table B-16 provides the format for this matrix.

Table B-16 Format for Requirements Validation Matrix		
Requirement	Method of Validation	Validation Status

Validation of the requirement occurs at the end of Phase 5 of the PPD process. In each design phase prior to validation, the team must perform analyses and/or POC testing to determine the probability that the device will meet the requirement during validation. These interim status findings need to be included in the phase design reviews.

B.2.7 A Simple Example: Requirements for a Travel Iron

The example travel iron project discussed previously will be used to illustrate how the engineering requirements development process can be used. This example is a team project to design and develop a compact iron for taking creases out of clothing when a person is traveling. Market research has determined the weighted customer needs shown in Table B-17. The next step is to assemble a set of engineering requirements that when fulfilled will result in customers' needs being met.

---

[1] The entire document can be accessed by visiting http://www.everyspec.com/MIL-SPECS/MIL-SPECS-MIL-S/MIL-S-19557D_38511/.

Table B-17 Customer Needs for a Portable Iron

Customer Need	Weighting (Scale 1–10)
Lightweight	10
Fits in carry-on bag	10
Operates on electric outlets in United States and Europe	8
Able to quickly press clothing	7
Able to adjust iron temperature	5
Visual signal that the iron is on	4
Easy to use on/off device	4
Does not require maintenance	6
Durable	7
Affordable	10

For each customer need, the team looked at the list of requirement categories previously presented in Table B-15 and then determined the requirements that, if met, would satisfy this need.

The design team created the list of travel iron requirements as shown in Table B-18.

Table B-18 Design Team's List of Travel Iron Engineering Requirements

Weight
Volume
Length
Input Power
Press Time
Iron Operating Temp. Range
Visual On Signal
On/Off Switch on Handle
Operating Life Reliability
Drop Test
Operating Life
Shelf Life
Unit Production Cost
Time to Production
Development/Tooling/Facility Cost
Initial Prototype Project Materials Cost

To make sure that all the customer needs were adequately covered by the engineering requirements, the needs and requirements were entered into a matrix as shown in Figure B-18.

Customer Need	Relative Weight	Weight	Weight	Volume	Length	Input power	Press time	Iron operating temp range	Visual on signal	On/off switch on handle	Operating life reliability	Drop test	Operating life	Shelf life	Unit production cost	Time to production	Development/Facilities/Tooling costs
								Engineering Requirements									
Lightweight	14	10															
Fits into carry-on bag	14	10															
Operates USA/European Power	11	8															
Able to quickly press clothes	10	7															
Able to adjust iron temperature	7	5															
Visual signal that iron is on	6	4															
Easy to turn on and off	6	4															
Does not require maintenance	8	6															
Durable	10	7															
Affordable	14	10															

FIGURE B-18 Example Customer Need vs. Engineering Requirements Matrix

In the referenced figure, it should be noted that a new column was added to convert the customer need weight to a percentage of relative weight. For example, because the weightings of the customer need items in Table B-17 add up to 71, the percentage relative weight for the lightweight need was (weighting/71) × 100 = (10/71) × 100 ≅ 14. Likewise, the percent relative weight of the durable need was (7/71) × 100 ≅ 10.

The importance of each engineering requirement to the customer needs was determined and entered into the matrix shown in the referenced figure. To do this, the following importance factors were used:

- 9–Engineering requirement is strongly important to meeting the customer need
- 3–Engineering requirement is moderately important to meeting the customer need.
- 1–Engineering requirement is weakly important to meeting the customer need.
- 0–Engineering requirement is not important to meeting the customer need.

Figure B-19 shows how the design team filled in the matrix for the travel iron.

LEGEND: Relationship between Engineering Requirement and Customer Need 9 = strongly important 3 = moderately important 1 = weakly important 0 = not important			Weight	Volume	Length	Input power	Press time	Iron operating temp range	Visual on signal	On/off switch on handle	Operating life reliability	Drop test	Operating life	Shelf life	Unit production cost	Time to production	Development/Facilities/Tooling costs
Customer Need	**Relative Weight**	**Weight**						Engineering Requirements									
Lightweight	14	10	9														
Fits into carry-on bag	14	10		9	3												
Operates USA/European Power	11	8				9											
Able to quickly press clothes	10	7					9	3									
Able to adjust iron temperature	7	5						9									
Visual signal that iron is on	6	4							9								
Easy to turn on and off	6	4								9							
Does not require maintenance	8	6									9						
Durable	10	7										3	9	3	9		
Affordable	14	10														9	3	9

FIGURE B-19 Customer Need vs. Requirements Matrix with Importance Ratings

Engineering requirements must be measurable. The next step for the team was to enter the metric and its target value for each engineering requirement. These metrics were placed below the engineering requirements matrix as shown in Figure B-20. For example, the travel iron weight must be less than 3 pounds.

Customer Need	Relative Weight	Weight	Weight	Volume	Length	Input power	Press time	Iron operating temp range	Visual on signal	On/off switch on handle	Operating life reliability	Drop test	Operating life	Shelf life	Unit production cost	Time to production	Development/Facilities/Tooling costs
								Engineering Requirements									
Lightweight	14	10	9														
Fits into carry-on bag	14	10		9	3												
Operates USA/European Power	11	8				9											
Able to quickly press clothes	10	7					9	3									
Able to adjust iron temperature	7	5						9									
Visual signal that iron is on	6	4							9								
Easy to turn on and off	6	4								9							
Does not require maintenance	8	6									9						
Durable	10	7										3	9	3	9			
Affordable	14	10														9	3	9
Engineering Requirements Target Values			<3.0 lbs	<100 cubic inch	<7.0 inch	110vac, 60Hz; 220vac, 50Hz	<2.0 minutes for pillow case	100-180F in 5 settings	Red on light on handle	On/off switch on handle	MCBF=2000 cycles	4 ft onto concrete floor	>200 cycles	5yrs (0-150F range)	<$4. (2020$s)	<2yrs	<$100K (2020$s)

FIGURE B-20 Travel Iron QFD Matrix with Requirement Target Values

Just as each customer need has a percentage relative weight, each engineering requirement needs a percentage relative weight. The weighting of each engineering requirement was determined by taking the sum of multiplying each customer need weight by the importance factor, as shown in the following equation:

Requirement weighting = Σ (customer need relative weighting)$_i$ x (importance factor)$_i$

For each requirement, this number is entered below the target requirement value. The relative weight of each engineering requirement is then calculated and listed in the following row. Figure B-21 shows the requirements weighting results for the travel iron.

Customer Need	Relative Weight	Weight	Weight	Volume	Length	Input power	Press time	Iron operating temp range	Visual on signal	On/off switch on handle	Operating life reliability	Drop test	Operating life	Shelf life	Unit production cost	Time to production	Development/Facilities/Tooling costs
								Engineering Requirements									
Lightweight	14	10	9														
Fits into carry-on bag	14	10		9	3												
Operates USA/European Power	11	8				9											
Able to quickly press clothes	10	7					9	3									
Able to adjust iron temperature	7	5						9									
Visual signal that iron is on	6	4							9								
Easy to turn on and off	6	4								9							
Does not require maintenance	8	6									9						
Durable	10	7									3	9	3	9			
Affordable	14	10													9	3	9
Engineering Requirements Target Values			<3.0 lbs	<100 cubic inch	<7.0 inch	110vac, 60Hz;220vac, 50Hz	<2.0 minutes for pillow case	100-180F in 5 settings	Red on light on handle	On/off switch on handle	MCBF=2000 cycles	4 ft onto concrete floor	>200 cycles	5yrs (0-150F range)	<$4. (2020$s)	<2yrs	<$100K (2020$s)
Weight= Need Relative Wheight * Matrix Relationship			126	126	42	99	90	93	54	54	102	90	30	90	126	42	126
% Relative Weight			10	10	3	8	7	7	4	4	8	7	2	7	10	3	10

FIGURE B-21 QFD for Travel Iron with Requirements Weighting Added (Example is highlighted in gray)

As shown in the example of Figure B-21 for the travel iron, the operating life reliability is important for two customer needs: (1) does not require maintenance and (2) durable. Table B-19 shows how the weighting for this requirement was determined.

Table B-19 Requirement Weighting Calculation for Operating Reliability Target Value

Customer Need	Customer Need Weighting	Importance Factor	Engineering Requirement Weighting Subtotal
Does not require maintenance	8	9	72
Durable	10	3	30

The steps of calculating the relative weight of the "operating life reliability" is summarized in the following lines:

- Engineering requirement weighting total = 72 + 30 = 102
- The total weight is the sum of the individual weights = (126 + 126 + 42 + 99 + 90 + 93 + 54 + 54 + 102 + 90 + 30 + 90 + 126 + 42 + 126) = 1290.
- The percentage relative weight = requirement weight/total weight = (102/1290) × 100 = 8 for the operating life reliability requirement.

The final step in creating the QFD house of quality is to show the relationships among the engineering requirements. This is done by adding a triangular matrix at the top of the QFD chart. This provides the HOQ with a roof. If there is a strong relationship between two requirements, a filled-in circle (•) is used. If the relationship is weak, an open circle (o) is used.

Figure B-22 shows the QFD roof for the travel iron example. In this case, the length is strongly related to volume. It is important to note that unit production cost is a function of all the other requirements.

The final QFD house of quality chart for the travel iron example is presented in Figure B-23.

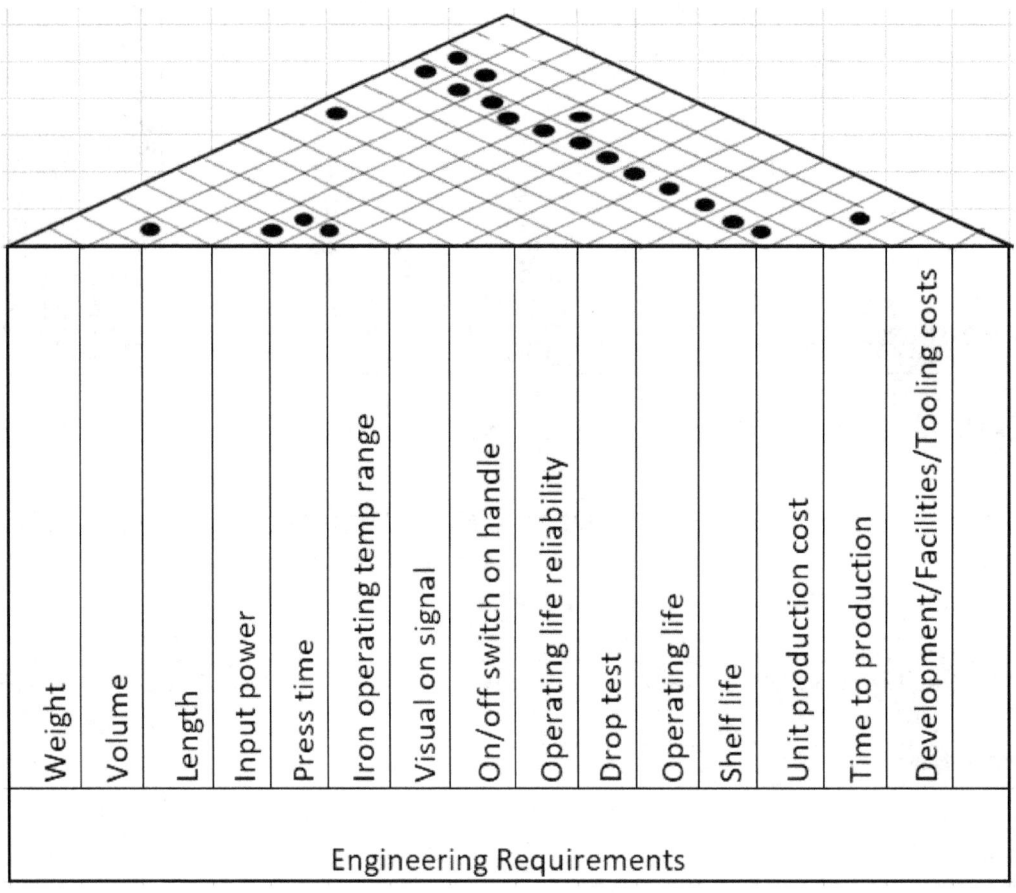

FIGURE B-22   QFD Requirements Matrix for Travel Iron Example

102 | Product Design and Development Handbook

Customer Need	Relative Weight	Weight	Weight	Volume	Length	Input power	Press time	Iron operating temp range	Visual on signal	On/off switch on handle	Operating life reliability	Drop test	Operating life	Shelf life	Unit production cost	Time to production	Development/Facilities/Tooling costs
								Engineering Requirements									
Lightweight	14	10	9														
Fits into carry-on bag	14	10		9	3												
Operates USA/European Power	11	8				9											
Able to quickly press clothes	10	7					9	3									
Able to adjust iron temperature	7	5						9									
Visual signal that iron is on	6	4							9								
Easy to turn on and off	6	4								9							
Does not require maintenance	8	6									9						
Durable	10	7										3	9	3	9			
Affordable	14	10													9	3	9
Engineering Requirements Target Values			<3.0 lbs	<100 cubic inch	<7.0 inch	110vac, 60Hz; 220vac, 50Hz	<2.0 minutes for pillow case	100-180F in 5 settings	Red on light on handle	On/off switch on handle	MCBF=2000 cycles	4 ft onto concrete floor	>200 cycles	5yrs (0-150F range)	<$4. (2020$s)	<2yrs	<$100K (2020$s)
Weight= Need Relative Wheight * Matrix Relationship			126	126	42	99	90	93	54	54	102	90	30	90	126	42	126
% Relative Weight			10	10	3	8	7	7	4	4	8	7	2	7	10	3	10

FIGURE B-23 HOQ for Travel Iron Example

B.2.8 Requirements Matrix and Goal Function

The next step for establishing the example project's engineering requirements is to create a requirements validation matrix, as shown in Table B-20. At the end of each subsequent development phase, the status of how the evolving design is meeting each engineering requirement will be entered into this matrix. The goal of the project is to have the final design be validated by the methods indicated in the matrix.

The travel iron team decided to use unit production cost as the goal function. This means that the production unit needs to cost no more than $4 (2020 dollars), but it is desirable to reduce the UPC as much as possible below this value. It is important to remember that the marketing strategy has imposed two key constraints on the design that may affect the UPC.

The first constraint is that the unit must be United States-made. The team is defining United States-made to be that the unit is assembled in the United States and that more than 90 percent of the UPC is made up of United States-made components and U.S. assembly labor costs.

Table B-20 Validation Matrix for the Travel Iron Example

Engineering Requirement and Target Value		Validation Method	Status
Weight	<1 lbs	Demonstration with scale	
Volume	< 100 cubic inches	Demonstration with scale	
Length	< 7.0 inches	Demonstration with scale	
Input Power	60 Hz 110-120 vac; 50 Hz 220-240 vac	Test by press cycle with power input and time	
Press Time	< 2.0 minutes for pillow case	Test and measure time	
Iron Operating Temp. Range	100-180 deg F in 5 settings	Test with thermocouples	
Visual On Signal	Red on-light located on handle	Inspection	
On/Off Switch on Handle	On/Off Switch on Handle	Inspection	
Operating Life Reliability	2000 cycles	Analysis—Based on FMEA failure modes	
Drop Test	4 feet onto concrete floor	Test by droping and then performing press test	
Operating Life	> 200 cycles	Test by cycling test rig	
Shelf Life	5 years (0-150 deg F range)	Analysis of materials	
Unit Production Cost	< $4.00 (2020 dollars)	Analysis based on production methods	
Time to Production	< 2 years	Analysis of similar projects	
Development/Tooling/Facility Cost	< $100k (2020 dollars)	Analysis of similar projects	
Initial Prototype Project Materials Cost	< $600 (2020 dollars)	Demonstration by tracking all manufacturing and assembly material and supplier costs	

Note: Avoid destructive testing, e.g., drop test, for single or few prototype item(s) when only limited resources are available. Use analysis for validation, when possible.

B.2.9 Preparing the Draft of Section 4 of the Final Report

As soon as the team completes the process of identifying the engineering requirements, Section 4: Engineering Requirements of the final report should be written. A detailed outline for this section is provided in Section C of this handbook.

B.2.10 Example Travel Iron Team Report for Week 6

The team completed their VOC research and identified the customer needs. The team then used the QFD process to translate customer needs into measurable engineering requirements. This work was documented in the project notebook. In addition, the team completed the draft of Section 4 of the final report. On Friday of Week 6, the team met with the sponsor to provide an update on the team's effort to finalize the engineering requirements.

MODULE 12

Exploring the Design Space

OVERVIEW
The team explores the design space and arrives at three candidate concepts. These are defined by configuration block diagrams, sketches and design narratives.

LEARNING OBJECTIVES
- Understand how to explore the design space.
- Demonstrate the ability to identify and define candidate designs.

PRE-LECTURE ASSIGNMENT
- Read sections B.2.11–B.2.15.
- Each team member prepares a paragraph on how the design space was explored.
- Each team member identifies and defines three candidate concepts.
- Each team member brings hard copy of the narrative and candidate concept descriptions to the lecture.

POST-LECTURE ASSIGNMENT
- Conduct team meetings to complete the following tasks: 1) Review each team member's method of exploring the design space and their candidate concepts and (2) determine if the design space has been adequately explored and, if not, what further tasks should be done. (3) Team members arrive at a final set of three candidate concepts and define them with sketches, block diagrams, and narratives.
- Conduct additional work and team meetings if the design space needs further exploration.

TEAM DELIVERABLES
- Final set of three candidate design concepts of the product. Define each candidate with sketches, block diagrams, and narratives.
- Minutes of the team meetings.
- Team activity as documented in the team notebook.

B.2.11 Updating the Preconcept Function Block Diagram

Once the product's engineering requirements are defined, the team can start the process of selecting a conceptual design. As discussed in Section A of the handbook, design is an iterative process. Many of the tools used in design are continually updated as the design becomes more refined. This is certainly the case with the function block diagram. The team needs to revisit the function block diagram prepared in the proposal phase and update it so that there are one or more functions to satisfy each of the engineering requirements. Figure B-24 illustrates this point by presenting a function block diagram for the portable travel iron example.

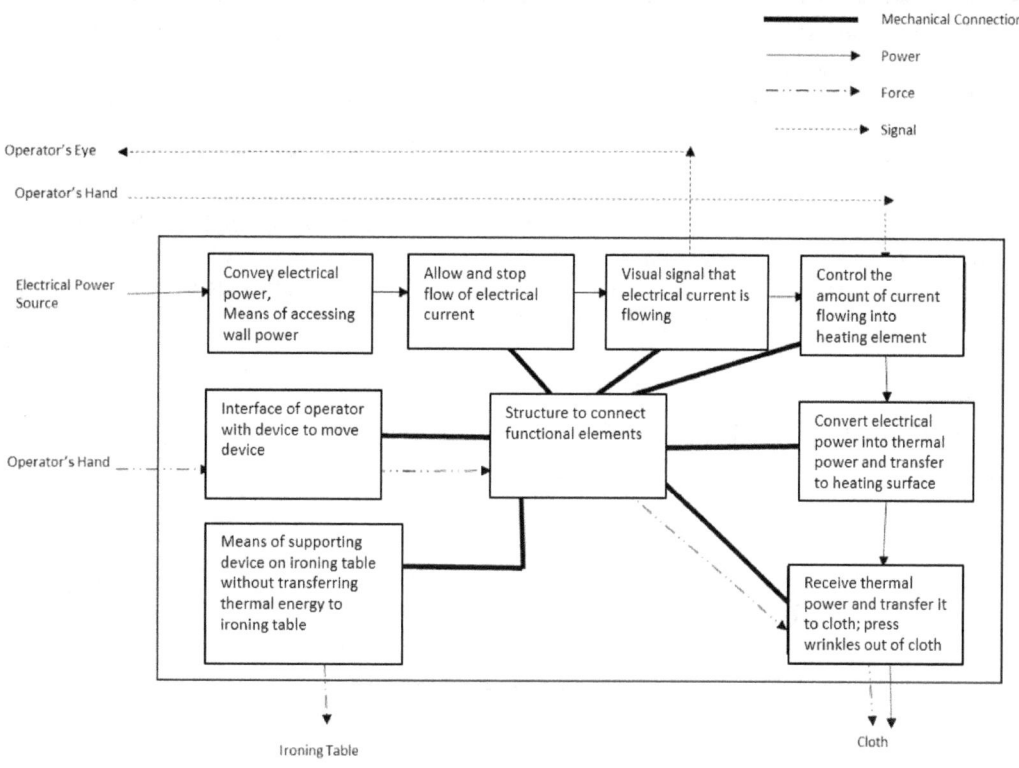

FIGURE B-24 Function Block Diagram for the Conceptual Design of Travel Iron

B.2.12 Exploring the Design Space

In the past, most companies were happy if the project team arrived at a design that met the customers' list of needs. However, the increased competition in today's markets have caused companies to avoid asking, what is a design that meets the customer's needs? Instead, these companies ask, what is the design that <u>best</u> meets the customer's needs? There is always the fear that even if the company has a good solution to the customer's need, there may be even a better solution in the design space, and a competing firm may

> **Creating Value Entrepreneurial Mindset:**
> *Curiosity*
>
> (EM)-(b)

find it. If the team is going to try to approach the optimum value proposition, they must be curious and explore the entire design space as much as possible within the given constraints of schedule and labor budget.

A key item that sponsors look for during conceptual design reviews is how well the projects have explored their design spaces. There is the tension between taking a lot of time to explore the design space and staying within the labor hours available for this task. The most important point here is to not embrace the first plausible concept, but instead to spend a concentrated effort on exploration. Table B-21 lists important questions that the team must answer as they explore the design space.

Table B-21 Exploring the Design Space Questions

- Did the team avoid just settling for the original concept presented in the proposal without exploring the design space?
- Did the team do a comprehensive literature search to understand what has already been done?
- Did the team conduct at least three separate brainstorming sessions to consider various conceptual approaches?
- Did the team allow adequate time between brainstorming sessions to allow the creative subconscious to identify potential approaches?
- Does the team have a written description of their design space exploration and a rationale for why it is adequate for the scope of the project?

It is important to remember that this exploration process is not linear. As shown in Figure B-25, the path will be circuitous. Often a filtering process will expand the number of concepts rather than reduce them. There will be blind alleys and frustrating periods of investigation without arriving at any new concepts. Discipline in following the process is required. Creativity is hard work! Eventually this process will reduce the number of concepts to a manageable number for detailed evaluation.

B.2.13 Using Morphological Analysis to Identify Candidate Conceptual Designs

A morphological analysis starts by listing the product functions. Then research is done to identify optional components for providing this function. These options are listed next to their respective function. Figure B-26 shows a generalized form of the morphological chart.

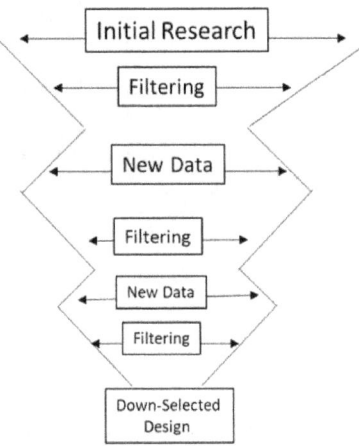

FIGURE B-25 The Conceptual Design Exploration and Filtering Process Is Not Linear

	Generalized Morphological Analysis Table		
Function	Options		
Function 1	Option 1	Option 2	
Function 2	Option 1	Option 2	
Function 3	Option 1	Option 2	Option 3
Function 4	Option 1	Option 2	
Function 5	Option 1	Option 2	
Function 6	Option 1	Option 2	Option 3

FIGURE B-26 Generalized Form of Morphological Chart

Once the options are identified, this chart can be used to map out different combinations of options that will make up the initial set of candidate product concepts. From the list of initial candidates, the team needs to select three or four final candidate concepts. Figure B-27 shows the entire morphological analysis process, including the selection of the final conceptual design using a Pugh comparison matrix. The figure shows that five candidate product concepts have been identified in the first round of the morphological analysis. A critical evaluation of the relative feasibility of the first-round candidates resulted in the discard of two first-round candidates to end up with three final-round candidates. The Pugh decision matrix method is used to compare the final-round candidates and select the final conceptual design.

The Pugh comparison chart will be described in the next section. But first, a morphological chart for the example travel iron will be discussed. Figure B-28 shows the morphological chart for the travel iron. For each function, the design team performs research into ways

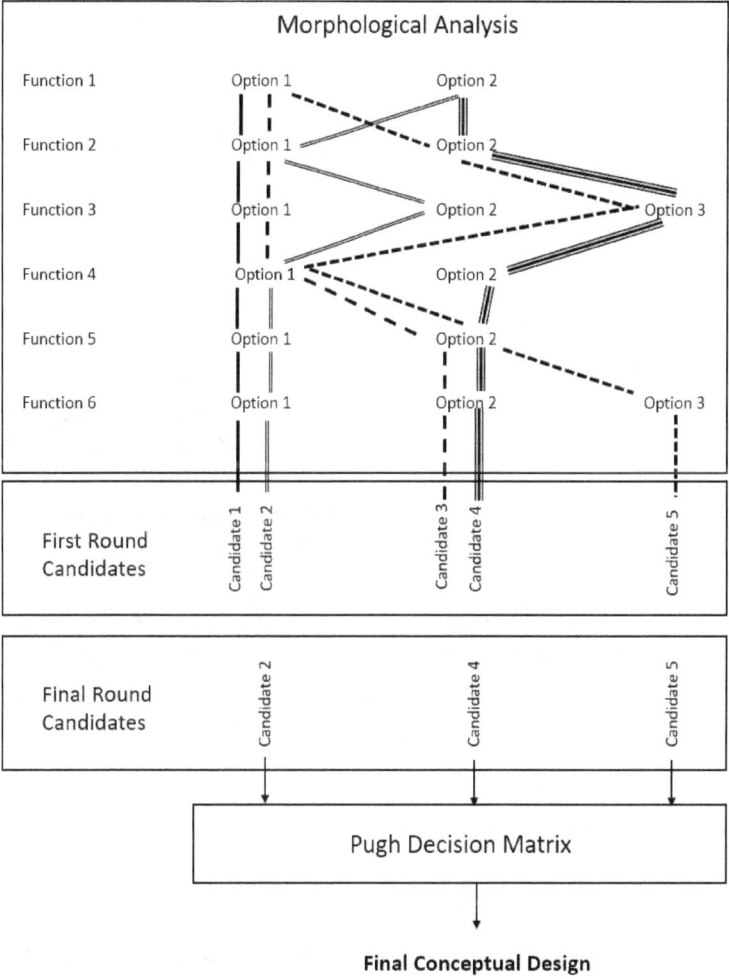

FIGURE B-27 Generalized Process for Selecting the Final Conceptual Design

Module 12 Exploring the Design Space | 109

of accomplishing this function. Doing adequate research—i.e., exploring the entire design space—is an essential part of this process.

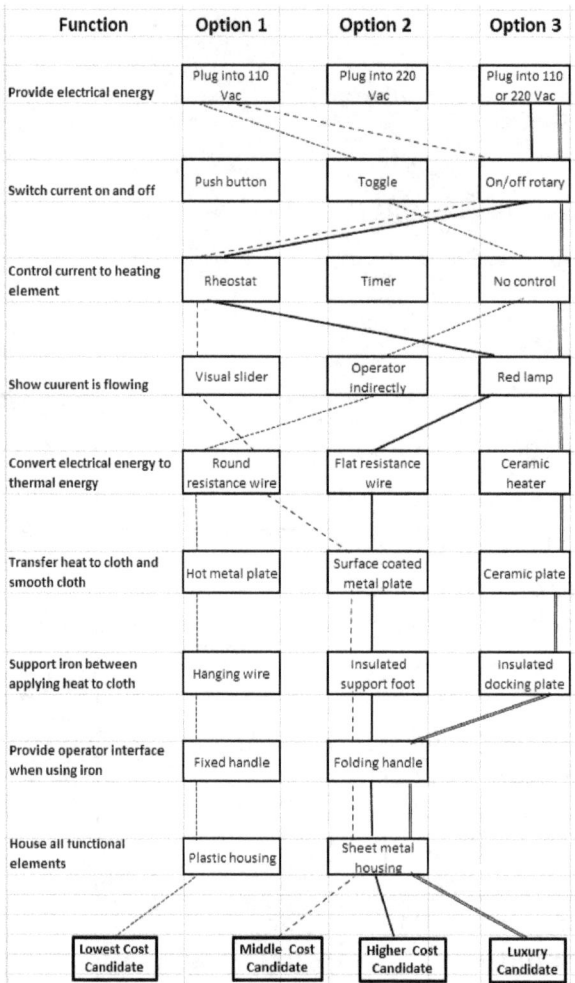

FIGURE B-28 Morphological Chart for the Travel Iron

Once the options for each function are identified, the team needs to decide how the different options should be combined to arrive at candidate product concepts. In the case of the travel iron, the team decided to identify candidate product concepts based on their production cost. The following four candidate configurations were selected:

- Candidate 1: Lowest-cost design
- Candidate 2: Middle-cost design
- Candidate 3: Higher-cost design
- Candidate 4: Luxury design

The lowest-cost design candidate selected the least cost component for each product function. The luxury design candidate selected the most costly components to both product and develop. The team thought that some potential customers would like the luxury nature of having a ceramic based iron because high-end customers have displayed an interest in ceramic knives as opposed to metal ones.

B.2.14 Selecting Candidate Conceptual Designs

The goal of the design exploration task is to arrive at three or four promising concepts for further evaluation. This process of going from a large number of initial concepts to just a few requires a filtering process by the team. For each potential concept, the team needs to consider how well it satisfies the functions identified in the updated function block diagram. There are several methods available for this filtering process. A simple voting system where each team member rates each concept from 1 to 10 is often used. The concepts (usually three or four) with the highest number of points among the team are then selected for further evaluation.

It is important that the concept exploration and filtering process be described in detail during the conceptual design review and also in the final report. The team must convince the reviewers that the team has done an adequate job of identifying the three or four most promising candidate concepts for further evaluation.

In the case of the travel iron, the team discussed the four initial candidate concepts coming out of the morphological analysis. They decided that the lowest-cost candidate although inexpensive, did not adequately address all the functions. The results of this discussion and decision was documented in the team meeting report and placed into the team project notebook.

B.2.15 Defining the Candidate Conceptual Designs

In order to adequately evaluate the candidate conceptual designs, each design needs to be further defined in terms of configuration and performance. As a minimum, the following items are needed:

- Configuration block diagram
- Sketch
- Narrative describing the concept and how it works

B.2.15.1 Configuration Block Diagram
The configuration block diagram replaces the functions in the function block diagram with specific hardware components that provide these functions. The relationship between components is the same as in the function block diagram. Energy, mass, and/or signals pass from one component block to another.

B.2.15.2 Product Conceptual Sketch
Most people are visual learners. A simple sketch is an effective means of defining a concept. Sketching is usually not part of the engineering curriculum, but it is a skill that is critical in the early phases of product design. It is a skill that every engineer should develop. Creating sketches during the capstone project is an important learning experience. Sketches are freehand drawings. Straight lines don't have to be perfectly straight. Approximation is all that is needed to convey the concept. However, sketches should be labeled, and key dimensions (at least overall dimensions) should be provided. Team members can improve their sketching skills using the vast engineering graphics free-hand sketching tutorials on the internet.

B.2.15.3 Concept Narrative

In addition to a sketch, the concept must be defined through a narrative that describes what the product looks like and how it performs. Any special features that enable it to meet the engineering requirements should also be discussed.

In a two-semester capstone course, there is not much time to perform analyses and POC on each of the candidate concepts. For projects with more resources, more analyses and POC testing should be accomplished to add more information to the concept narrative.

MODULE 13

Selecting and Analyzing the Conceptual Design

OVERVIEW

The use of a Pugh matrix to compare candidate concepts is introduced. This is followed by discussions of how analyses and/or testing should be used to determine how well the selected final conceptual design meets the engineering requirements. The difference between the production prototype and the capstone prototype is presented.

LEARNING OBJECTIVES

- Understand and apply the Pugh matrix to compare candidate concepts.
- Understand the general guidelines for conducting analyses and testing.
- Demonstrate the ability to follow the guidelines as analyses and testing are accomplished to evaluate the candidate concepts and arrive at a final conceptual design.

PRE-LECTURE ASSIGNMENT

- Read sections B.2.16–B.2.20 and appendices BA3 and BA4.
- Each team member prepares a Pugh matrix for evaluating the candidate concepts and brings a hard copy to the lecture. Analyses and testing may be required.
- Each team member prepares a written plan for analyzing and/or testing the final project concept to determine how well the concept meets the engineering requirements. Hard copies of these plans are due at the beginning of the lecture.

POST-LECTURE ASSIGNMENT

- Conduct a team meeting to 1) review team member inputs and then prepare a team Pugh matrix for comparing the project candidate concepts and (2) prepare team analysis and testing plans for the final concept.
- Each team member executes their part of the analysis and testing plans.
- Conduct a second team meeting to discuss the analyses and testing results and prepare a requirements matrix for the final conceptual design.

TEAM DELIVERABLES

- Team final conceptual design selected from the three candidate concepts. Define the final conceptual design with sketches, block diagrams, and narratives.
- Team analysis and testing plans for the final concept.
- Minutes of the team meetings.
- Team activity as documented in the team notebook.

B.2.16 Selecting the Final Conceptual Design

Once the candidate conceptual designs have been chosen and defined, it is time to compare these candidates in order to select a final conceptual design. The comparison factors are the engineering requirements with the weighting factors from the QFD house of quality analysis. The comparison (often called a *trade study*) uses a Pugh matrix in which one of the concepts is selected as a baseline.

The travel iron example can be used to demonstrate the proper use of the Pugh matrix method. As shown in Figure B-29, the engineering requirements for the travel iron are listed in the left column. The candidates are then listed in columns to the right of the requirements. For each candidate, there are two columns. The first is a ranking of how well the candidate meets each of the requirements. The baseline candidate scores a 3 for all requirements. The other candidates are scored from 1 to 5, where 1 is strongly does not meet the requirement compared to the baseline candidate and 5 is strongly exceeds the requirement as compared to the baseline candidate. The column next to the ranking is the weighted ranking, which is the product of the requirement weight and ranking. For example, for the UPC requirement, the rankings for candidates A, B, and C are 3, 2.5, and 1, respectively. The requirement weighting is 10, so the weighted rankings for candidates A, B, and C are 30, 25, and 10, respectively.

The rationale for each ranking must be a team decision and be recorded in the team project notebook. For example, the rationale for Candidate B being 2.5 is that this design has an electrical component that will work on either 110 Vac (alternating current voltage) or 220 Vac as compared to Candidate A, which has a lower-cost component that only works on 110 Vac. Candidate C scores only a 1 because it has a number of components more expensive than those used in Candidate A. For example, the ceramic components are more expensive.

The candidate with the highest total weighted rankings is the preferred candidate. In the case of the travel iron, Candidate B is the preferred one. Of course, the team has the opportunity of continuing to refine the final conceptual design if the team feels this is necessary.

In the case of the travel iron, the team did indeed need to do more work to arrive at a final conceptual design. In parallel with doing the Pugh analysis, some of the team members were putting together an initial UPC estimate. As discussed in more detail in Section B.2.20, the team concluded that their UPC requirement was too optimistic. The UPC analysis for Candidate B resulted in a UPC of greater than $10 per unit. As shown in Figure B-30, the team decided to use the following lower-cost items from the eliminated low-cost candidate in the final conceptual design:

- Round heating element
- Fixed handle
- Plastic housing

By making these changes, the travel iron team believes it is possible to achieve a UPC requirement of not $4 per unit but instead $10 per unit.

		Candidate A Baseline-Middle Cost		Candidate B Higher Cost		Candidate Luxury	
Engineering Requirement	Weighting*	Ranking	Weighted Ranking	Ranking	Weighted Ranking	Ranking	Weighted Ranking
weight<3.0 lbs	10.0	3	30.0	2.7	27.0	1	10.0
volume<100 cubic inch	10.0	3	30.0	2.7	27.0	2.5	25.0
length<7.0 inch	3.0	3	9.0	3	9.0	3	9.0
110vac, 60Hz;220vac, 50Hz	8.0	3	24.0	5	40.0	5	40.0
press time<2.0 minutes for pillow case	7.0	3	21.0	3	21.0	2	14.0
temperature range: 100-180F in 5 settings	7.0	3	21.0	3	21.0	2	14.0
Indication of current	4.0	3	12.0	5	20.0	5	20.0
On/off switch on handle	4.0	3	12.0	3	12.0	3	12.0
MCBF=2000 cycles	8.0	3	24.0	2.5	20.0	2	16.0
drop test: 4 ft onto concrete floor	7.0	3	21.0	3	21.0	1	7.0
service life >200 cycles	2.0	3	6.0	3	6.0	3	6.0
shelf life: 5yrs (0-150F range)	7.0	3	21.0	3	21.0	3	21.0
unit production cost<$4. (2020$s	10.0	3	30.0	2.5	25.0	1	10.0
development time<2yrs	3.0	3	9.0	3	9.0	1	3.0
dev/facility/tooling cost<$100K (2020$s)	10.0	3	30.0	3	30.0	1	10.0
Total Weighted Ranking			**300.0**		**309.0**		**217.0**
Total	100.0						
* Weighting from QFD analysis							

FIGURE B-29 Concept Trade Study for Travel Iron Example Using a Pugh Matrix

Eliminated Candidate Lowest Cost	Candidate A Middle Cost	Candidate B Higher Cost	Candidate C Luxury Candidate
110 Vac	110 Vac	110/220 Vac	110/220 Vac
Toggle	On/off rotary	On/off rotary	On/off rotary
No control	Rheostat	Rheostat	Rheostat
Operator senses heat	Current visual slider	Red lamp on indicator	Red lamp on indicator
Round heating element	Round heating element	Flat heating element	Ceramic heating element
Metal sole plate	Coated sole plate	Coated sole plate	Ceramic sole plate
Hanging wire	Insulated support foot	Insulated support foot	Insulated docking plate
Fixed handle	Folding handle	Folding handle	Folding handle
Plastic housing	Sheet metal housing	Sheet metal housing	Sheet metal housing
	Pugh Score = 300	Pugh Score = 309	Pugh Score = 217

FIGURE B-30 Travel Iron Final Conceptual Design Components Are Circled

B.2.17 Conceptual Design Analyses and Testing

In order to determine how well the candidate concepts are meeting the engineering requirements, a combination of analysis and POC testing is required. A general discussion

EM-(h)

of how to use engineering analyses in the design process is provided in Subsection B.2.18. In a like manner, Subsection B.2.19 provides a general discussion of how to conduct testing.

B.2.18 General Guidance on Conducting Design Analyses

A major engineering skill set is the ability to do design analyses. The level of analysis spans from back-of-the-envelope stress calculations to computerized performance models of large systems. It is vital that the design engineer not only know how to perform a wide range of analyses but also be able to select the proper level of analysis for the design issue being addressed. When confronted with a design issue that requires analysis, the engineer should first ask and answer the questions listed in Table B-22. Table B-23 lists the key points the team should consider as they plan the analyses to be performed during each PPD process phase. Table B-24 presents the documentation format for analyses that should be used. A more detailed discussion of design analysis is provided in Appendix BA3.

Table B-22 Questions to Answer When Initiating a Design Analysis

1) Is there an overall analysis plan that considers all the analyses required, and have the proper available resources been allocated for each of these analyses?
2) Has the engineering issue been reduced to a specific problem statement with listed assumptions?
2) Has the minimum level of acceptable answer precision been determined?
3) Is the scope of the answer only a point estimate, or does it require a sensitivity study of the point estimate to variations in some or all the parameters involved?
3) Is the level of analysis appropriate for the level of answer precision required?
4) Have the procedures used, including software programs, been validated?
5) Has the analysis undergone a "sanity check" using an alternate analysis method?
6) Is there a plan to document the analysis in such a way that another engineer unfamiliar with the issue could replicate the analysis?

Table B-23 Key Points Concerning Design Analyses for the Capstone Project

- As a minimum, analyses should address each of the engineering requirements.
- The lowest level of analysis that answers the design question should be used to conserve resources.
- At the beginning of each PPD phase, an analyses plan should be developed by the team to find the best way to use the limited project resources to adequately cover all the analysis needs.
- A graphic should be included in the analysis problem statement.
- Analysis documentation is vital in design work. All analyses should use the format given in Table B-24.
- All analyses should be placed in the project notebook.
- Results of each analysis should be summarized in the requirements validation matrix (this is a living document that resides in the project notebook).

> **Table B-24 Analysis Documentation Format**
>
> - Issue: The issue is described with a graphic included.
> - Problem statement: Proper problem-solving techniques are used to arrive at a well-articulated problem statement that effectively addresses the issue, with a graphic included.
> - Approach: The approach to solving the problem in terms of simplifying assumptions and equations to be used should be covered. The approach also describes what kind of answer is needed (i.e., a ballpark solution, a precise and accurate "home plate" answer, or something in between). The method of analysis, such as hand calculations, computer-aided analysis, etc., should be included. A flowchart may be helpful to show the analysis steps.
> - Defining Equations: The defining equations are listed, and each variable is defined.
> - Assumptions: The assumptions are to be listed, with the rationale for each assumption included.
> - Calculations: The calculations must be documented. If several calculations use the same equations, the top-level documentation should include a set of sample calculations. The remaining calculations should be archived in an appendix or the project notebook.
> - Results: The results must be displayed in one or more tables or graphs. These graphics must be clearly and completely explained.
> - Conclusions: The conclusions drawn from the results are listed and explained.
> - Solution: The solution to the problem statement is based on the conclusions.
> - Recommendations: What should the team do as a result of the analysis? Should the solution be implemented? If so, how and when?

B.2.19 General Guidance on Conducting Testing

Because so much of engineering tool development has been in the area of computer simulation, many students be under the impression that testing now plays a small role in a product's design. In general, this is not the case. Testing should be prominent during all phases of design. In the early stages, there are often times when the designer wants to integrate a new concept into the design, but its viability is in question. Rather than just performing analyses and waiting until the prototype testing in Phase 5P, the designer should conduct simple POC tests during all project phases. These POC tests are an excellent method of retiring development schedule and budget risk. For even simple tests, the engineer needs to follow proper testing protocol, including test planning, test procedures, test data collection, test data analysis, and test documentation. During Phase 4P: Prototype Detailed Design, a formal plan for prototype testing in Phase 5P should be prepared and presented at the Phase 4P design review. A more detailed discussion of testing is provided in Appendix BA4.

B.2.20 Example Travel Iron Team Report for Week 7

The team explored the design space and identified a number of design features and component types for the travel iron. The team then used a morphological chart to identify four candidate designs. The team eliminated the lowest-cost design because it did not address the other requirements as well as the remaining three candidates did. These candidates were evaluated using a Pugh matrix. Candidate B was selected because it satisfied all the requirements and had a UPC lower than Candidate C, the luxury design. However, a parallel effort by other team members performing a UPC analysis concluded that the UPC requirement of $4 per unit needed to be changed to $10 per unit. The analysis also concluded that lower UPC elements from the low-cost candidate needed to be used to keep the UPC at or below the new $10-per-unit requirement. Figure B-31 summarizes the UPC analysis.

Initial UPC Summary
1. Used online injection molding calculator to estimate costs for top shell using ABS plastic (https://www.custompartnet.com/estimate/injection-molding/).
        a. Material and production cost is approximately $2.00
        b. Tooling is approximately $11,000
2. Other costs based on quoted prices on Alibaba internet site and engineering judgement.
3. Initial UPC estimate is $10.00/unit as given in the table below.

Item	Qty	Unit $	Qty $s
Top shell	1	2.00	2.00
Sole plate	1	1.00	1.00
Heating element	1	1.00	1.00
Insulation plate	1	0.40	0.40
Dual voltage device	1	1.50	1.50
Thermostat	1	0.30	0.30
Power cord	1	0.32	0.31
Misc. Hardware: well cover, screw covers, well potting material, LED washer, LED screw, shell screws and glue	1 LOT		0.20
Misc. Electrical: LED harness, wire nuts, wiring, flux and solder	1 LOT		0.19
Misc. plastic parts: window and temperature knob	1 LOT		0.10
TOTAL Components			7.00
Assembly labor			2.50
Packaging			0.50
TOTAL UPC			10.00

FIGURE B-31 Initial UPC Analysis Summary for Travel Iron

As part of this conceptual design process, the team accomplished numerous analyses following the guidance given in the handbook. Likewise, the team evaluated several irons by conducting simple POC ironing tests following the testing guidance provided in the handbook. Figure B-32 presents the requirements validation matrix with updated results in the status column. It should be noted that the weight requirement of <3 pounds was changed to <1 pound based on more recent VOC research.

Engineering Requirement		Validation Method	Status
Weight	<1 lbs	Demonstration with scale	No weight analysis yet
Volume	< 100 cubic inches	Demonstration with scale	Vol=5x3x4x80%=48 cu in
Length	< 7.0 inches	Demonstration with scale	Sketch is 5 inches
Input Power	60 Hz 110-120 vac; 50 Hz 220-240 vac	Test by press cycle with power input and time	Assuming device that switches from 110 to 220 Vac within size and weight limits
Press Time	< 2.0 minutes for pillow case	Test and measure time	Tests with existing irons pass
Iron Operating Temp. Range	Three settings: 275 deg F, 375 deg F, 400 deg F	Test rig A with IR thermometer	Thermostat allows temperature settings
Visual On Signal	Red on-light	Inspection	Not selected yet
On/Off Switch	On/Off Switch	Inspection	Part of thermostat
Operating Life Reliability	MCBF > 2000 cycles	Test by cycling test rig	Not analyzed yet
Drop Test	4 feet onto concrete floor	Test	Not analyzed yet
Operating Life	> 200 cycles	Test	By similarity with existing iron designs, components should be structurally adequate for desired operating life.
Shelf Life	5 years (0-150 deg F ambient temperature range)	Analysis	Materials being considered should have properties for suitable for shelf life.
Unit Production Cost	< $4.00 (2020 dollars)	Analysis	Initial UPC analysis results indicate UPC target should be $10.00 or less.
Time to Production	< 2 years	Analysis	Initial analysis by similarity to other small appliances, time to production is estimated to be one year.
Development /Tooling/ Facility Cost	< $100k (2020 dollars)	Analysis	Cursory analysis suggests cost may be higher than $100k.
Initial Prototype Project Materials Cost	< $600 (2020 dollars)	Demonstration by tracking all manufacturing and assembly material and supplier costs	Rigs and prototype estimated to be less than $600 based on Internet prices

FIGURE B-32 Updated Travel Iron Requirements Validation Matrix

MODULE 14

Completing Phase 2: Conceptual Design

OVERVIEW

In Module 14, the method of defining the final concept is presented. The commercialization topic of defining the concept's value is presented. Phase 2 is finalized.

LEARNING OBJECTIVES

- Understand and demonstrate the ability to define a conceptual design.
- Understand the concept of product value and apply that concept to the final conceptual design.
- Understand how the PPD process for product development fits into the larger scope of the Goldsmith Commercialization Model.
- Apply communications skills to produce a high-quality draft of Section 5 of the project final report.
- Analyze the work done in Phase 2 in order to complete the Phase 2 exit checklists.

PRE-LECTURE ASSIGNMENT

- Read sections B.2.21–B.2.31.
- Bring hard copies of the following to the lecture for your team meeting:
    - A list of the value streams provided by the final concept
    - A list of items not already in the project notebook
    - Based on prior research and engineering judgment, a break-even analysis of the conceptual product and a list of the assumptions

POST-LECTURE ASSIGNMENT

- Conduct a team meeting to 1) define the conceptual design in detail, 2) prepare the Phase 2 design review presentation, and 3) assign tasks to complete Section 5 of the project final report and complete the Phase 2 exit criteria checklists.
- Each team member conducts 360-degree team member peer evaluation.
- Team reconvenes to finalize the Phase 2 deliverables.

TEAM DELIVERABLES

- Final team conceptual design including the items in Table B-25
- Commercialization and value proposition for the customer and sponsor
- Section 5: Conceptual Design of the project final report
- DR2 presentation slides
- Phase 2 exit criteria checklists
- Minutes of the team meetings
- Team activity as documented in the team notebook

B.2.21 Defining the Final Conceptual Design

Once the final capstone conceptual design has been chosen, it needs to be clearly defined with the items listed in Table B-25.

Table B-25 Items to Define the Final Conceptual Design
• Narrative description   • Table of features and benefits   • Labeled sketch with overall dimensions   • Configuration block diagram   • Performance tables (weight, power output, UPC, etc.)   • Requirements validation matrix (include results from analyses and POC testing)

B.2.22 Commercialization

Commercialization of a product is the entire process of identifying a customer need, designing a product that meets that need, and profitably bringing that product to market in production quantities. As discussed in Section A of the handbook, the authors have selected the Goldsmith Commercialization Model to provide structure to the commercialization process. Commercialization of a product requires a multidisciplinary team of engineers, marketing experts, and those with business acumen. This handbook primarily focuses on the technical aspects of product development that can be covered in a two-semester capstone course. In industry, the capstone team would be part of a larger commercialization team that would have more resources to fully conduct the various commercialization tasks.

EM-(i)

EM-(j)

EM-(k)

Conceptual design, which is covered in Phase 2, fits in the investigation stage, which is the first stage of the Goldsmith model. The capstone team has identified a market need and used primary and secondary research to determine the VOC. Part of that research should provide an initial estimate of the market size and amount of money customers are willing to pay for the product. Based on these estimates, the team can use their engineering judgement to estimate the UPC, cost to develop the product, and cost to build and equip a production factory. Using these estimates, the team can begin to see whether the product has a good profitability value (i.e. the product has a reasonable payback time). These estimates will be refined during Phase 3: Preliminary Design.

EM-(l)

EM-(q)

Creating value is one of the key elements of the entrepreneurial mindset. As discussed previously, there are two sets of value to consider. Set 1 is value to the customer. The customer must find enough value in the product offering to purchase and be delighted with the product. Value is more than just an affordable selling price. It also includes other items that customers value, such as being environmentally friendly and being produced in a factory that is safe for its workers. Set 2 is value for the business venture.

Entrepreneurial Mindset: *Creating value*

The business venture must find enough value in developing and producing the product to meet its profitability, business growth, and sustainability goals.

B.2.23 Marketing Basics

EM-(m) This section provides the following key marketing concepts: definitions, types, segmentation, mix, and strategies.

B.2.23.1 Marketing Definitions

The following definitions are provided by the American Marketing Association (AMA)[1]:

Marketing is the activity, set of institutions, and processes for creating, communicating, delivering, and exchanging offerings that have value for customers, clients, partners, and society at large (approved 2017).

Marketing research is the function that links the consumer and public to the marketer through information—information used to identify and define marketing opportunities and problems; generate, refine, and evaluate marketing actions; monitor marketing performance; and improve understanding of marketing as a process. Marketing research specifies the information required to address these issues, designs the method for collecting information, manages and implements the data collection process, analyzes the results, and communicates the findings and their implications (approved 2017).

Brand is a name, term, design, symbol, or any other feature that identifies one seller's goods or service as distinct from those of other sellers. The [International Organization for Standardization] brand standards add that a brand "is an intangible asset" that is intended to create "distinctive images and associations in the minds of stakeholders, thereby generating economic benefit/values."[2]

B.2.23.2 Marketing Types

The AMA lists six types of marketing:

- **Influencer**—using individuals to drive a brand message
- **Relationship**—segmenting customers to build loyalty
- **Viral**—people pass along a marketing message
- **Green**—environmentally friendly products and/or production processes
- **Keyword**—placement of an ad when certain words are searched
- **Guerilla**—unconventional strategy to get maximum results from a minimum of resources

---

1 American Marketing Association. n.d. *Definitions of Marketing.* https://www.ama.org/the-definition-of-marketing-what-is-marketing/.

2 International Organization for Standardization, n.d. *Brand Evaluation – Principles and Fundamentals.* https://www.iso.org/obp/ui/#iso:std:iso:20671:dis:ed-1:v1:en

B.2.23.3 Marketing Mix

In 1960, E. J. McCarthy, in his book *Basic Marketing: A Managerial Approach*, introduced the 4P's of the marketing mix.[3] A brief discussion of each of these elements follows:

- **Product** is a group of functions and benefits in the form of a good and/or service that is capable of exchange or use.
- **Price** is the quantity of money needed to acquire a given quantity of goods and/or services.
- **Place** or distribution is the process of marketing and transferring products from producers to customers.
- **Promotion** is providing information to potential and existing customers about the product, brand, and company.

B.2.23.4 Marketing Strategy

Michael Porter states that there are the following five competitive forces that shape marketing strategy[4]:

- **Competitive rivalry** forces the organization to differentiate themselves from the competition. Some of these differentiators include selling price, quality, customer loyalty, etc.
- **Threat of substitution** forces the organization to consider other products that meet the customers' need. For example, the aircraft firm Boeing considers Zoom to be a threat to air travel for business meetings that can now be held face to face online.
- **Buyer Power** exists when there is more supply than demand. When buyer demand in a market decreases, a producer with a fixed production output must gain more market share to remain profitable. Strategies such as reducing production costs to lower the selling price or differentiating on quality or performance should be considered.
- **Supplier power** exists when the producer is dependent on a sole supplier for material or labor needed for production. Having more than one supply source and having the ability to substitute are strategies for resisting this force.
- **Threat of new entry** exists when a new competitor enters the market. Barriers to market entry include economies of scale, technology protection, special knowledge, and entry costs. For the project seeking to enter an existing market, the challenge is to overcome these barriers.

B.2.24 Business Opportunity Development

For the capstone project to evolve into a commercialized product, the team needs to consider how their product design will be produced. There are a number of options. The capstone team could start their own company, sell their product design to another firm, or license the right of a company to produce the product, with the team being paid a royalty for each unit sold.

It takes resources to turn a product idea into a commercialized product. If the project team decides to develop the product themselves, they will need working capital. This can come from individual savings, a loan from a bank, and/or a partnership with a joint venture firm.

---

3   McCarthy, E. Jerome. 1960. *Basic Marketing: A Managerial Approach*. Homewood, IL: Richard D. Irwin, Inc.
4   Porter, Michael E. 2008. The Five Competitive Forces That Shape Strategy. *Harvard Business Review* 86, no. 1 (January): 78–93.

Sometimes, a project team will take the product close enough to production that they can either sell or license the product along with the rest of the product development to another firm. Whatever entity ends up commercializing the product, they will want a clear business case that establishes the potential rewards for the risks they take in investing in the product.

B.2.25 Simple Business Model

EM-(o)

Figure B-33 presents a simple business model. In this model, the business takes resources such as raw materials and labor and converts them into a product that is sold into a market. This part of the process results in sales revenue. Some of this revenue is used to pay for the resource expenses and taxes. What is left is called *profit*. The money invested in the business for business startup, facilities, and equipment is called *capital*. Capital is provided by investors who are rewarded by receiving the profits. In simple terms, the return on investment is the profit minus the investment divided by the investment. The figure also shows that the business not only produces marketable products, but it also produces waste products. These can have a negative effect on the environment. The business operates in other aspects of the environment, such as government regulations. It should be noted that this model is dynamic. The market, environment, resources, and available capital are always in a state of flux, so the business must be nimble enough to adjust to these changing conditions.

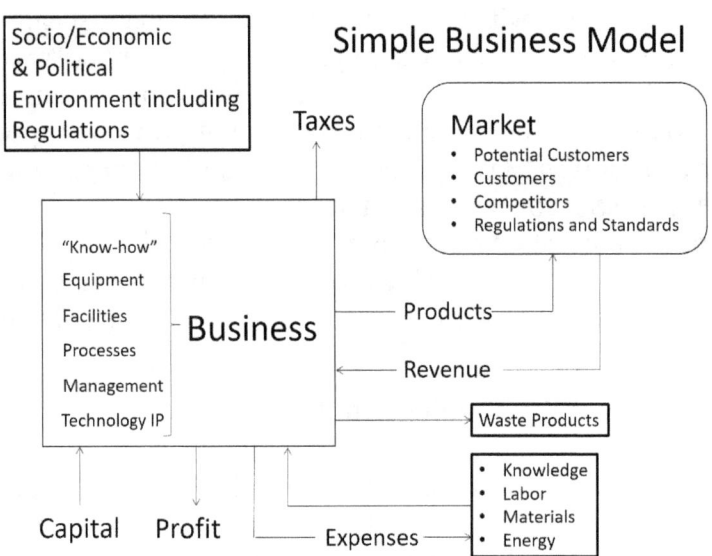

FIGURE B-33 Simple Business Model

B.2.26 Business Case

The British Association for Project Management uses the following definition: "A business case provides justification for undertaking a project, programme or portfolio. It evaluates the benefit, cost and risk of alternative options and provides a rationale for the preferred solution."[5]

In addition to addressing the market conditions and the strategy to enter the market, the business case describes the value streams that result from the project. A major value stream is the acceptable return to the investors. Examples of other value streams are customers buying an environmentally friendly product or customers seeing this product as an extension of an existing brand product line (customer loyalty).

B.2.27 Return-on-Investment Calculations

There are a number of methods financial analysts use to evaluate the investment potential of a proposed project. A major method involves determining the project's internal rate of return (IRR). This is done by creating a cash flow diagram that shows the investments and profits as a function of time. The IRR is the discount rate that yields a net present value (NPV) of these cash flows to be 0. The defining equation for this is given below:

$$NPV = 0 = \sum \frac{CF_i}{(1+IRR)n_i}$$

Where

NPV = net present value

CF_i = cash flow for period i

n_i = period i

IRR = internal rate of return

To illustrate how this equation is used, the following simple example is provided.

Company ABC produces and sells 100,000 units per year at a price of $100 per unit. The cost of production, overheads, and plant depreciation is $60 per unit. Taxes are $20 per unit. So the profit per unit for the company is equal to $100 - $60 - $20 = $20 per unit. The company's business case projects sales over three years of 20,000, 30,000, and 20,000 for years 1, 2, and 3, respectively. The company projects that it will take one year at a cost of $1,200,000 to design and develop the product, build the factory, hire and train the workers, and equip the factory. At the end of the third year of sales, the business is sold for a salvage value of $500,000. The IRR analysis for this scenario is given in Figure B-34. In this analysis, all cash flows are assumed to

---

5   Association for Project Management. n.d. *What Is a Business Case?* https://www.apm.org.uk/resources/what-is-project-management/what-is-a-business-case/.

occur on the last day of the year. The analysis shows that with the model input data given above, the IRR is about 21 percent. For many investors, this level of return for a project of this level of risk would be acceptable.

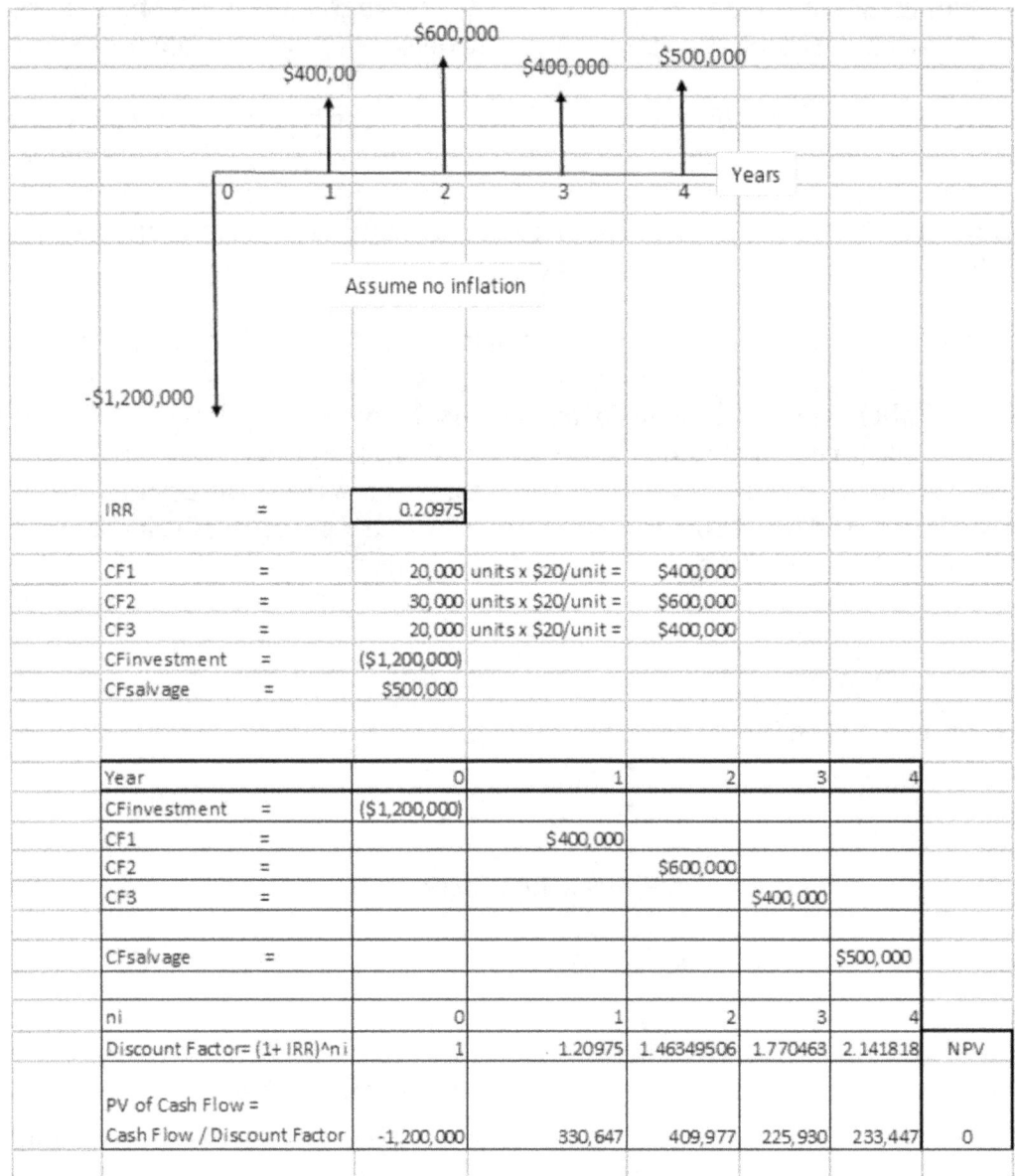

FIGURE B-34 Example Problem IRR Analysis

B.2.28 Phase 2 Conceptual Design Review (DR2)

The conceptual design review is a key activity. It has two main purposes: (1) It gives the team the opportunity to present their work to a group of experts who can identify areas of design improvement, and (2) it provides the project sponsors with the information they need to decide if they want to proceed with the project and fund the preliminary design phase.

The conceptual design review members should include the team, sponsors, and design experts. The design experts should be knowledgeable about the types of design problems associated with the project. Invitations to the design review along with an agenda should be provided to the review members several weeks before the review. About a week before the review, the team should send out a rough draft of the design review slides to help the reviewers prepare for the review.

It should be noted that for the capstone course, it may be difficult to have outside design experts attend the review. If this is the case, the course instructor serves as both the design expert and the project sponsor.

At the beginning of the design review, the team should hand out printed copies of the design review slides to the reviewers so that it is easy for them to take notes. The team should dress appropriately for the review. Every member of the team should participate in the presentation. It is important that the team practices the presentation to ensure that they stay within the time limits of the agenda. Table B-26 provides an outline for the conceptual design review slides.

Table B-26 Slides for Phase 2 Conceptual Design Review

1. Title slide—name of project, list of team members, picture of team
2. Customer need—scope of research, weighted list of customer needs
3. Engineering requirements—weighted list, from QFD and HOQ
4. Explore the design space—scope, brainstorming results, filtering
5. Candidate concepts—morphological chart to arrive at three candidate concepts of the production unit
6. Concepts trade study—Pugh matrix table used to select one final concept; any other considerations that result in even a better final selection
7. Final production concept—labeled and neat sketches to describe the final production concept
8. Capstone prototype concept—labeled sketches, changes in the production unit requirements
9. Commercialization—investigation phase, technical, market and business case
10. Analyses and testing—key results used to update the production unit requirements matrix
11. Overall project schedule
12. Detailed Phase 2 project schedule
13. Project labor chart—updated project labor chart
14. Go-forward plan—issues and preliminary design approach

The amount of material on a presentation slide is limited by the size of the room and audience. For large audiences, the slides need to be in large fonts to keep them easy to read even from the back of the room. For small group presentations, one can use smaller fonts; however, too much information on a slide may cause the audience to lose sight of the slide's main point.

For example, two methods of presenting the selection of the travel iron final concept using the Pugh matrix chart method are provided in Figure B-35. Assume there is time enough only for one slide on this subject. In Slide A, the team has put the entire matrix on the slide. Note how hard it is to read. There is so much information that the audience is studying the slide rather than listening to the key points being covered by the presenter. Slide B, on the other hand, is much easier to read and quickly understand. There is more time for the audience to listen to the speaker and read the slide. There is more information provided in Slide A, but the audience will actually process more information using Slide B. In addition, Slide B contains more types of information about the selection process. It provides a better overall story if only one slide can be used.

Slide A
- Hard to read
- Only shows Pugh scoring detail

Final Conceptual Design Selection

Engineering Requirement	Weighting*	Candidate A Baseline-Middle Cost		Candidate B Higher Cost		Candidate Luxury	
		Ranking	Weighted Ranking	Ranking	Weighted Ranking	c	Weighted Ranking
weight<3.0 lbs	10.0	3	30.0	2.7	27.0	1	10.0
volume<100 cubic inch	10.0	3	30.0	2.7	27.0	2.5	25.0
length<7.0 inch	3.0	3	9.0	3	9.0	3	9.0
110vac, 60Hz;220vac, 50Hz	8.0	3	24.0	5	40.0	5	40.0
press time<2.0 minutes for pillow case	7.0	3	21.0	3	21.0	2	14.0
temperature range: 100-180F in 5 settings	7.0	3	21.0	3	21.0	2	14.0
indication of current	4.0	3	12.0	5	20.0	5	20.0
On/off switch on handle	4.0	3	12.0	3	12.0	3	12.0
MCBF=2000 cycles	8.0	3	24.0	2.5	20.0	2	16.0
drop test: 4 ft onto concrete floor	7.0	3	21.0	3	21.0	1	7.0
service life >200 cycles	2.0	3	6.0	3	6.0	3	6.0
shelf life: 5yrs (0-150F range)	7.0	3	21.0	3	21.0	3	21.0
unit production cost<$4. (2020$s)	10.0	3	30.0	2.5	25.0	1	10.0
development time <2yrs	3.0	3	9.0	3	9.0	1	3.0
dev/facility/tooling cost<$100K (2020$s)	10.0	3	30.0	3	30.0	1	10.0
Total Weighted Ranking			**300.0**		**309.0**		**217.0**
Total	100.0						

* Weighting from QFD analysis

Slide B
- Easy to read
- Summarizes Pugh scores
- Shows candidate components
- Shows selected components

Final Conceptual Design Selection

Eliminated Candidate Lowest Cost	Candidate A Middle Cost	Candidate B Higher Cost	Candidate C Luxury Candidate
110 Vac	110 Vac	110/220 Vac	110/220 Vac
Toggle	On/off rotary	On/off rotary	On/off rotary
No control	Rheostat	Rheostat	Rheostat
Operator senses heat	Current visual slider	Red lamp on indicator	Red lamp on indicator
Round heating element	Round heating element	Flat heating element	Ceramic heating element
Metal sole plate	Coated sole plate	Coated sole plate	Ceramic sole plate
Hanging wire	Insulated support foot	Insulated support foot	Insulated docking plate
Fixed handle	Folding handle	Folding handle	Folding handle
Plastic housing	Sheet metal housing	Sheet metal housing	Sheet metal housing
	Pugh Score = 300	Pugh Score = 309	Pugh Score = 217

◯ Selected Component

FIGURE B-35 Example of How to Display Slide Data

Sometimes less is more. This is especially true when showing schedules, tables, or written explanations. When possible, use a picture instead of words. However, label or caption the picture to make sure the message is clear to the audience.

B.2.29 Documenting the Conceptual Design

It is important for the team to document the design process as it progresses. The best way to do this is to complete the appropriate sections of the final report during the applicable project phase.

B.2.30 Example Travel Iron Team Report for Week 7.5

The team was very busy this past half-week in completing Phase 2. The major efforts are described below.

B.2.30.1 Description of Final Conceptual Design

As shown in Figure B-36, the exterior design of the travel iron remained similar to the one presented in the proposal. The team focused on packaging the selected final internal conceptual design components. As shown in Figure B-37, the heating element is placed in a well that is part of the sole plate casting. The heating element is placed in the bottom of the well, and thermally conductive potting material is poured around it. A thin metal cover is used to cover the potting material. A plastic insulation cover is used to separate the outer shell from the sole plate. At this point, the team has not determined the actual 110/220 Vac switching device, but it was decided to locate it in the back of the iron on the plastic insulating cover. The team decided to locate the red lamp window near the front of the iron, where it is glued into the outside shell. A sketch of the sole plate casting is given in Figure B-38. A detailed weight analysis has not been completed.

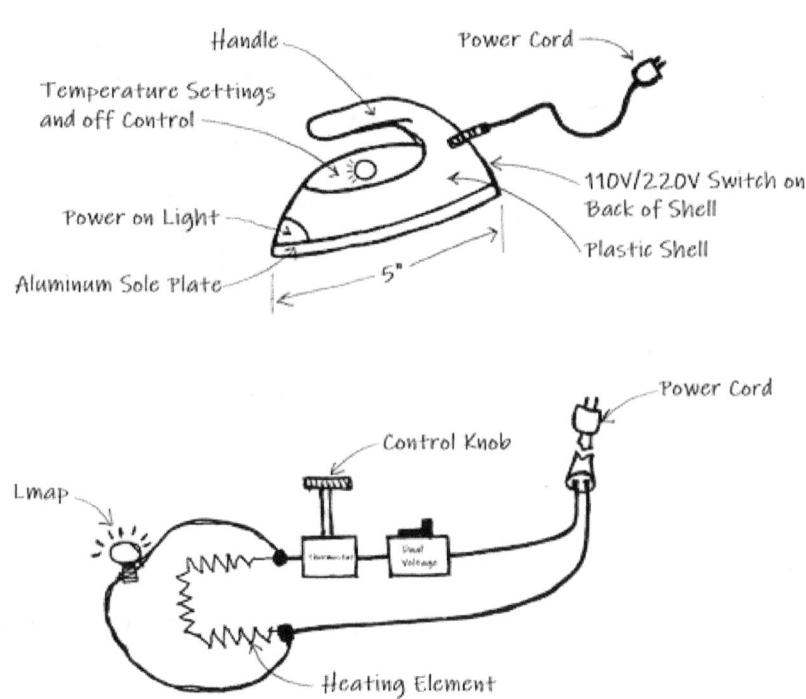

FIGURE B-36 Travel Iron External View and Wiring Schematic

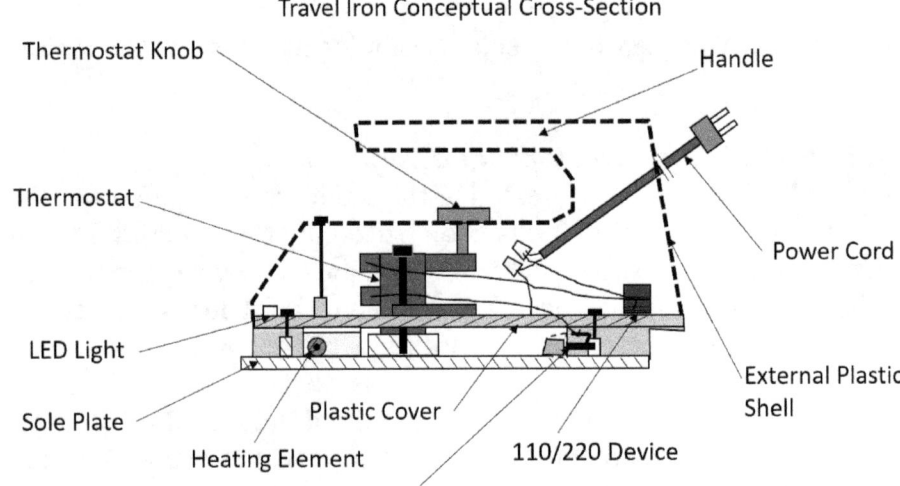

FIGURE B-37 Cross-Section of Travel Iron Final Conceptual Design

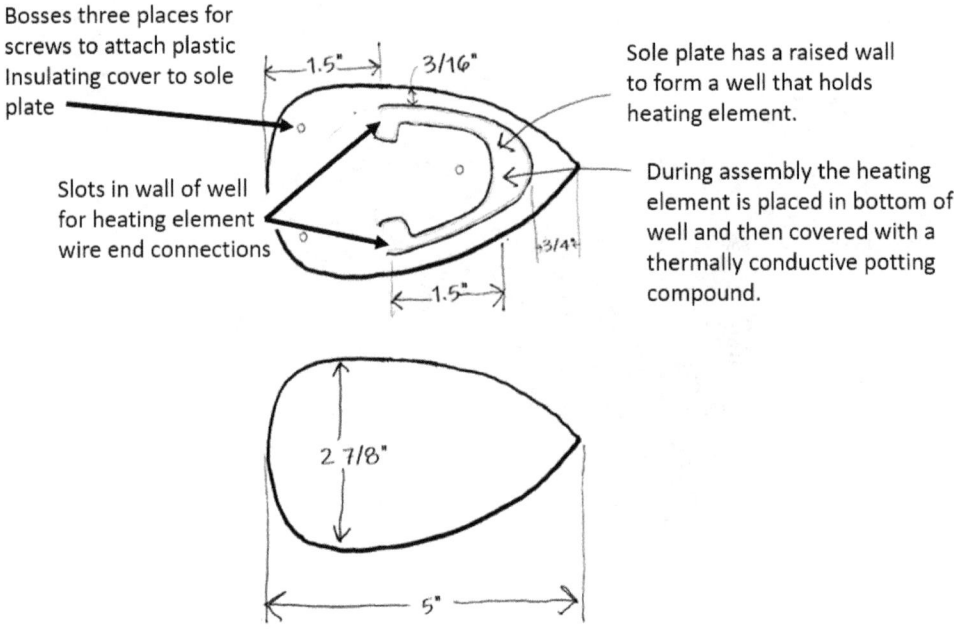

FIGURE B-38 Sketch of Sole Plate Casting

An exploded view sketch of the travel iron is given in Figure B-39. The thermostat will be calibrated to provide specific operating temperatures to accommodate different types of clothing material. The thermostat also serves as the on/off switch.

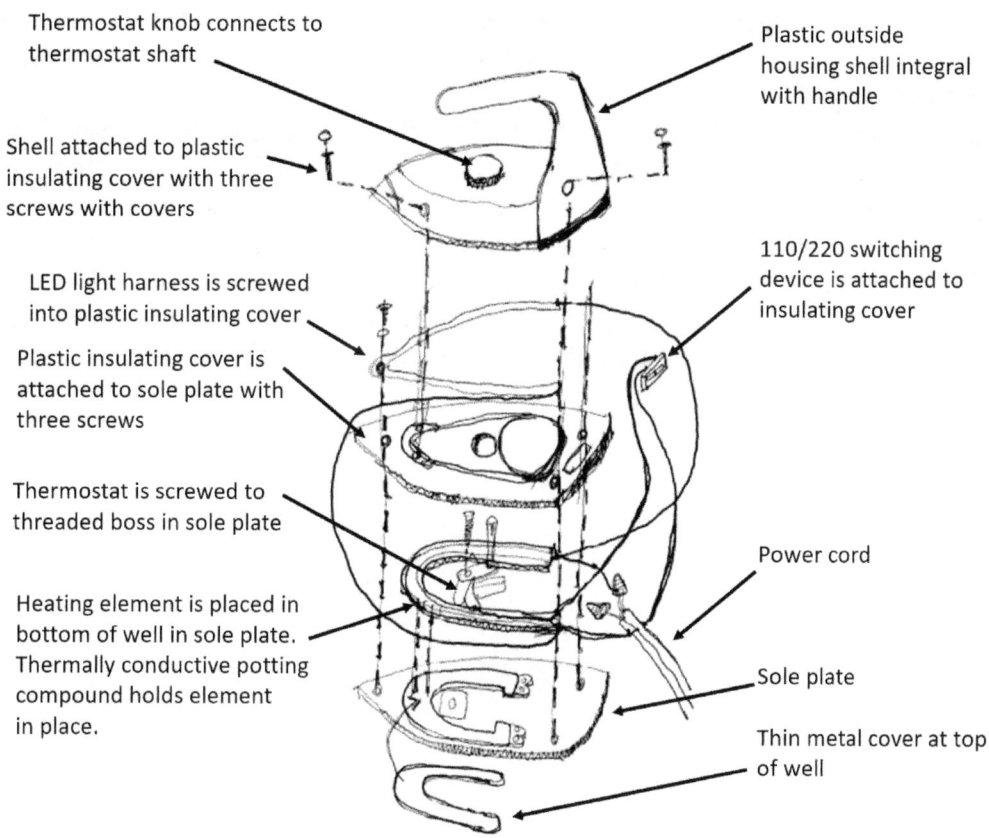

FIGURE B-39 Travel Iron Final Conceptual Design Exploded View

B.2.30.2 Analyses and Testing

Top-level analyses and simple POC testing were accomplished to establish the likelihood that when turned into the Phase 5 production prototype, the conceptual design will meet all the engineering requirements. These early analyses and tests indicate that there is a high probability that when it is fully developed, the travel iron will be able to meet all of its engineering requirements. It should be noted, however, that the UPC target of $4 per unit had to be changed to $10 per unit based on the team's initial UPC analyses. As discussed in the commercialization part of this weekly report, the team still believes that the travel iron can be commercialized with a UPC of $10 per unit.

B.2.30.3 Commercialization

There are numerous large companies involved in the production and marketing of travel irons across the globe. The goal of this project is to focus on the underserved niche market of travelers who want a travel iron that is locally made and environmentally friendly. These customers are willing to pay a 10 to 30 percent price premium in order to have a product that is based on the above characteristics.

The team believes a relatively small but profitable and sustainable business can be built around the production and marketing of their travel iron. Several types of potential investors exist. One type of investor may be an established appliance company that wants to gain experience in producing locally made and environmentally friendly products. Another type

EM-(n)

of producer under consideration is a relatively small operation that provides rewarding employment for persons who have a significant disability but are still able to do travel iron production tasks.

The team has investigated this second type of producer situation for a production rate of 15,000 units per year for three years. It is assumed that it will take $200,000 of capital and one year to develop the product, production facility, workforce, and market. It is assumed that the product would yield an after-tax profit margin of $6.50 per unit. As shown in Figure B-40, this results in an IRR of 23 percent. The team believes this rate would be acceptable to a variety of investors and that financing would not be a major concern. The team is planning more detailed commercialization planning in the project's remaining development phases.

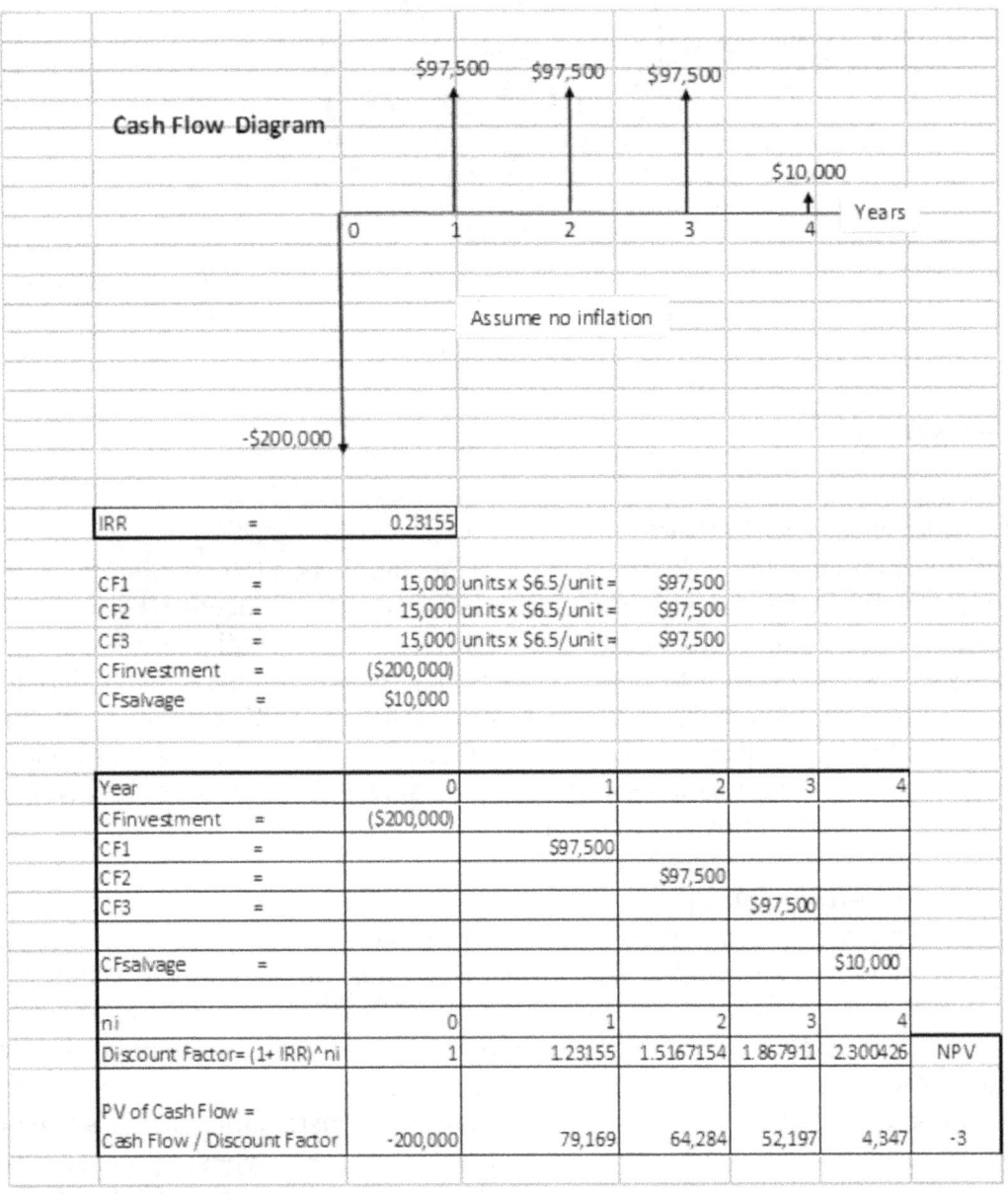

FIGURE B-40 Travel Iron Initial IRR Analysis

B.2.30.4 Team Assessments

The team performed team member 360-degree reviews and discussed the results during the team meeting. During the meeting, the team discussed whether they were a high-performing team and whether they were playing to their strengths. Based on their ability to stay on schedule and labor budget and their resilience to certain problems with getting adequate customer feedback, they decided that the team was becoming high-performing but still needed to improve.

EM-(p)

B.2.31 Phase 2 Criteria Checklist

The checklist for this phase is provided on the next page.

Exit Criteria Checklist: Phase 2: Conceptual Design

Team Name: _____ Team Leader: _____

Members: _____ _____

_____ _____

_____

Date Started: _____ Date Completed: _____

Done		
	1. Team agreement	
	2. Detailed Phase 2 project plan	
	3. Conduct secondary and primary VOC research	
	4. VOC-based needs converted to a requirements list and validation matrix for the production unit	
	5. Design space fully explored	
	6. At least three concepts defined in detail and compared in a weighted-criteria matrix table	
	7. Analysis plan and test plan (if applicable) prepared to support conceptual design trade studies.	
	8. Final production concept defined	
	9. Final capstone prototype defined	
	10. Commercialization plan updated	
	11. Conceptual design review completed and approved	
	12. Final report draft chapters 1, 3, 4, and 5 completed and approved	
	13. Notebook checked and approved	

Yes or No	Entrepreneurial Mindset Indicators Exit Criteria		Questions Related to Criteria
	a)	Critically observes surroundings to recognize opportunity	Did the team do research to identify customer needs?
	b)	Explores multiple solution paths	Did the team explore the design space to identify candidate concepts?
	c)	Gathers data to support and refute ideas	Did the team do research to identify customer needs?
	d)	Suspends initial judgment on new ideas	During the brainstorming process for identifying concepts, did team members suspend judgment?
	e)	Observes trends about the changing world with a future-focused orientation/perspective	Did the team use a future-focused orientation when listing the customer needs?
	f)	Collects feedback and data from many customers and customer segments	Did the team do primary customer research? Did the research cover the key customer segments?
	g)	Applies technical skills/ knowledge to the development of a technology/product	Did the team properly use QFD to arrive at the requirements? Did the team properly conduct analyses and testing (if applicable)?
	h)	Modifies an idea/product based on feedback	Did the team integrate the sponsor's proposal feedback into the Phase 2 planning process?
	i)	Focuses on understanding the value proposition of a discovery	Did the team clearly state the value proposition for both the production unit and the capstone prototype?
	j)	Describes how a discovery could be scaled and/or sustained, using elements such as revenue streams, key partners, costs, and key resources	Did the team develop a simple financial return model for delivering the product to the customer?

(continued)

	k)	Defines a market and market opportunities	Did the team define the market and estimate the price point and sales volume so that a business case could be developed for the investigative phase of commercialization?
	l)	Engages in actions with the understanding that they have the potential to lead to both gains and losses	Did the team use initial analyses and testing to estimate the probability of the final concept developing into a production unit that meets the customers' needs?
	m)	Articulates the idea to diverse audiences	Was the team successful in conducting their conceptual design review?
	n)	Persuades why a discovery adds value from multiple perspectives (technological, societal, financial, environmental, etc.)	Did the team include multiple value perspectives in their commercialization assessment for this phase of the project?
	o)	Understands how elements of an ecosystem are connected	Did the team define an ecosystem around the product, including waste products and resource utilization?
	p)	Identifies and works with individuals with complementary skill sets, expertise, etc.	Did the team demonstrate that they were a high-performing team? Did they conduct team member assessments?
	q)	Integrates and synthesizes different kinds of knowledge	Did the team improve their marketing and business acumen to properly integrate these considerations into the prototypes' technical design?
Approved by	Name and Title		Completion Date

Section B

Phase 3: Preliminary Design

MODULE 15

Introduction to Phase 3: Preliminary Design and Planning

OVERVIEW
Module 15 focuses on detailed planning of the Phase 3 preliminary design activities. The design team will review feedback from the Phase 2 conceptual design review and complete the Phase 3 detailed planning. Each team member must complete the reading prior to the class lecture. After the lecture, the team will divide up the Phase 3 detailed planning tasks among its members. Each member is responsible for completing the first draft of their assigned planning tasks. The team then meets again to complete the plan prior to the beginning of Module 16.

LEARNING OBJECTIVES
- Understand the preliminary design phase goals and processes.
- Apply project management principles to create a detailed Phase 3 plan.

PRE-LECTURE ASSIGNMENT
- Read and study Section B.3.
- Each team member makes a list of the Phase 3 tasks and identifies which ones they would like to take the lead in planning.

POST-LECTURE ASSIGNMENT
- Conduct a team meeting to 1) review the feedback from the Phase 2 conceptual design review and 2) create an outline for the Phase 3 detailed plan and assign tasks to each member.
- Conduct a second team meeting to compile the individual write-ups into a complete and edited final Phase 3 detailed plan.

TEAM DELIVERABLES
- Detailed Phase 3 plan
- Minutes of the team meetings
- Team activity as documented in the team notebook

B.3 Phase 3: Preliminary Design

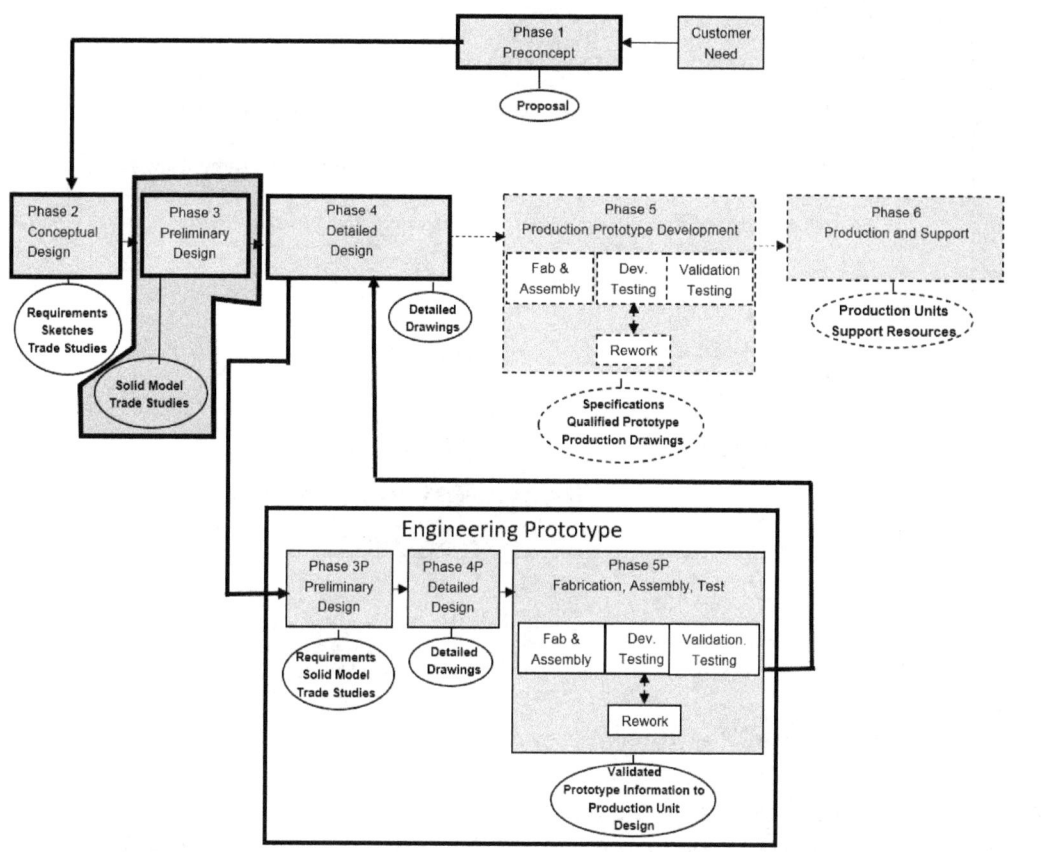

B.3.1 Overview

As described in Section B.2, during the conceptual design phase, the team arrives at their best idea of how the product can meet the engineering requirements that are based on the voice of the customer. Also during this phase, the team selects a goal function that is composed of one or more engineering requirements for which the goal is to not only meet but continue to improve upon the requirement. This phase's deliverables include a product sketch, a list of the product's major components, a configuration block diagram showing how these components interact to provide the required product functions, and a collection of analyses and proof-of-concept (POC) tests that indicate that the conceptual design has a reasonable probability of meeting the engineering requirements when fully designed. The components are either commercial-off-the-shelf (COTS) or fabricated items.

Figure B-41 graphically shows the flow of the preliminary design process. In preliminary design, also called *embodiment design*, the team takes the conceptual design and turns it into a specific product design with the goal of it being the best configuration to meet the engineering requirements and approach an optimum goal function as established during Phase 2. The deliverables from this preliminary design phase include a specific list of components that has each part specified and its source of manufacture identified, as well as an assembly plan and a collection of analyses and POC tests that show with a high level of confidence that the final preliminary design meets all of the production unit's engineering requirements. Moreover, the design has achieved a near optimum value of the project's goal function.

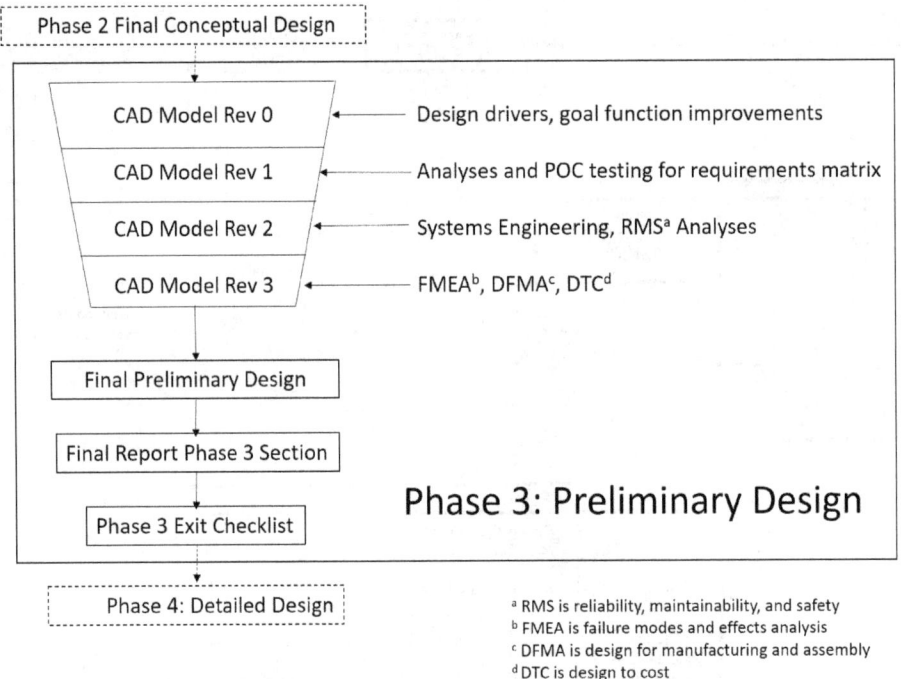

FIGURE B-41 Phase 3 Preliminary Design Flowchart

In addition to the specific engineering requirements, the team has used a system engineering approach to make sure that the design is based on good engineering practice relative to reliability, maintainability, safety, materials selection, manufacturing methods, and design-to-cost goals. At the end of Phase 3, the preliminary design is ready to enter detailed design, during which a complete production drawing package is created. A few analyses may be carried on into detailed design, but at least 90 percent of the design analyses should be completed prior to the start of this phase.

Figure B-42 summarizes the key objectives for Phase 3.

1. Update final conceptual design based on Phase 2 design review feedback.
2. Prepare a detailed go-forward plan for Phase 3.
3. Continue to use an Entrepreneurial Mindset in achieving the other Phase 3 objectives.
4. Conduct the preliminary design of the product by converting the final conceptual design into a defined production design configuration that meets all the production design requirements and approaches an optimum value for the design goal function.
5. Successfully conduct a preliminary design review
6. Document the work accomplished during this phase in the team project database and prepare the Preliminary Design potion of the Final Report.
7. Complete the Phase Exit Criteria Checklist and obtain the required approval signatures

FIGURE B-42 Phase 3 Key Objectives

Experienced design teams tend to conduct many of the preliminary design tasks in parallel. Because most capstone design teams are still discovering how to accomplish these duties, they will first learn about the task and then perform it in a sequential manner, as shown in the abovementioned Figure B-41.

Each task area will be discussed in general terms first, and then the travel iron example will be used to show how the general information can be applied to a specific capstone project. For the practicing engineer using this handbook, the information presented is a helpful review of the preliminary design process.

B.3.2 Continue to Build Team Commitment

It is vital in Phase 3 to ensure once more that all team members understand their roles and responsibilities on the team and make a commitment to be an active team member.

EM-(p)

B.3.3 Phase 3 Planning and Management

There are two basic types of companies that perform product development. The first are established companies with substantial existing processes and knowledge about their products. Engineers in these companies tend to follow established analysis processes and use existing computer models of their products. These computer models were developed over a substantial amount of time and have been validated with field data. Most engineering courses emphasize this type of detailed analysis.

The second type of company is involved in the development of new products for which existing computer models do not exist. Often these companies, like the capstone course, have very limited calendar time and labor resources to develop the new product. In these cases, there is not adequate time to develop sophisticated computer modeling tools. The team must rely on established knowledge for similar products, simple POC testing, and back-of-the-envelope analyses that are grounded in basic physics and engineering judgment.

For a two-semester capstone product development course, there are usually only seven or eight weeks available for preliminary design. When these courses are taught in summer school, this calendar time is cut in half. Often there is not enough time to create sophisticated computer models.

A key task in preliminary design is deciding what, how, when, and who will accomplish the tasks needed to arrive at a product design that meets all engineering requirements in a way that improves the goal function and meets the required calendar time constraints. The team starts the planning process by having each member read Section B.3. Following this, the team must think through and document how the preliminary design will be conducted. Only then is the team ready to create a plan that includes task descriptions, team assignments, a Gantt chart schedule, and a labor chart.

B.3.3.1 *Preliminary Design Tasks, Schedule, and Labor Budget*

At the beginning of the project, the team created a top-level project plan. Upon entering Phase 3, the team must prepare a more detailed plan for accomplishing all Phase 3 tasks with the resources available. Table B-27 presents a typical capstone design team's planning chart for Phase 3. The chart lists the tasks to be performed and indicates when each will be conducted. Further, it shows how the team's labor resources will be allocated among the tasks on

Table B-27 Example Phase 3 Planning Chart

Task/Weeks/Modules		8	9	10	11	12	13	14	15	
		15	16	17	18	19	20	21	22	23	24	25	26	27	28	
A	Feedback Update	3														
B	Plan Phase 3	27														
C	Design Drivers COTS		40													
D	Identify COTS Sources		10													
E	Size and Spec COTS		10													
F	Size and Spec Fab Components			30												
G	Initial Assembly Plan			10												
H	Prepare CAD Rev 0			20												
I	Requirements Matrix Analyses & POC Testing				40											
J	Prepare CAD Rev 1				20											
K	RMS Analyses					25										
L	Prepare CAD Rev 2					5										
M	FMEA/Lower RPN*						20	10	10							
N	Manufacturing and Materials Studies						9	19	5							
O	DTC (design to cost) Studies						1	1	5							
P	Prepare CAD Rev 3								10							
Q	PDR** Preparation									10	10					
R	Preliminary Design Review											In class				
S	Write Final Report Sections											16	16	20		
T	Phase Exit Checklists											4	4	10		
	Total hours per week	30	60	60	60	60	60			60	60	30				

Note: This table assumes a team of six for which each has pledged to devote at least 10 hours per week.

* RPN = risk priority number

** PDR = preliminary design review

142 | Product Design and Development Handbook

a weekly basis. The chart presented is for a team of six members who have committed to each member spending at least 10 hours per week outside of class time on their capstone project. The schedule that goes with the planning chart is shown in Figure B-43.

It is important to remember that the Phase 3 plan, like any plan, must be monitored and modified when necessary to adapt to unforeseen developments. It is good practice for the team at the beginning of each team meeting to review the team's plan and make any needed modifications to keep the project progressing with the available resources and within schedule constraints.

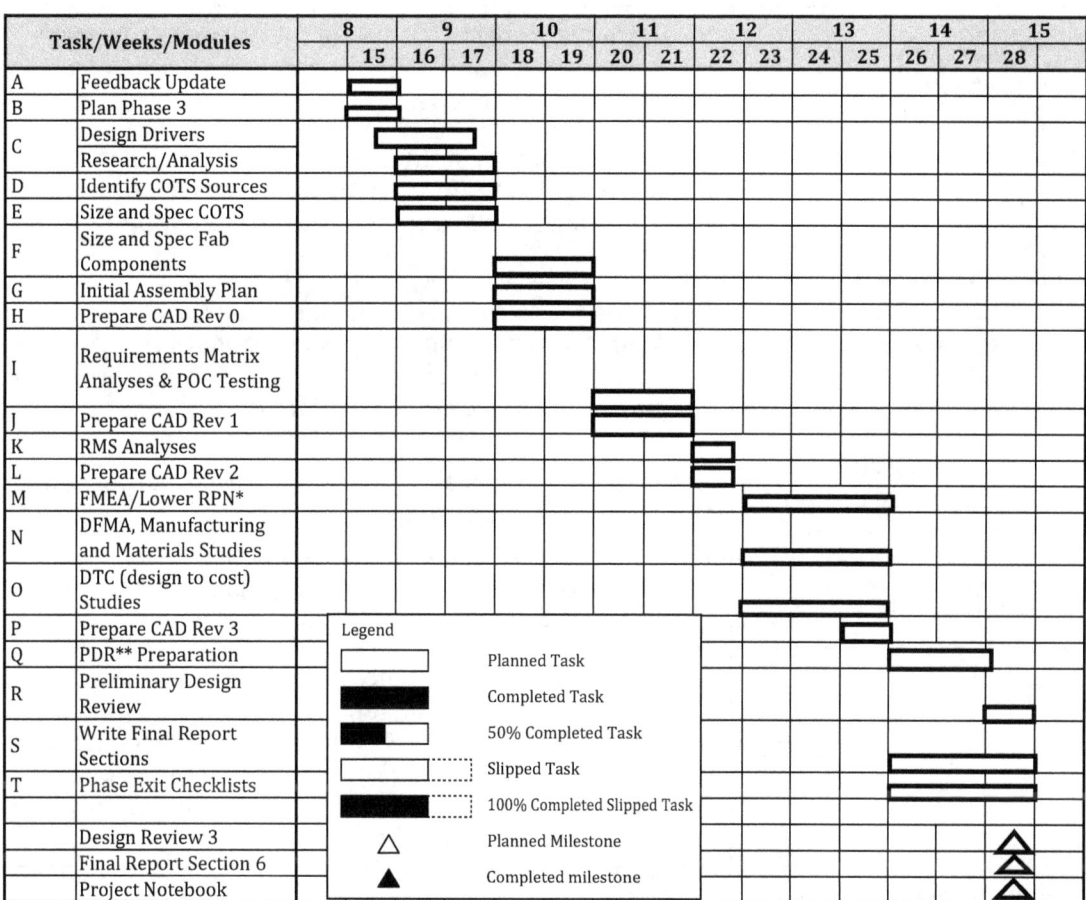

FIGURE B-43 Phase 3 Detailed Schedule

B.3.3.2 Phase 3 Task Dictionary

In parallel with preparing the Phase 3 schedule, the team must create a task dictionary that describes each task in terms of what is to be done, by whom, when, and its deliverables. Table B-28 provides a generic Phase 3 task dictionary.

Table B-28 Task Dictionary for Phase 3 Tasks

No.	Task	Description
A	Feedback update	Updates the conceptual design and the project plan based on the sponsor's feedback on Phase 2 design review.
B	Plan Phase 3	All members read Section B.3 and research design topics, divide up planning among team, and meet to finalize task assignments and scope of work for each task.
C	Design drivers research/analysis	Provide a set of goal functions for the project against which designs can be measured. Use research/analysis to identify key performance parameters.
D	Identify COTS sources	Identify the sources of the product's commercial off-the-shelf (COTS) items. These are ready-made and available products, such as a pump, electric motor, bearings, engine block, and flanges.
E	Size and specifications of COTS	Decide on the sizes (geometry, dimensions, and configuration) and technical specifications of the COTS items. Ensure that the materials and specifications of the COTS parts will stand the service environment of the product.
F	Size and specification of fabricated components	Use modelling and analysis for sizing special-purpose parts that must be manufactured.
G	Initial assembly plan	Create a layout of the initial assembly plan of the product from all standard and special-purpose parts and modules.
H	Prepare CAD Rev 0	Start by making hand sketches and drawings of the COTS and special-purpose parts, including their geometry, dimensions, and configurations. Using SolidWorks or similar computer-aided design (CAD) software, prepare an initial set of drawings for the parts, modules, and complete assembly of the product from the COTS and special-purpose parts.
I	Requirements matrix analyses and POC testing	Identify the analysis and POC testing plans for validating each of the requirements matrix items.
J	Prepare CAD Rev 1	Based on the results of item I above, prepare an updated CAD Rev 1 solid model, including part, subassembly, and assembly drawings.
K	RMS analyses	Conduct reliability, maintainability, and safety (RMS) analyses of the product.
L	Prepare CAD Rev 2	Based on the results of item K above, prepare an updated CAD Rev 2 solid model, including part, subassembly, and assembly drawings.

(continued)

No.	Task	Description
M	FMEA/Lower RPN	Conduct failure mode and effect analysis (FMEA) to lower the risk priority number (RPN).
N	Manufacturing and materials studies	Select the material and manufacturing processes for each of the special-purpose items. Ensure the material will stand the service environment of the parts.
QP	Prepare CAD Rev 3	Based on the results of items M, N, and O above, prepare an updated CAD Rev 3 solid model including parts subassemblies, and assembly drawings.
RQ	PDR prep	Prepare for Phase 3 design review (PDR).
SR	Preliminary design review	Prepare final report section for Phase 3.
TS	Write final report sections	Team members share writing the final report sections and then edit them as a group and compile them.
VT	Phase-exit checklists	Prepare exit checklist.

B.3.3.3 Phase 3 Management

The team needs to start each Phase 3 team meeting by reviewing the detailed Phase 3 schedule and making the necessary changes in the phase plan so that the project remains on schedule and within budget. This often requires some team members to help others with their assignments. In some cases, the scope of certain tasks must be reduced in order to stay within the project resource constraints. Results of these management discussions should be recorded in the team meeting minutes and placed in the team notebook.

EM-(g)

B.3.4 Example Travel Iron Team Report for Week 8

To illustrate the Phase 3 processes, the experiences of a capstone team conducting the preliminary design of the travel iron discussed in Phase 2 are provided. These experiences are divided into seven weekly reports, with each placed in the handbook so that it correlates with the appropriate modules. The first of these weekly reports is provided below.

Initially, the six-person travel iron project team met to start planning their Phase 3 activities. They realized that with only seven weeks to do all the preliminary design work, they would need to be clear on the scope of each task, its deliverables, and the time available. The team started by each member reading the entire Section B.3 in the textbook and listing their ideas on what should be done. They agreed to use the generic schedule provided that listed the tasks and indicated what needed to be done each week. Table B-29 shows what the team eventually decided on for their plan.

The team further agreed that each member would spend at least 10 hours outside of class hours each week on the project. Each member also agreed to email the other members their completed assignments before each team meeting so the other members would have time to review it prior to the meeting.

The team's design strategy was to use information about existing electric irons and tailor this information to meet the need of a compact and affordable travel iron. They also agreed to use COTS components as often as possible to reduce production costs.

It is important for the capstone team to understand that the capstone course requires an accelerated schedule to cover all the new course material. Therefore, there is only a small amount of time available for tasks such as creating the failure mode and effects analysis (FMEA). An important goal in the capstone course is for the team to have some experience of preparing parts of an FMEA. In regular engineering practice, the design team would spend more time conducting the FMEA and many of the other Phase 3 tasks.

Table B-29 Phase 3 Task Dictionary for the Travel Iron

Task	Description	Wk	Hrs**	Persons
A. Feedback update	Update conceptual design during team meeting	8	1	all
B. Phase 3 plan	All members read Sec. B.3 and research design topics, divide up planning among team, meet to finalize task assignments and scope of work for each task.	8	29	all
C. Design drivers	Subteam 1: POC testing to select minimum sole plate size	9	20	1,2
	Subteam 2: Determine power required	9	6	3,4
	Subteam 3: Make dual voltage schematic	9	14	5,6
D. Identify COTS sources	Subteam 2: Do COTS internet research	9	10	3,4
E. Size/spec COTS	Subteam 1: Analysis of fabricated components	9	10	1,2
F. Size/spec fab comp.	Subteam 2: Analysis of COTS components	10	30	3,4
G. Initial assembly plan	Subteam 3: List components in order of assembly	10	10	5,6
H. Prepare CAD Rev 0	Subteam 4: Make first CAD drawing with packaging	10	20	5,6
I. Reqts analyses/testing	Subteam 1: Split up requirements for analysis and then review	11	50	1,2,3,4,5,6
J. Prepare CAD Rev 1	Subteam 2: Incorporate changes to meet requirements	11	10	5,6
K. RMS Analyses	Subteam 1: Reliability Analyses	12	10	1,2
	Subteam 2: Maintainability Analyses—No sched or repair	12	0	
	Subteam 3: Safety Analyses	12	10	3,4
L. Prepare CAD Rev 2	Subteam 4: Incorporate needed RMS changes	12	10	5,6
M. FMEA –Initial	Team divides up components and several discussion meetings	12	20	1,2,3,4,5,6
N. Mfg/Matl's Studies	Subteam 2: Part 1 of study	12	8	1,2,3,4,5,6
O. DTC Studies	Subteam 3: Part 1 of study	12	2	1,2,3,4,5,6
M. FMEA–Lower RPN	Subteam 1: Design changes to lower RPN and document	13	20	1,2
N. Mfg/Matl's Studies	Subteam 2: Part 2 of study	13	24	3,4
O. DTC Studies	Subteam 3: Part 2 of study	13	6	4,5
P. Prepare CAD Rev 3	Person 6 incorporates changes from FMEA, Mfg/Matl's, DTC	13	10	6
Q. PDR Preparations	Team divides up slides, meets to finalize and practice	14	25	1,2,3,4,5,6
R. Prel Dsn Review	Held during class	14		
S. Write Final Report Sections	Team splits up sections and then edits as a group	14	23	1,2,3,4,5,6
T. Phase Exit Cklists	Team completes together using notebook information	14	12	1,2,3,4,5,6
S. Write Final Report Sections	Team splits up sections and then edits as a group	15	23	1,2,3,4,5,6
T. Phase Exit Cklists	Team completes together using notebook information	15	12	1,2,3,4,5,6

*Done in initial Phase 3 team meeting; ** Hours outside of classroom meetings*

MODULE 16

Design Drivers and Commercial off the Shelf (COTS) Components
Part 1

OVERVIEW

In this and the following modules, the team focuses on (1) identifying and analyzing design aspects that largely affect the overall product design and (2) specifying and sourcing the design components that are already available from suppliers or manufacturers.

LEARNING OBJECTIVES

- Understand how to identify the design drivers for a product.
- Demonstrate the ability to improve the overall product's design by focusing on improving the product's design drivers.
- Demonstrate the ability to use the results of the design driver analyses to specify the other parts of the product that can be COTS items.

PRE-LECTURE ASSIGNMENT

- Reread and study sections B.3.4–B.3.6. (Note that Section B.3.6 is the travel iron project report for Week 9 located in Module 17.)
- Each team member makes a list of what they consider to be the product's two or three major design drivers and how these should be analyzed to start the preliminary design process.
- Each team member makes a list of the items that should be in a specification for a COTS part used in the product's design.

POST-LECTURE ASSIGNMENT

- Conduct a team meeting to 1) agree on the top two or three product design drivers and how they should be analyzed, 2) agree on a generic specification format for the product's COTS items, and 3) divide the team into three or four subteams to do the following:
  - Subteam 1: Analyze Design Driver 1.
  - Subteam 2: Analyze Design Driver 2.
  - Subteam 3: Analyze Design Driver 3 (if needed).
  - Subteam 4: Size, specify, and identify the sources for the COTS items used in the design.
- Each subteam completes at least 50 percent of their assigned work and sends the results to the other team members by the beginning of Module 17.

TEAM DELIVERABLES

- First 50 percent of product design drivers analyses
- Minutes of the team meetings
- Team activity as documented in the team notebook

B.3.4 Design Drivers Research and Analysis

B.3.4.1 Types of Drivers

A good way to start preliminary design is to look at the final conceptual design and ask the question, what engineering requirements drive the design? There are two types of drivers.

The first type of driver is a requirement that is especially difficult to meet. For example, the item must weigh less than X pounds (lb), and the team knows that without concentrated design work, this requirement cannot be met.

The second type of driver is a goal function for which there is a need to meet some minimum threshold and then continue to improve the design to exceed the threshold requirement as much as possible. For example, the item must be as light as possible, but it cannot exceed X lbs. This second type of design driver sometimes has additional constraints such that an optimum goal function can be determined.

B.3.4.2 Examples of Design Drivers

To illustrate the idea of identifying design drivers early in the preliminary design process, three different scenarios are presented below. In each scenario, an aircraft designer or producer identifies the need for an additional pneumatic control valve in the wing of an aircraft being designed after most of the spaces in the wing have been taken by other components.

In Scenario 1, the late addition of the valve means that the valve must meet all its engineering requirements and fit in an odd-shaped volume. The unit must fit in the specified volume, but if the valve fits in the space, there is no additional benefit for being smaller. This design driver is of the "go, no-go" type. In this case, the design is complete when it meets all the engineering requirements and fits in the specified space. One of those requirements is that it can weigh no more than 5 lbs.

In Scenario 2, the valve must still fit in the odd-shaped space in the wing and meet all engineering requirements. But in this scenario, the overall aircraft design is overweight. The aircraft designer or producer is offering the valve supplier a weight reduction incentive by paying an additional $400–per-pound reduction per unit for every pound and fraction of a pound of reduction from the original requirement of 5 lbs of valve weight. In this scenario, the valve designer keeps trying to reduce the valve's weight during the design process to obtain the most monetary incentive per unit sold to the aircraft designer or producer.

In Scenario 3, the valve needs to fit in the specified space, there is the same weight reduction incentive as given in Scenario 2, and the aircraft designer has found that the weight of the valve's actuation motor can be reduced if it is allowed to consume more electrical power than the original requirement of 150 watts (W). The aircraft designer or producer will accept a valve that uses more than 150 W to reduce the valve's weight, but there is a penalty of $5 per W for every watt over 150 W. The incentive and penalty equations are given below:

$$\text{Incentive dollars/unit} = \$400 \times \text{weight reduction in lbs per unit}$$

$$\text{Penalty dollars/unit} = \$5 \times \text{number of watts over 150 W required by the valve}$$

In Scenario 3, the valve design team did some research and arrived at two conclusions: (1) The valve will probably be near the weight limit of 5 lbs when it meets all its engineering requirements, including fitting in the odd-shaped volume specified by the aircraft designer, and (2) by selecting different motors, they can achieve different degrees of weight reduction as a function of the additional power required in watts. This relationship is given by the following equation:

$$\text{Additional power in watts} = 35 \times (\text{weight reduction in pounds})^2$$

These three equations can now be combined to form the following goal function for scenario 3:

$$\text{Incentive dollars/unit} = 400\,X - 175\,X^2, \text{ where } X \text{ is the weight reduction in pounds.}$$

Figure B-44 shows that this goal function can be optimized by selecting the motor that yields a weight reduction of 1.15 lbs. This weight reduction design uses a motor that requires almost 50 W more power than the original specification of 150 W. This combination of weight reduction and power increase results in a maximized net incentive of $228.56 per unit.

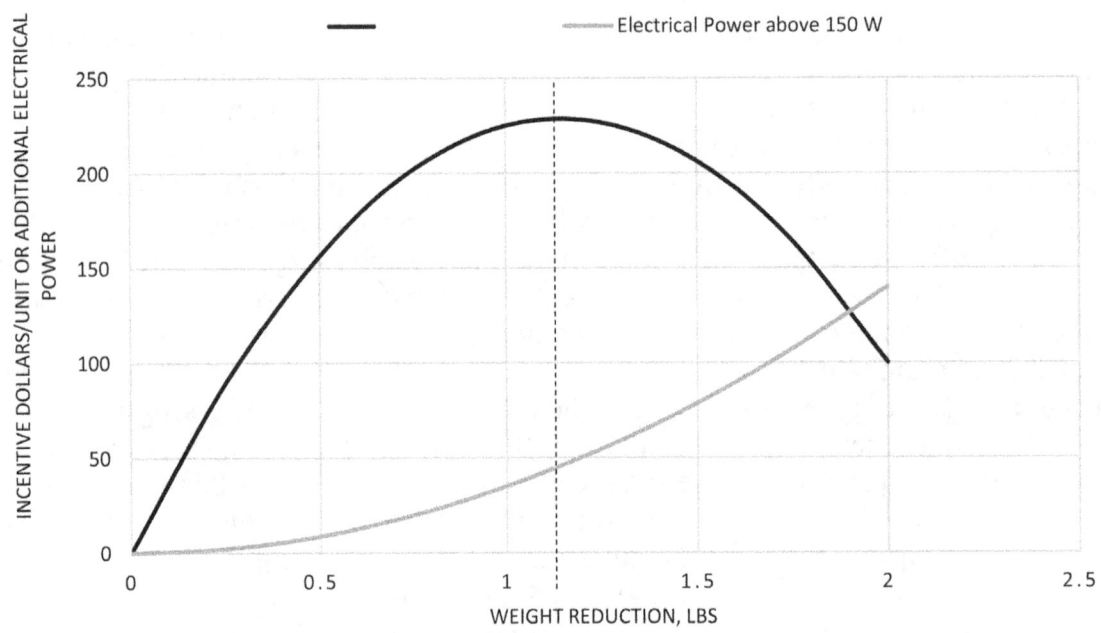

FIGURE B-44 Example: Optimization of Valve Weight Reduction for Maximum Incentive Dollars

It is important to remember that this weight optimization analysis assumes that before the new motor is incorporated into the design, (1) the valve design meets all the engineering requirements, (2) the valve fits in the odd-shaped space in the aircraft wing, (3) the valve weighs exactly 5 lbs before being optimized, and (4) the valve has a power draw of 150 W before being optimized. Further, when the new motor is integrated into the design, the valve will still meet all its requirements, except for the electrical power, and it will fit in the wing. However, it will be lighter and have a larger power draw. (It should be noted that the actual valve may not be able to be sized for exactly the 1.15-lb weight reduction due to available component sizes. In this case, the design should be as close to the 1.15-lb weight reduction as possible.)

B.3.5 Selecting the Commercial off the Shelf (COTS) Components

B.3.5.1 Process for Selecting COTS Items

Once the design drivers have been identified and a strategy for satisfying them has been formulated, the team can take the final conceptual design, list the components, and decide whether each of them is (1) a COTS part or (2) a fabricated part. It should be noted that the fabricated part can be either provided by a supplier or manufactured in house.

EM-(b)

This section describes how the team should procure COTS items. Procuring fabricated items is covered in Section B.3.7.

The process of selecting a supplier for a COTS item starts with the designer creating a desired specification for the item. This specification should include the dimensions, weight, and other required technical data. Then the designer searches for a supplier that can provide a part that meets that specification. If such a part does not exist, then the designer needs to find a COTS item that is similar and determine if their design specification can be modified to meet the COTS item's specification. If there is a match, then the supplier should be contacted to obtain a quote on the price of the part based on a specific procurement rate. It should be noted that the selected COTS item probably has more specification data than the original designer's list of specification data. All of the actual COTS specification data should be recorded and placed in the project notebook. Eventually, these data will be turned into a purchase part control drawing. This will be discussed in Section B.4: Detailed Design.

To select the COTS parts, the team can use the internet or hard-copy sources to find manufacturers' and distributors' product catalogues. The path for procuring the COTS components is given in Table B-30.

Table B-30 Process for Selecting COTS Items

1. Define the technical specifications and features of each of the COTS items.
2. Discover the potential suppliers from suppliers/manufacturers catalogues. A good starting point is using the internet to search for the potential suppliers. For example, Thomas[1] served for 120 years as North America's number one industrial sourcing platform. Others such as McMaster-Carr[2] have proven to be a good source for capstone design projects.
3. Use the technical information in item 1 to select the COTS components from relevant suppliers' catalogues and submit a request for information and quote for the price to the supplier.
4. Document the specifications, features, and product line description of the selected COTS component. Ensure that these matches or exceed the component's required specifications.
5. Obtain from supplier catalogue any additional item details such as images, downloadable specification sheets, schematics, and three-dimensional computer-aided design (CAD) models to enable adequate mounting and assembly of the item into the product.

---

1 https://www.thomasnet.com.
2 https://www.mcmaster.com.

B.3.5.2 *Example of How to Select COTS Items*

A team needed to select a flange to connect two 3-inch outside-diameter drinking water plastic pipes. The water pressure in the pipe is 140 pounds per square inch. The team decided to choose plastic pipe and pipe fittings because plastic is lighter in weight than steel pipe and plastic is more corrosion-resistant compared to steel.

The team opened the McMaster-Carr link and entered the search word *flanges*. A page with various types of flanges and fittings opened. The application allows the user to enter certain parameters to narrow the search and final selection. To select the required flange, the team selected *3"* for pipe size, *flange* for type of fitting, *flanged* for the connection style, *plastic* for material, *flat* for flanged connection surface, *4* for number of bolt holes, *PVC plastic* for pipe material, fixed for flange type, *7"* for flange OD, and *white* for color. This resulted in the selection in Figure B-35 from McMaster-Carr.[3]

The final technical information for the flange selection includes specifications, price, temperature range, and connection method. It has been customary for vendors such as McMaster-Carr to provide buyers with technical drawings and three-dimensional models to assist in incorporating the selected item in the drawing package of the product. The images also show important technical details, such as the number of bolts, their diameter, circle diameter, and holes diameter, which are needed for the flanges' assembly.

---

[3] McMaster-Carr. n.d. *Aboveground Standard-Wall PVC Pipe Flanges for Drain, Waste, and Vent.* https://www.mcmaster.com/flanges/pipe-size~3/type~flange/connection-style~flanged/material~plastic/flanged-connection-surface~flat/number-of-bolt-holes~4/for-pipe-material~pvc-plastic/flange-od~7inches/flange-type~fixed/color~white/.

MODULE 17

Design Drivers and COTS Components Selection
Part 2

OVERVIEW

This is a continuation of the work started in Module 16. In this module, the team completes their analyses of the product design drivers and the specification and selection of the product parts that are COTS items.

LEARNING OBJECTIVES

- Gain more insight into how to evaluate design drivers by reading about two more case studies.
- Demonstrate the ability to use the results of the design driver analyses to specify the other parts of the product that can be COTS items.

PRE-LECTURE ASSIGNMENT

- Each subteam prepares a progress report describing what was done to complete 50% of their assigned tasks relative to either design drivers or COTS.
- For most students, the preliminary design process involves several new concepts. Each team member should make a list of at least three questions they have regarding the preliminary design process and bring the list to class for a discussion.

POST-LECTURE ASSIGNMENT

- Conduct a team meeting to complete the following tasks: 1) Review the work done so far by the subteams regarding design drivers and COTS parts specifications and selection. 2) The team agrees on what each subteam should do to complete their assignments.
- Each subteam completes their assigned work and sends the results to the other team members by the beginning of the second team meeting.
- Conduct a second team meeting to finalize the design drivers analyses and selection and specification of each COTS item for the production unit.

TEAM DELIVERABLES

- Completed analyses of the product design drivers
- Completed specification and selection of each of the product parts that are COTS items
- Minutes of the team meetings
- Team activity as documented in the team notebook

B.3.6 Example Travel Iron Team Report for Week 9

These reports for the example travel iron are provided to help the reader see how the general information provided in the handbook can be applied to a specific design project.

B.3.6.1 Team Planning

At the beginning of Week 2, the team met to review the week's plan. The team decided that there were two major design drivers for the travel iron: (1) the size of the sole plate and (2) the mechanism for providing dual-voltage operation. The size of the sole plate is important because it sets the plan dimensions for the product. The dual-voltage mechanism is a design driver because the team was not familiar with how this could be done, and the size of the required mechanism would have a major influence on the iron's height dimension.

The team also identified the need to determine which parts would be COTS and which would be fabricated. They also realized that they needed to specify and identify sources for the COTS parts.

B.3.6.2 Sole Plate POC Testing

Subteam 1, consisting of team members 1 and 2, determined how small the sole plate could be by making four wooden sole plates of varying sizes. They attached a wooden handle to each sole plate to simulate an entire electric iron. The POC test was to see how long it took to simulate the pressing of a dress shirt with each of the simulated irons. The smallest sole plate was only 5 inches long and about 3 inches at its widest point. This small unit was able to simulate the pressing of the shirt in the reasonable amount of time of less than two minutes. Based on this simple test, the team decided to use these dimensions for the sole plate.

B.3.6.3 Temperature Settings and Electrical Power Requirement

Subteam 2, consisting of team members 3 and 4, spent the early part of the week researching the power required by a full-sized dry electric iron for pressing various fabrics. The team decided that these temperatures were independent of the size of the iron's sole plate. They concluded that the minimum number of temperature settings for their travel iron was three, as shown in Table B-31. They agreed on the following settings: 275, 375, and 400 °F. They also compared the surface area of the sole plate selected by Subteam 1 with the surface area of a full-sized iron and determined that the travel iron needed only 400 W of power. Subteam 2 then performed an internet search to identify sources for the COTS components needed by the system.

Table B-31 Travel Iron Temperature Settings

Recommended Temperatures*		Travel Iron	
Material	Temperature, deg F	Setting No.	Temperature, deg F
Acrylic	275	1	275
Lycra/Spandex	275	1	275
Nylon	275	1	275
Acetate	290	1	275
Wool	300	1	275
Polyester	300	1	275
Silk	300	1	275
Rayon	375	2	375
Triacetate	390	2	375
Cotton	400	3	400
Linen	445	3	400

Source: Leverette, Mary M. 2020, April 29. **Select the right setting for ironing any fabric.** The Spruce. https://www.thespruce.com/select-correct-ironing-temperature-for-fabrics-2146186.

B.3.6.4 *Specifying and Selecting COTS Components*

In the latter part of the week, subteam 2 (team members 3 and 4) used the product configuration schematic given in Figure B-40 to make a list of the components and decide whether they would be COTS or fabricated items, as shown in Table B-32. Subteam 2 then started the task of specifying each COTS item and identifying potential suppliers. Figure B-45 illustrates this process for the thermostat. In addition to the thermostat's technical specifications, the supplier also provides a drawing of the part to show its shape, dimensions, and mounting screw hole diameters and locations. This information helps the design team integrate the thermostat in the electric iron assembly drawings. (It should be noted that at this point, the team is selecting a thermostat manufactured offshore. Later in the design process, the team will have to locate a local manufacturer for this component in order to meet one of the product's marketing goals, which is that it be locally made.)

Table B-32 List of Travel Iron COTS vs. Fabricated Components

No.	Component	COTS	Fabricated	Comments
1	Sole plate		X	Aluminum casting with PTFE coating, threaded bosses
2	Outer heater element		X	Have supplier design and produce to spec
3	Inner heater element		X	Have supplier design and produce to spec
4	Well cover		X	Thin steel stamping
5	Well filler	X		Thermal insulating epoxy
6	Thermostat	X		Source from China
7	Thermostat screw	X		
8	Black plastic cover		X	Injection-molded part
9	Light-emitting diode (LED) harness	X	X	LED light connected to wires; LED is COTS
10	LED harness washer	X		
11	LED harness screw	X		
12	Dual-voltage switch	X		
13	Power cord	X		
14	Wire nut (2)	X		
15	Wiring	X		
16	Wiring	X		
17	Wiring	X		
18	Wiring	X		
19	Top shell		X	Thermoplastic plastic injection molded
20	Temp/Off knob	X		
21	Covers (4)		X	Rubber
22	Screws (4)	X		
23	Clear plastic window		X	Glues into top shell

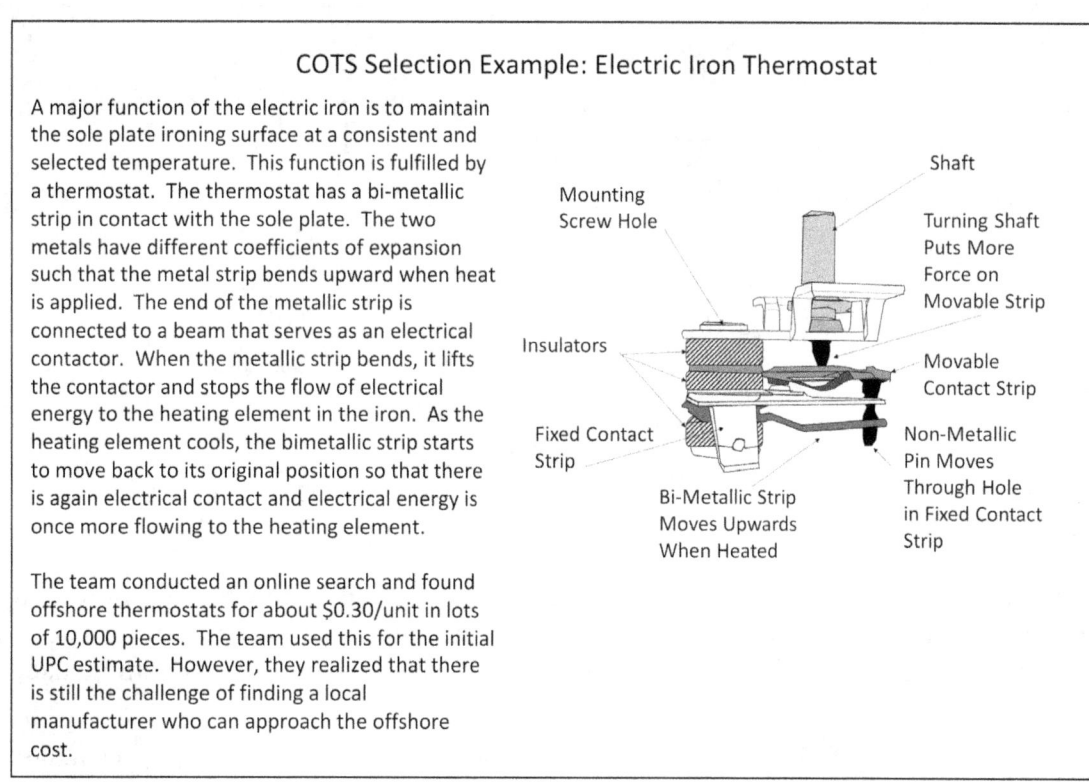

FIGURE B-45 COTS Component Selection Example—Electric Travel Iron Thermostat

B.3.6.5 Dual-Voltage Capability

Subteam 3 (team members 5 and 6) researched how an iron can be designed to operate on the dual voltage of nominally 120 voltage alternating current (VAC) or nominally 240 VAC. The initial guess was that there was a component that received either voltage as input and put out a common output voltage to the heating element. The subteam suspended judgment that this was the best way to achieve dual-voltage capability and was open to other ideas. The team learned that there is no specific component for dual-voltage capability. Instead, the subteam discovered that dualvoltage irons require two heating elements and a switching mechanism. Team Member 5 found an example on the internet in which two heating elements with the same resistance were put in parallel for 120 VAC and series for 240 VAC. Each configuration produced the same heating thermal power. This concept was applied to the travel iron, as shown in Figure B-46. However, this configuration requires a double-pole-double-throw (DPDT) switch, as shown in Figure B-47.

EM-(b)

EM-(c)

EM-(d)

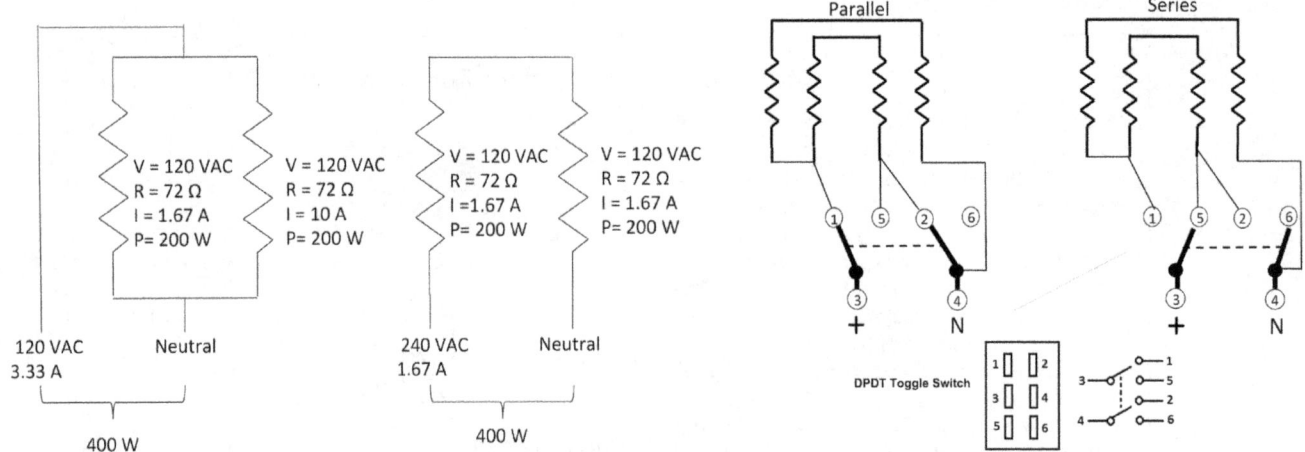

FIGURE B-46 Dual-Voltage Travel Iron Idea A: Use Same Heating Elements

FIGURE B-47 Same Heating Element Requires DPDT Switch

The subteam wanted to find a wiring configuration that required only a single-pole-double-throw (SPDT) switch. As shown in Figure B-48, two configurations are possible. A separate heating element for each voltage can be used, or a single heating element can be used for

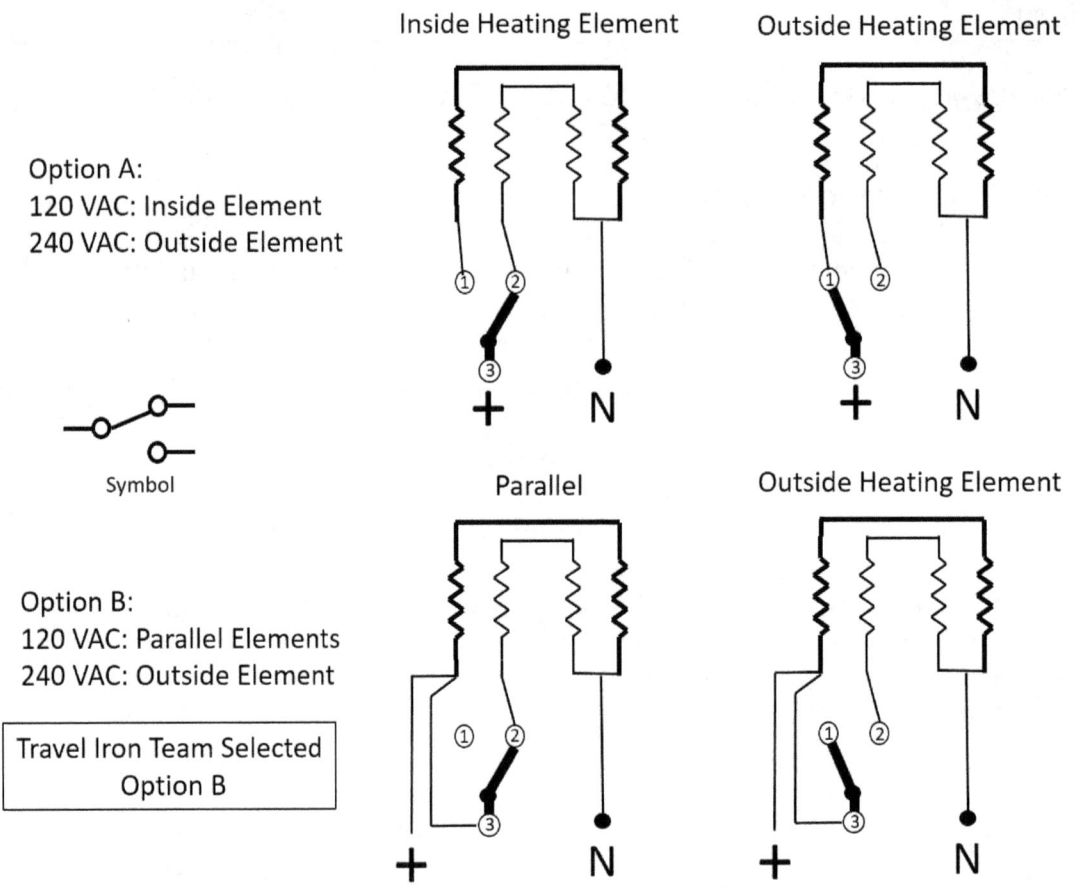

FIGURE B-48 Travel Iron Heating Element Options for Using a SPDT Switch

158 | Product Design and Development Handbook

240 VAC. Then, this element, along with the second heating element of the proper resistance, can operate in parallel for the 120 VAC case. The team decided to use this second configuration. In this configuration, the switch could actually be just a single-pole-single-throw device. The subteam then calculated the resistance needed by each heating element, as shown in Figure B-49.

Travel Iron Calculations for Resistances of Heating Elements

Issue: The heating element resistances for travel iron design shown as Option B in Figure B-36 need to be selected.

Problem Statement: Select heating element resistances for travel iron such that one element operates at 240 VAC and that element in parallel with a second element of different resistance operates at 120 VAC. In both cases the iron produces 400 W of thermal energy.

Approach: Neglect the fact that 120 VAC and 240 VAC are only nominal voltages and there are acceptable tolerances allowed by the electricity provider. Assume the heating element can be procured for any resistance needed. Start by sizing the heating element for the 240 VAC case. Then, find the second heating element resistance.

Defining Equations:
AC Voltage = $V = V_{RMS}$ AC Current = $I = I_{RMS}$ Resistance = V/I
Electrical Power = VI = Thermal Power = $I^2 R = V^2/R$
For Travel Iron: V = 120 VAC, 240 VAC P = 400 W   Inner, outer heating element resistances = R1, R2

Calculation Results:
1) For 240 VAC use R2:
a) Find the current: $I = P/V$ = 400 W/240 V = 1.67 A
b) Find heating element resistance: $R2 = P/I^2$ = 400 W / (1.67 A)^2 = $\boxed{144\ \text{Ohm} = R2}$
c) Calculation Check: P = 400 W = $(I)^2 * R2$ = (1.67 A)^2 * 144 Ohm = 400 W <u>Check</u>
2) For 120 VAC use R2 and R1 in parallel:
a) R2 has V = 120 VAC = I2 R2 → I2 = V/R2 = 120 V / 144 Ohm = 0.833 A
b) P2 = $V^2/R2$ = (120 V)^2 / 144 A = 100 W
c) Total current = I1 + I2 = I = P/V = 400 W / 120 V = 3.333 A
d) I1 = I – I2 = (3.333 – 0.833) A = 2.50 A
e) R1 = V1/I1 = 120 V / 2.5 A = $\boxed{48\ \text{Ohm} = R1}$
f) P1 = $V^2/R1$ = (120 V)^2 / 48 A = 300 W
g) Calculation Check: P = 400 W = P1 + P2 = 300 W + 100 W = 400 W <u>Check</u>

Conclusions/Solution:
Travel Iron inner heating element resistance = 48 Ohm
Travel Iron outer heating element resistance = 144 Ohm

Recommendation:
Team should design the travel iron to use outer heating element when operating at 240 VAC. Team should design the travel iron to use the outer heating element in parallel with the inner heating element when the iron operates at 120 VAC.

FIGURE B-49 Travel Iron Calculations for Resistances of Heating Elements

B.3.6.5 Configuration Schematic and Wiring Diagram

The team ended the week by creating a configuration schematic of the product, as shown in Figure B-50. The team also drew a pictorial wiring schematic, as shown in Figure B-51.

1. Sole plate
2. Outer heating element
3. Inner heating element
4. Well cover
5. Well red thermal filler
6. Thermostat
7. Thermostat screw
8. Black plastic cover
9. LED harness
10. LED harness washer
11. LED harness screw
12. Dual voltage switch
13. Power cord
14. Wire nut (2 pcs)
15. Wire from power cord to right heating elements
16. LED harness wire from right heating element terminals to left outer heating element terminal.
17. Wire from inner left heating element terminal to dual voltage switch
18. Wire from outer left heating element terminal to dual voltage switch
19. Top shell (plastic)
20. Temperature/Off knob
21. Covers over screw holes (4)
22. Shell-to-sole plate screws (4)
23. Clear plastic window

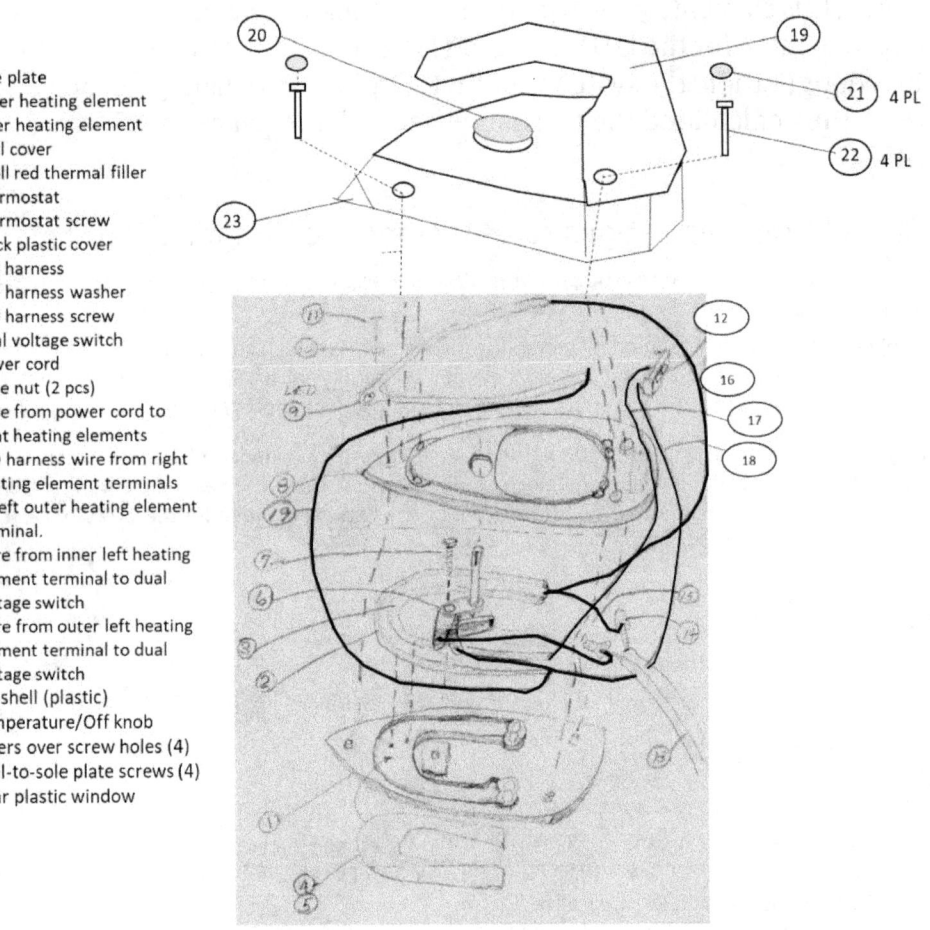

FIGURE B-50 Configuration Schematic of Travel Iron

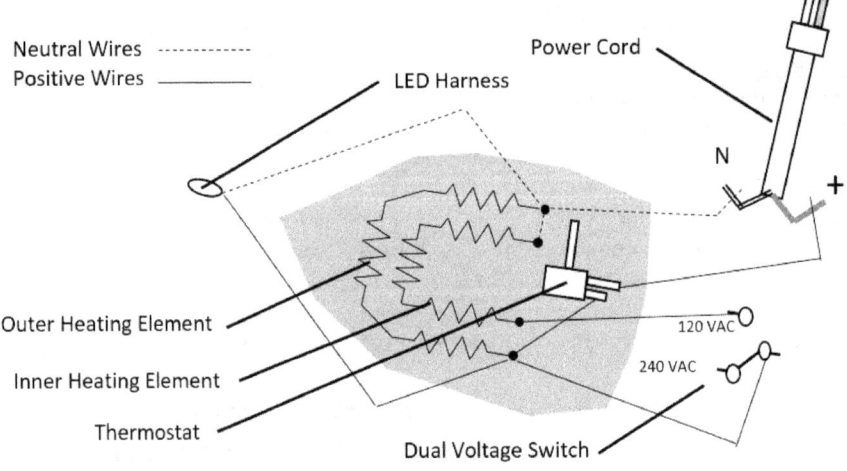

FIGURE B-51 Pictorial Wiring Schematic of Travel Iron

MODULE 18

Product Baseline Preliminary Design
Part 1

OVERVIEW

Modules 18 and 19 cover the remaining work necessary to establish the product's baseline preliminary design in the form of a CAD solid model with nominal dimensions. The COTS parts have already been specified and selected as part of Module 17. The tasks to be completed during modules 18 and 19 are the following: 1) Size and specify the components that need to be fabricated, 2) prepare an initial assembly plan, 3) draw parts-and-assembly sketches, and 4) prepare the Rev 0 version of the product CAD solid model.

LEARNING OBJECTIVES

- Gain knowledge and demonstrate the ability to complete all of the tasks necessary to establish the baseline preliminary design of a product.

PRE-LECTURE ASSIGNMENT

- Reread sections B.3.7 and B.3.8 and come to the lecture with issues you would like to discuss.

POST-LECTURE ASSIGNMENT

- Conduct a team meeting to divide the team into subteams and prepare task assignments for the following:
  - Subteam 1: Size and prepare specifications for all product components that need to be fabricated. Prepare a sketch of each of these parts including nominal dimensions and initial thoughts regarding materials and fabrication methods.
  - Subteam 2: Prepare an initial product assembly plan that includes sketches of how these and the COTS parts will be assembled.
- Each subteam completes at least 50 percent of their assigned work and sends the results to the other team members by the beginning of Module 19.

TEAM DELIVERABLES

- First 50 percent of the task to size and specify all components that need to be fabricated
- First 50 percent of task to prepare an initial assembly plan
- First 50 percent of initial CAD model, i.e., Rev 0 version, of the product using a software (such as SolidWorks)
- Minutes of the team meetings
- Team activity as documented in the team notebook

B.3.7 Size and Specify Fabricated Components

The first logical choice for sourcing of the product components is to buy them if they meet the required specifications and are affordable. However, some of the product parts must be fabricated because either they are not readily available from suppliers or the cost exceeds the target production cost.

In preliminary design, the fabricated parts need to be defined by hand sketches or simple CAD drawings to the point where all features are present and key nominal dimensions have been established. It is desirable at this time for the team to document their initial thoughts regarding the materials of construction and the methods of manufacture for each of the fabricated items. For key part interface dimensions, the tolerances for those dimensions should also be defined. The remaining part tolerances, final decisions on manufacturing methods, and final materials selection will be established during Phase 4: Detailed Design.

Most students are not used to making hand sketches; however, this skill is widely used in industry, so each team member should practice making some part sketches. These sketches may be rough, but with practice the students will become proficient at this skill. During detailed design, the drawings will be prepared to specific drawing standards. At the beginning of preliminary design, students should not worry about making perfect sketches. The important point is that they communicate the design intent. As students gain experience with engineering drawing standards (especially in detailed design), they will start incorporating the correct procedures into their hand sketches as well. Figure B-52 shows some examples of part hand sketches suitable for preliminary design.

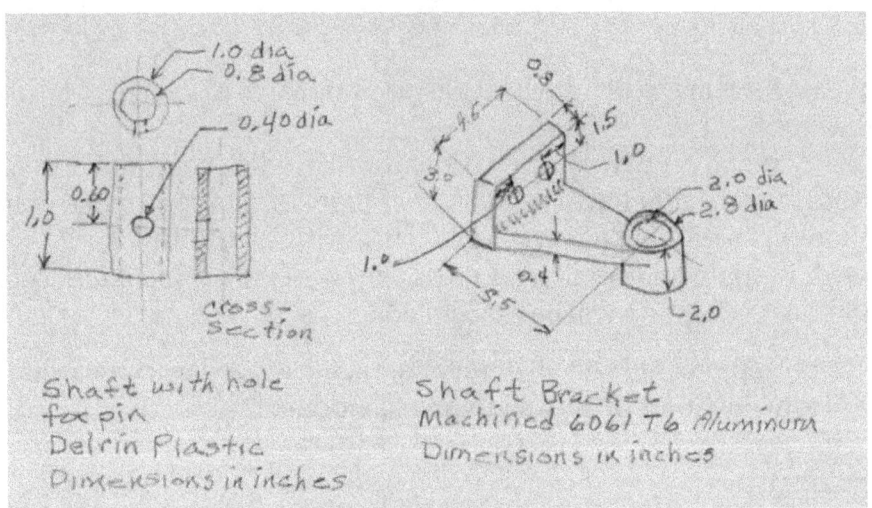

FIGURE B-52 Examples of Hand Sketches Suitable for Preliminary Design

For fabricated parts, the preliminary design analyses should ensure that when installed in the product, each component will enable it to meet all the engineering requirements. For fabricated items such as housings, brackets, and other structural members, it is important to verify that the part has been sized to withstand the loadings applied to it when installed in the product. For stress and fatigue analyses, it is especially important to state the source of the material properties and for the team to establish proper safety factors.

B.3.8 Initial Assembly Plan

At this stage, the design team has a list of all COTS components and their specifications as well as the specifications and working sketches of the components that need to be fabricated. The next step is to prepare an initial plan for assembling the components into the product configuration. A more detailed assembly plan will be addressed later in Phase 3 in the section on design for manufacturing and assembly (DFMA). The purpose of this initial plan is to provide the CAD designer on the team with enough information to construct the initial CAD model of the product using a software such as SolidWorks.

EM-(c)

The plan is a two- or three-page written document. It should address the questions listed in Table B-33. Based on the answers to these questions, the team prepares a list of sequential assembly tasks. This list is provided to the CAD designer to help with the product's overall layout.

Table B-33 Questions to Be Answered in the Initial Assembly Plan

- What is the production rate?
- Will there be an assembly line, or will each unit be assembled by the same person?
- Are any subassemblies or final assembly accomplished with automation?
- Are there any quality inspections during the assembly process?
- In what type of facility will the product be assembled?
- Are there any special tools or fixtures required for assembly?
- Are there any performance tests conducted during assembly?
- Are there any subassemblies, and why is there a need for subassemblies?
  - Some of the replacement parts are sold as subassemblies.
  - Certain subassemblies are needed to conduct subassembly tests prior to the product's complete assembly.
- Are there any safety concerns related to the product's assembly?

MODULE 19

Product Baseline Preliminary Design
Part 2

OVERVIEW
Module 19 covers the remaining work necessary to establish the product's baseline preliminary design in the form of a CAD solid model with nominal dimensions.

LEARNING OBJECTIVES
- Gain knowledge and demonstrate the ability to complete all the tasks necessary to establish the baseline preliminary design of a product.

PRE-LECTURE ASSIGNMENT
- Read sections B.3.7–B.3.10 and come to the lecture with issues you would like to discuss.

POST-LECTURE ASSIGNMENT
- Conduct a team meeting to complete the following tasks: 1) Discuss the work done by subteams 1 and 2 and agree on what each team should do to complete their assigned tasks before the CAD model is completed; 2) select the members of subteam 3 and give them the assignment of preparing the CAD solid model Rev 0 of the production unit configuration.
- Each subteam completes their assigned work and sends the results to the other team members prior to a second team meeting.
- Conduct a second team meeting before starting Module 20. During this meeting, review all the baseline preliminary design work and make any necessary changes.

TEAM DELIVERABLES
- Completed task of sizing and specifying components that need to be fabricated
- Completed task of preparing initial assembly plan
- Completed task of preparing initial CAD (i.e., Rev 0 version of the product using a software such as SolidWorks)
- Minutes of the team meetings
- Team activity as documented in the team notebook

B.3.9 Completing the Product Baseline Preliminary Design CAD Solid Model Rev 0

In Module 19, each design subteam completes their assigned work and sends the results to the other team members prior to a second team meeting to discuss and approve the product baseline preliminary design. The team completes preparing Rev 0 version of the product solid model, which includes all the parts with nominal dimensions. If required, certain parts are arranged into subassemblies. The model is constructed to follow the initial product assembly plan. Once the model is prepared, it is used to make (1) an exploded drawing of the product with all parts labeled and (2) a product outline drawing (this can be an isometric or have three orthogonal views) with the major dimensions shown. For more clarity, the design team may need to prepare sectioned assembly and subassembly drawings to show internal parts and/or details.

B.3.10 Example Travel Iron Team Report for Week 10

This week the travel iron team is divided into four stubteams. The activities of these teams are provided in this report.

B.3.10.1 COTS Specification and Supplier Identification

Subteam 1 (team members 1 and 2) specified the COTS components and identified suppliers for these items.

B.3.10.2 Dimensioned Sketches of Fabricated Components

Subteam 2 (team members 3 and 4) made dimensioned sketches of the components to be fabricated. They also selected initial material and manufacturing processes. One of these sketches is shown in Figure B-53.

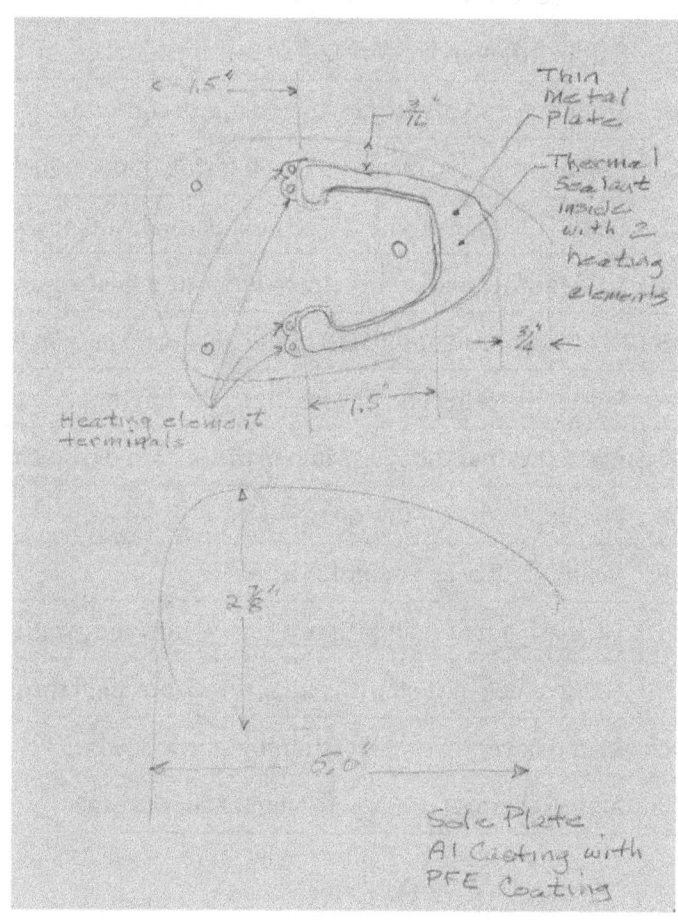

FIGURE B-53 Fabrication Sketch for Sole Plate

B.3.10.3 Assembly Plan

Subteam 3 (team members 5 and 6) started the week by making an initial assembly plan that included an assembly order list, as shown in Table B-34.

	Table B-34 Assembly Process for Travel Iron
1.	Sole plate coated with PFE
2.	Place heating elements in sole plate well with terminal connections outside well
3.	Pour thermal sealing epoxy to top of well
4.	Place well cover into well and let epoxy seal cover to well
5.	Install thermostat into sole plate with screw
6.	Install plastic cover and secure front of cover to sole plate with a screw
7.	Solder ends of LED harness to the outer heating element terminals.
8.	Install power cord sheath into cover shell and bring cord wires into bottom of shell
9.	Connect one end of wire (24) to power cord positive wire with wire nut
10.	Solder other end of wire (24) to one terminal of the thermostat
11.	Soler one end of wire (25) to end of other thermostat terminal
12.	Solder other end of wire (25) to left side of outer heating element terminal
13.	Solder one end of wire (18) to pole of dual voltage switch
14.	Solder other end of wire (18) to left, outer heating element terminal
15.	Solder one end of wire (17) to the dual voltage switch
16.	Solder the other end of wire (17) to the left inner heating element terminal
17.	Place the dual voltage switch in the pocket of the plastic cover
18.	Install plastic window on to shell
19.	Install shell over assembled iron
20.	Secure front of shell with two screws into the plastic cover bosses
21.	Secure back of shell with two screws from shell through plastic cover to sole plate
22.	Install covers over shell retaining screw heads
23.	Attach thermostat knob to shaft of thermostat

B.3.10.4 CAD Model Rev 0

During the latter part of the week, subteam 4 (team members 5 and 6) took the information from the other team members and created the initial CAD computer model of the product, as shown in Figure B-54. They tried several packaging approaches in order to find the best configuration for ease of assembly and a compact shape.

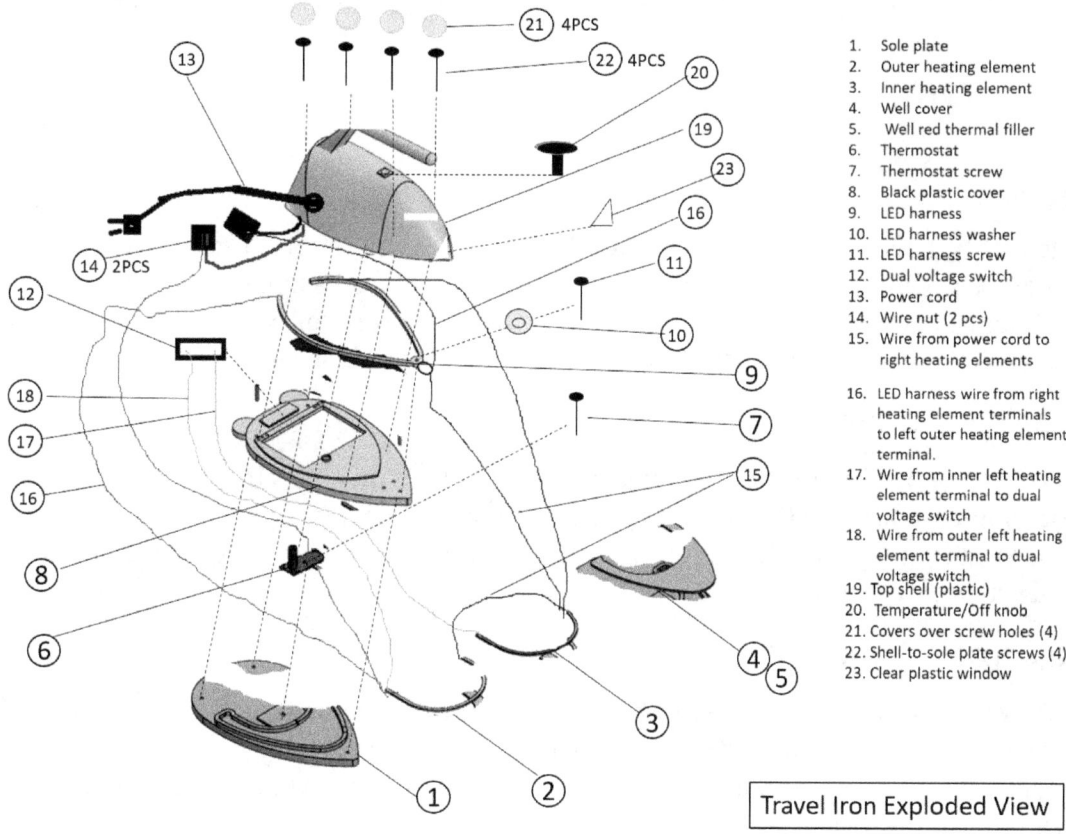

1. Sole plate
2. Outer heating element
3. Inner heating element
4. Well cover
5. Well red thermal filler
6. Thermostat
7. Thermostat screw
8. Black plastic cover
9. LED harness
10. LED harness washer
11. LED harness screw
12. Dual voltage switch
13. Power cord
14. Wire nut (2 pcs)
15. Wire from power cord to right heating elements
16. LED harness wire from right heating element terminals to left outer heating element terminal.
17. Wire from inner left heating element terminal to dual voltage switch
18. Wire from outer left heating element terminal to dual voltage switch
19. Top shell (plastic)
20. Temperature/Off knob
21. Covers over screw holes (4)
22. Shell-to-sole plate screws (4)
23. Clear plastic window

Travel Iron Exploded View

FIGURE B-54 Travel Iron CAD Model Rev 0 Exploded View

Module 19 Product Baseline Preliminary Design | **167**

MODULE 20

Updating the Requirements Matrix
Part 1

OVERVIEW

Modules 20 and 21 cover the additional analyses and testing required to update the requirement matrix to show that there is a high probability that the baseline preliminary design of the product will meet all engineering requirements. Changes required based on these analyses and tests will be included in an update of the CAD solid model to Rev 1.

LEARNING OBJECTIVES

- Gain experience in going from back-of-the-envelope analyses in conceptual design to more sophisticated analyses of the product in preliminary design.
- Gain experience in designing and conducting more involved POC tests of selected product components.

PRE-LECTURE ASSIGNMENT

- Review the status of the product's requirement matrix and make a list of what additional work is necessary to complete/validate the requirements matrix.
- Reread Section B.3.11 and appendices BA3 and BA4 to review the process steps for performing analyses and testing.
- Reflect on the engineering sciences courses taken so far in your engineering career and identify what analyses are applicable to addressing the requirements associated with the current product design.

POST-LECTURE ASSIGNMENT

- Conduct a team meeting to 1) identify the additional analyses and POC testing necessary to update the requirement matrix and 2) assign each team member with specific analyses and/or POC testing relative to the product requirements (including what needs to be accomplished before the start of Module 21).
- Each team member completes the first part of their assignments and sends the results to the other team members prior to the start of Module 21.

TEAM DELIVERABLES

- First 50 percent of the analyses and testing required to update the requirement matrix to show that there is a high probability that the updated baseline preliminary design (i.e., CAD solid model Rev 1 will meet all engineering requirements)
- First 50 percent of CAD solid model Rev 1
- Minutes of the team meetings
- Team activity as documented in the team notebook

B.3.11 Updating the Requirements Matrix and CAD Model

Up to this point in preliminary design, the team has taken the conceptual design from Phase 2 and has done the following: (1) identified the design drivers and analyzed them in order to establish the design's key characteristics, such as overall size, weight, cost, and/or performance; (2) created specifications for the product's components and established how these components are configured in the product; (3) prepared an initial assembly plan; and (4) captured these Phase 3 changes in a baseline preliminary design CAD solid model Rev 0.

The next step is to take the Phase 2 requirements matrix and update it in two ways. First, the requirements analysis must consider the changes in the design as captured in the CAD model Rev 0. Second, the analyses and POC testing—to establish the probability that the design will meet each of the engineering requirements—must be conducted at a more detailed level than that accomplished in Phase 2. The analyses and testing must be documented in the project notebook and summarized in the updated requirements matrix.

To accomplish this task in a relatively short amount of time, the team needs to divide into subteams and assign specific work to each subteam. Each subteam's results need to be shared and discussed with the whole team before the updated requirements matrix is approved by the team.

The process of analyses and testing will identify design changes that are needed to increase the probability that the design will meet all its engineering requirements when it enters Phase 5 development. These design changes need to be incorporated into an updated CAD model Rev 1.

MODULE 21

Updating the Requirements Matrix
Part 2

OVERVIEW

Modules 20 and 21 cover the additional analyses and testing required to update the requirement matrix to show that there is a high probability that the product's baseline preliminary design will meet all engineering requirements. Changes required based on these analyses and tests will be included in an update of the CAD solid model from Rev 0 to Rev 1.

LEARNING OBJECTIVES

- Gain experience in 1) going from back-of-the-envelope analyses in conceptual design to more sophisticated analyses of the product in preliminary design, 2) designing and conducting more involved POC tests of selected product components, and 3) reviewing the engineering work of others and suggesting ways of improving the quality of the reviewed analyses and/or tests.

PRE-LECTURE ASSIGNMENT

- Reread Section B.3.12.
- Review 1) other team members work relative to updating the product's requirements matrix and send suggested changes back to them, 2) the suggestions from team members on how you can improve your assigned analyses and/or testing tasks, and 3) the team's schedule and labor chart and identify a strategy for the team completing all the required work due by the end of Module 21.

POST-LECTURE ASSIGNMENT

- Conduct a team meeting to 1) discuss the work done on updating the requirements matrix, 2) finalize a strategy for completing all the work on the updated requirements matrix before the second team meeting, and 3) assign to one of the team members the task of updating the CAD model after the second team meeting but before the start of Module 22.
- Conduct a second team meeting to 1) review the work of the team members relative to updating the requirements matrix, 2) agree on when all inputs to the updated CAD model must be provided to the team member doing the model updating, and 3) discuss the progress of the team to date and discuss ways in which the team can be more effective.
- Update the CAD model from Rev 0 to Rev 1.

TEAM DELIVERABLES

- Completed analyses and testing required to update the requirement matrix to show that there is a high probability that the updated baseline preliminary design (i.e., CAD solid model Rev 1 will meet all engineering requirements)
- Completed CAD solid model Rev 1
- Minutes of the team meetings
- Team activity as documented in the team notebook

B.3.12 Example Travel Iron Team Report for Week 11

At the beginning of Week 11, the team split up the product engineering requirements and performed analyses and/or POC tests to determine how well the product's initial preliminary design would meet all engineering requirements. Product design changes were made as necessary to meet the requirements. This information was documented in the team notebook and summarized in the updated product engineering requirements matrix chart. Near the end of the week, team members 5 and 6 took the updated design information and incorporated it into the updated CAD model Rev 1 configuration.

MODULE 22

Systems Engineering and RMS Analyses

OVERVIEW

Preliminary design is an iterative process. Initial design steps identify the components required to perform the needed functions. These components and their assembly into an initial product configuration are accomplished to meet specific engineering requirements. The team is now ready to analyze this design relative to good engineering practice in the areas of product reliability, maintainability, and safety (RMS). This module, which falls under systems engineering topics, describes the RMS procedures and applies them to the team's product design. Changes resulting from these analyses are captured in the updated version of the CAD model.

LEARNING OBJECTIVES

- Understand the key processes involved in addressing the design areas of RMS.
- Gain experience applying the RMS design procedures to the team's product design.

PRE-LECTURE ASSIGNMENT

- Review the other team members' final work relative to updating the product's requirements matrix and the CAD model to the Rev 1 configuration.
- Read sections B.3.13–B.3.19.
- Prepare a list of key RMS tasks that need to be accomplished by the team.

POST-LECTURE ASSIGNMENT

- Conduct a team meeting to 1) discuss the team strategy for completing the needed RMS design activities within the project's resource constraints, 2) agree on a time to hold a second team meeting prior to the beginning of Module 23 to review the RMS analyses accomplished by the team members and the resulting changes to the product preliminary design in the form of CAD model Rev 2, and 3) form the following subteams:
  - Subteam 1: Reliability Analyses
  - Subteam 2: Maintainability Analyses
  - Subteam 3: Safety Analyses
  - Subteam 4: Update CAD Model to Rev 2
- All teams complete their assignments prior to the second team meeting.
- Conduct a second team meeting to complete and approve all the RMS activities accomplished by the team prior to the start of Module 23.

TEAM DELIVERABLES

- RMS analyses results
- Design changes in the CAD solid model that are needed to increase the probability that the design will meet all its engineering requirements based on the RMS analyses
- Updated CAD model Rev 2, based on the design changes dictated by the RMS analyses
- Minutes of the team meetings
- Team activity as documented in the team notebook

B.3.13 Systems Engineering

Thus far in Phase 3 preliminary design, the focus has been on designing the product to meet the specific engineering requirements developed in Phase 2. Substantial analyses and POC testing have been done to ensure that the product design has a high probability of meeting the engineering requirements when it enters Phase 5 product development. Now, the design team needs to review the Phase 3 design to make sure that it meets not only the specific engineering requirements, but also the best practices of engineering design.

EM-(q)

To do this, the team needs to learn more about the broad discipline of systems engineering. Systems engineering looks at the design from a functional point of view. It asks the question, what should the design be? Once the design is built and tested, systems engineering then asks the question, is the product what was wanted?

In a large engineering organization, systems engineering experts usually work with project design teams to make sure that all the necessary product functions are identified and addressed during the design process. Small companies generally rely on everyone in their engineering staff to participate in the systems engineering tasks. In the capstone project, each team member is expected to be knowledgeable and skilled at being a systems engineer as well as a design engineer.

So what is a system? Table B-35 provides the definition that appears in the National Aeronautics and Space Administration's (NASA) *Systems Engineering Handbook*. This definition is very broad. The components can be more than pieces of hardware, and an engineering system can include hardware, packaging, software, training, and manuals. An example of a large system is the Boeing 787 Dreamliner. This includes the aircraft itself, spare parts, manuals, methods of transportation, test equipment, and warranties. For the capstone design project, the system includes the prototype hardware, test rigs, software, and any documentation necessary to define, manufacture, operate, and maintain the system.

Table B-35 NASA's Definition of a System

A "system" is the combination of elements that function together to produce the capability required to meet a need. The elements include all hardware, software, equipment, facilities, personnel, processes, and procedures needed for this purpose—that is, all things required to produce system-level results. The results include system-level qualities, properties, characteristics, functions, behavior, and performance.

Source: NASA. 2020. NASA Systems Engineering Handbook Revision 2. NASA.

Figure B-55 presents the V model that is often used to present the systems engineering concept. The figure shows that systems engineers are interested in the design's functional aspects, whereas design engineers are interested in the design's physical configuration.

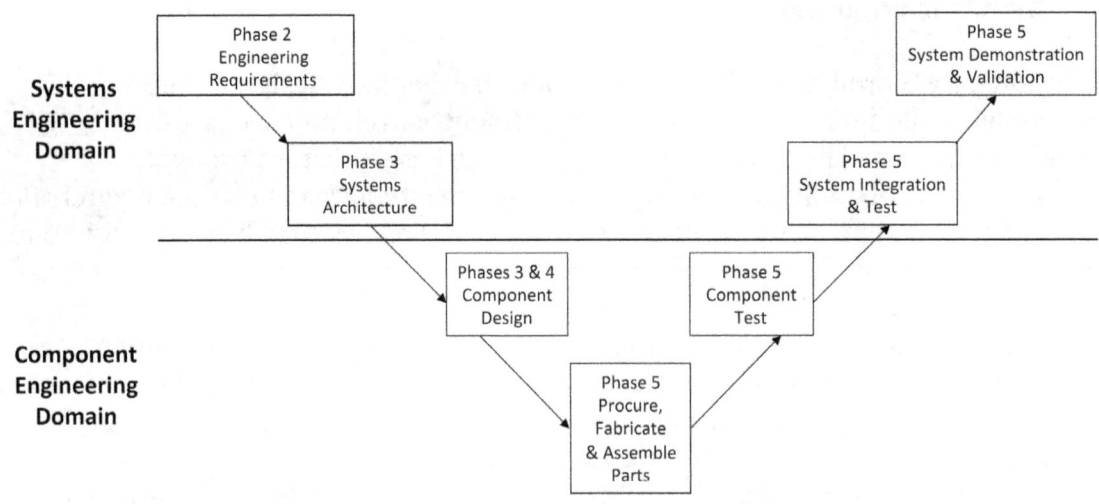

FIGURE B-55 The Vee Model of System Development Applied to the Phased Product Development Process

A large number of topics fall under the category of systems engineering, as shown in Table B-36. Some of these topics have already been introduced in prior sections of the handbook, while others are discussed in the following subsections.

Table B-36 System Engineering Topics Covered in This Handbook
Engineering requirements
Requirements validation matrix
Function block diagrams
Trade studies
Product improvement with goal function
Reliability
Maintainability
Logistics support
Safety
FMEA
DTC
Requirements validation
Commercialization

B.3.14 Trade Studies

The systems engineer oversees design trade studies to make sure that adequate alternatives have been considered and evaluated relative to the project's goals, requirements, and commitment to robust engineering. The *NASA Systems Engineering Handbook* is an excellent source of information on this subject. From this referenced document, Figure B-56 provides a flowchart for conducting trade studies.

EM-(b)

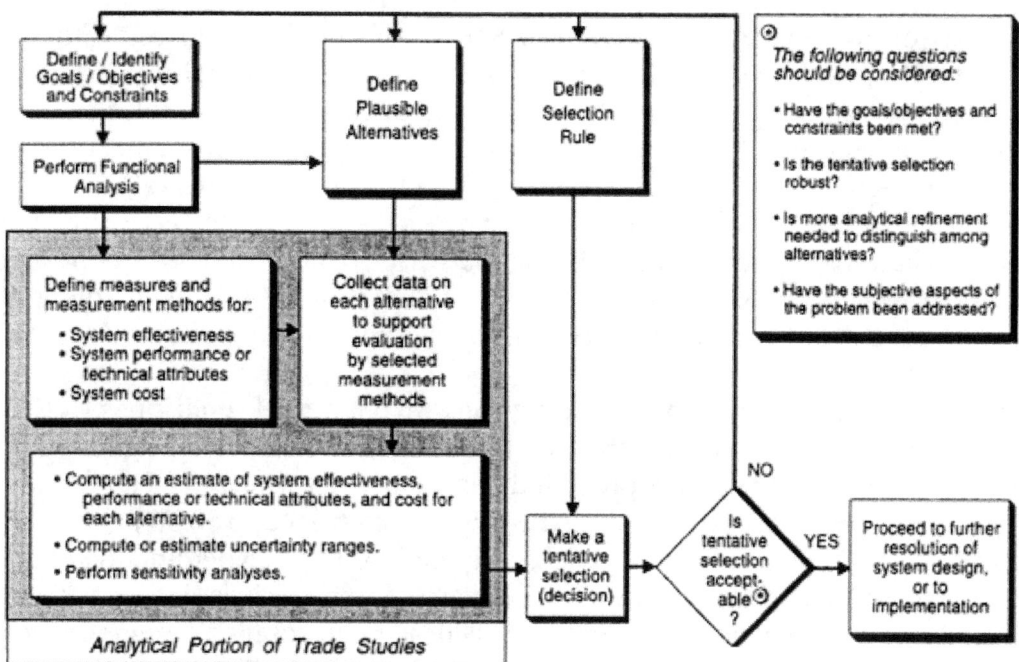

FIGURE B-56 Flowchart for Conducting Trade Studies

B.3.15 Product Design Improvement with Goal Function

A key system engineering task is performing design improvement analyses. Design improvement is a decision-making process. The designer must determine the options, what attributes each option has, and how customers value these attributes. In most cases, the designer has limited resources to identify these options, determine the customer value preferences, and assess each option to determine which presents the best overall value for gaining the largest market share.

The constraint of limited resources means that the designer cannot explore all possible options and hence it is not possible to conduct an optimization analysis where the best possible option is identifiable. However, it is possible to follow a strategy that narrows the set of potential options. For example, the design team can state that it is now known that the product's unit production cost (UPC) can be at least this low a value.

B.3.15.1 Example of Goal Function Improvement Limited by Project Resources
In the Phase 2 section of this handbook, the decision-making process involved identifying three or four specific concepts and then evaluating them using a Pugh matrix with the criteria

based on the engineering requirements. In the case of the travel iron, the Pugh analysis did not result in the definitive selection of one of the candidates, but instead it helped the team identify a final concept that was a combination of attributes from the other candidates.

The point here is that once the final concept was selected, the decision-making process changed. In the case of the travel iron, the design criteria for improvement was to just meet all the engineering requirements except for the UPC. The UPC was then selected as the design goal function, for which it is desirable to achieve as low a UPC as the team can design within resource limits.

If the travel iron project had more resources in terms of labor hours, material budget, and calendar time, the goal function could have included other attributes besides the UPC. In this case, there would need to be a weighting assigned to each of the variables.

For capstone projects in general, a good strategy is to arrive at a set of engineering requirements where all but one are valued as being met, but there is no additional value in exceeding the requirement. For example, the device must survive a 4-foot drop onto a concrete floor. This requirement is based on a technician dropping the device from their workbench. Surviving a 6-foot drop has no more value, because the workbench is only 4 feet high. This approach leaves only one requirement that benefits from further improvement beyond meeting the threshold requirement value.

For example, a team is designing a product for which the goal function is UPC. In this case, the threshold goals are to have the UPC less than $5 per unit and be as low as possible. As Figure B-57 shows, the team's proposed design had an estimated UPC of $5 per unit. However, by the end of conceptual design, the UPC had increased to $6.50 per unit. During preliminary design, the team was able to decrease the UPC to $5.25 per unit by redesigning the unit's housing. The UPC was further decreased to $4 per unit by going to a COTS shaft. Unfortunately, a finite element analysis of the housing early in detailed design resulted in an increase of UPC to $4.50 per unit. This value is still below the threshold goal, but the team had enough resources left in detailed design to redesign the unit's controller and thus decrease the UPC to about $3.50 per unit.

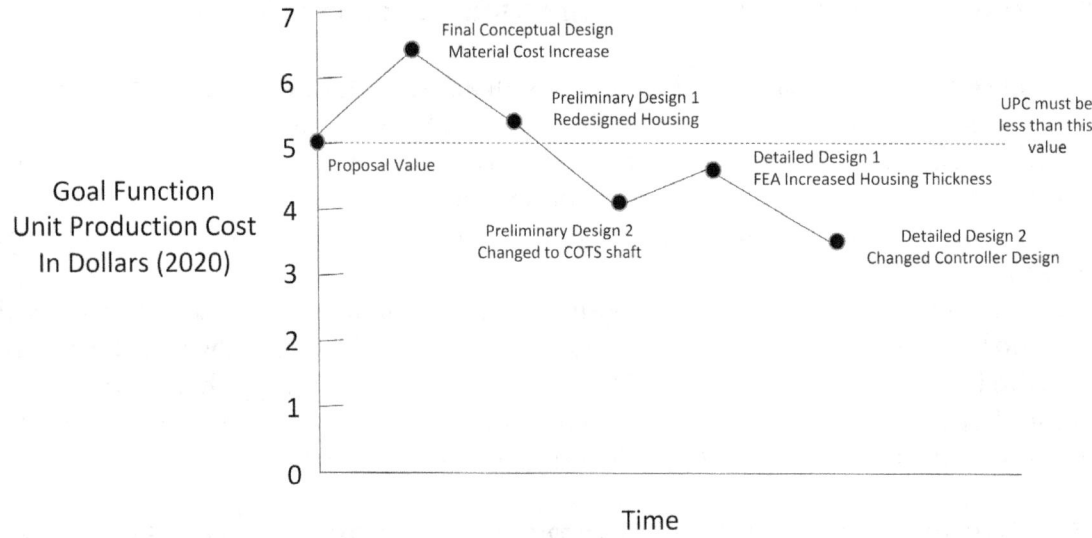

FIGURE B-57   General Format for Reporting Goal Function Progress

B.3.15.2 Example of Optimizing the Goal Function

Sometimes a team has enough knowledge about the design situation to actually identify an optimum design. An example of this is given below.

A 10-kilowatt-electric engine and generator set is being designed for remote communications towers. The market entry strategy is to offer a product that is as reliable as existing sets but more efficient such that it has a cost of electricity (COE) less than that of its nearest competitor. The analysis given in Figure B-58 shows that engine and generator set is optimized relative to COE with efficiency features that result in a sell price of $7,400.

Issue: Customers need a 10-kWe diesel engine/generator set for remote communications towers. Diesel fuel delivered to this remote site is expensive at $6.00 per gallon, so the engine needs to be efficient. The engine-generator set operates about 20% of the time and always at its rated output power. The designer wants to know what efficiency features to put in the engine so that the COE is optimized to the lowest value possible. The designer has the graph given below that shows the effect of engine improvements on the sell price and efficiency of the engine/generator set.

Problem Statement: Identify the engine configuration that optimizes the customer's COE.

Approach: Use the generalized equations for COE with the following assumptions:

- All engine configurations have an operations and maintenance (O&M) cost of $0.020/kWhe
- The yearly fixed charge rate (FCR) is 20%. This includes capital repayment, insurance, etc.
- The rated capacity (RC) of the unit is 10 kWe.
- The capacity factor (CF) is 15%.
- The fuel cost (FCo) is $2.80/gallon.
- The thermal energy content of diesel fuel is 0.02456 gallons/kWht.

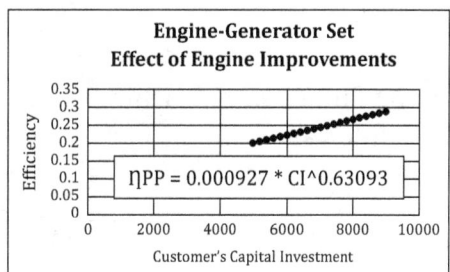

Defining Equations:

Where ηPP = power plant efficiency in kWe/kWt and CI = Capital Investment = Engine-Generator Set Consumer Cost

$$COE = \text{fixed charges} + \text{fuel charges} + \text{O\&M costs}$$

$$= \frac{(FCR)(CI)}{(RC)(CF)(8760 \text{ hrs/yr})} + \frac{(FCo)(0.02456 \text{ gallons/kWht})}{\eta PP} + \text{O\&M costs}$$

$$= \left[\frac{(0.20) CI}{(10)(0.15)(8760)} + \frac{(2.80)(0.02456)}{\eta PP} + 0.020 \right] \$/kWhe$$

Calculation Results: See graph at right

Conclusions: The engine/generator set should be designed with the improvements that yield at capital investment to the customer of $7400. This results in a COE minimum of $0.457/kWhe.

Recommendation: Use the engine improvements that result in engine/generator set customer capital investment of $7400. The efficiency for this unit will be 25.6 percent.

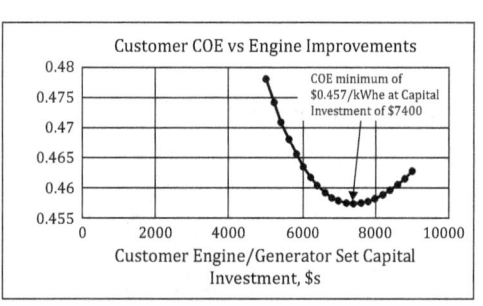

FIGURE B-58 Example of Goal Function Optimization

B.3.16 Reliability Engineering

B.3.16.1 Basic Concepts

System reliability is an important discipline within systems engineering. A good place to start in determining how to design reliability into a product is to use the definition of reliability given below.

> Reliability: The probability that the product will function properly over a specified amount of operating time or number of operating cycles.

An effective method of converting the above definition into mathematical terms is provided by Bazovsky[1] as follows: A test program is conducted where a fixed number of units, N_o, are repeatedly tested. After any time, t, N_s units have survived and N_f units have failed. Therefore, reliability, R, is defined as

$$R = \frac{N_s}{N_o} = 1 - N_f / N_o$$

Furthermore, the failure rate, λ, which is the instantaneous probability of failure for one component is defined as

$$\lambda = \frac{1}{N_s} dN_f / dt$$

Taking the derivative with respect to t for the equation for R results in

$$\frac{dR}{dt} = -\frac{1}{N_o} \frac{dN_f}{dt}$$

Rearranging this equation becomes

$$\frac{dN_f}{dt} = -N_o \frac{dR}{dt}$$

Substituting this value for $\frac{dN_f}{dt}$ into the equation for λ and noting that $\frac{N_s}{N_o} = R$ results in the following:

$$\lambda = -\frac{1}{R} \frac{dR}{dt}$$

By rearranging and integrating the above equation, the general formula for reliability becomes

$$\text{General Equation for Reliability: } R = e^{-\int_0^t \lambda \, dt}$$

As indicated in the above equation, the time-varying relationship of the failure rate, must be determined before the integral can be evaluated and the reliability determined.

B.3.16.2 Describing Failure Rates with the Bathtub Curve

To illustrate the general way in which units fail, the following example is given. Component A is tested by taking 785 units and testing each one to failure. The last unit fails after 985 hours of operation. The data are then put into 10 sequential bins of 100 operating hours for each bin. The derivative of the number of units failing at time t, $Nf(t)$, is then estimated by calculating the number of failures per hour for each bin and assigning this value to the mid-point time of that bin. The value of the number of surviving units at time t, $Ns(t)$, is estimated by subtracting half of the failures occurring in the bin from the number of surviving units at the beginning of the bin's timeline. The results of this analysis are shown in the spreadsheet of Table B-37.

Figure B-59 shows the failure rates from Table B-37 plotted versus operating time. This curve is typical of most components. They have high and decreasing failure rates at the beginning of life, relatively constant failure rates during their useful life, and increasing failure rates as they wear or age out. This is referred to in the reliability literature as the *bathtub curve*. An explanation of each of these three regions follows:

- <u>Infant Mortality</u>. This region of failure occurs at the beginning of the unit's operating life. It is caused by a lack of proper production quality control. The units with these failure modes should be removed during production and prior to the unit leaving the production facility. These units can be removed through inspection that detects the failure. The unit

Table B-37 Component A Test Data Failure Rates

Bin	1	2	3	4	5	6	7	8	9	10
Operating Time Frame, hrs	0–100	100–200	200–300	300–400	400–500	500–600	600–700	700–800	800–900	900–1000
Operating Mid-point Time, hrs	50	150	250	350	450	550	650	750	850	950
Bin Failures	400	47	42	37	32	29	25	22	81	70
Units at Beginning of Bin	785	385	338	296	259	227	198	173	151	70
Units at Bin Mid-Point	585	361.5	317	277.5	243	212.5	185.5	162	110.5	35
Failure Rate, Failures/hr	0.0068	0.0013	0.0013	0.0013	0.0013	0.0014	0.0013	0.0014	0.0073	0.0200
Failure Rate, Failures/1000 hrs	6.8	1.3	1.3	1.3	1.3	1.4	1.3	1.4	7.3	20.0

can also be operated for a brief period of time to uncover these early failure modes. This process of operating the unit at the factory is called *burn-in*. Specific procedures for the burn-in of electronic units is provided in the military standard MIL-STD-883C.
- <u>Wear-Out and Aging</u>. This region shows the failures increasing with time. This is due to the failure modes of wear-out and aging, such as material fatigue, corrosion, embrittlement, and contact wear.
- <u>Normal Operation</u>. The middle region of the *bathtub curve* has a relatively constant failure rate versus operating time. The product's useful operating life should occur in this region.

Figure B-60 shows why failures in the normal operating region tend to be random. The operating environmental stresses tend to vary normally. Likewise, each unit has some variability in its strength due to material property and manufacturing variations. The variation also tends to be normally distributed. The unit is designed to have the product's mean strength be substantially greater than the anticipated mean environmental stress. However, there is always some overlap of the stress and strength distributions. The region of overlap represents the product's random failures. These failures can be reduced by increasing the product's mean strength and/or reducing its variability. Likewise, the failures can be reduced by decreasing the mean environmental stress and/or reducing its variability.

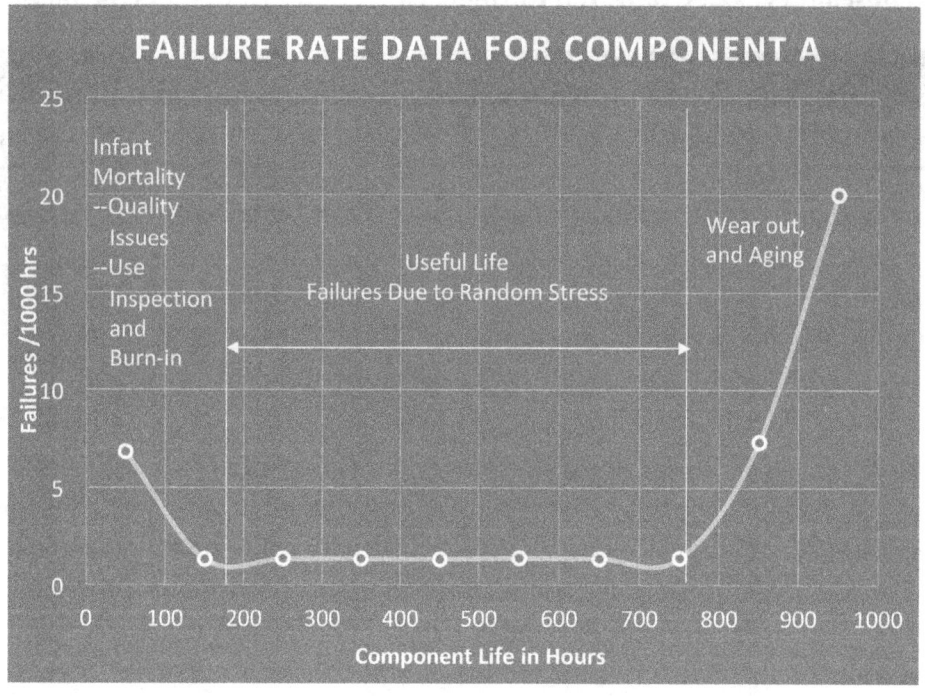

FIGURE B-59 Failure Rate vs. Operating Time for Component A Example

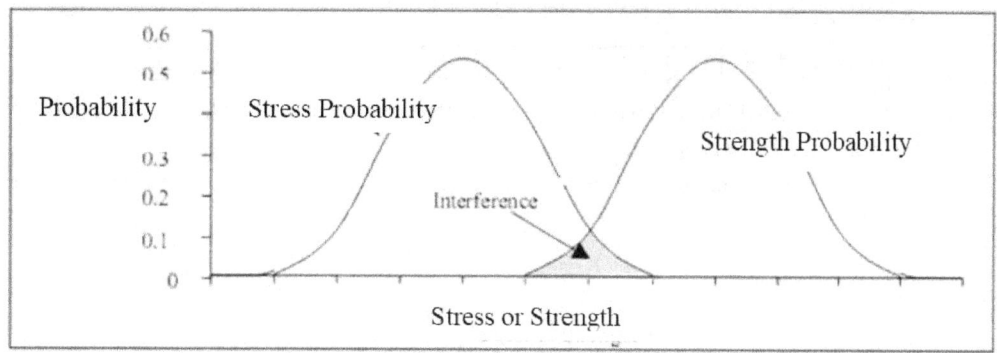

FIGURE B-60 Random Failures Occur When Stress and Strength Probabilities Intersect

B.3.16.3 Reliability for Constant Failure Rate Components

The first step the team must take to *design reliability into the product* is to make sure that the product's useful life does not include any infant mortality or wear-out/aging failure modes. This enables the designer to assume that the failure rates are constant during the operating life.

The next step is to minimize the useful life random failures by making sure that each component has the strength probability distribution to keep interference with the environmental probability distribution to a minimum. Too often components are selected or designed without taking into account all of the environmental conditions that will be encountered by the product during its useful life.

If the failure rate, λ, is constant, then the general reliability equation can be simplified to the following:

> Constant Failure Rate Equation for Reliability: $R = e^{-\lambda t}$

To illustrate, if the failure rate is 200 failures/million hours and the operating time is 100 hours, then the probability that the unit will operate without failure is 0.98.

The term *mean time between failures* (MTBF) often appears in a list of engineering requirements. When the failure rate is constant (not a function of time), the MTBF is the reciprocal of the failure rate, λ:

$$MTBF = 1/\lambda$$

When designing a system, it is usually assumed that all the components must operate properly in order for the system to function. Therefore, when the components are in a series arrangement, as shown in Figure B-61, the system reliability is the product of the component reliabilities, and the system failure rate is the sum of the component failure rates.

System with Components in Series

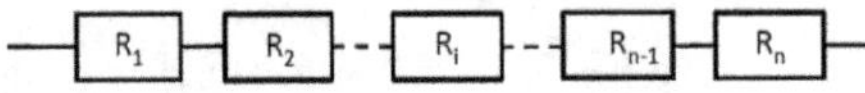

The reliability of the n components system, R_s, is:

$$R_s = R_1 \times R_1 \times R_1 \times .... \times R_n$$

For the exponential distribution model with $R_i(t) = e^{-\lambda_i t}$

$$R_s = e^{-\lambda_1 t} \times e^{-\lambda_2 t} \times ... \times e^{-\lambda_n t} = \exp\left[-\left(\sum_{i=1}^{n} \lambda_i\right) t\right]$$

∴ The equivalent system failure rate is $\lambda_s = \sum_{i=1}^{n} \lambda_i$

∴ The failure rate of the system is the sum of the failure rates of its components.

The system reliability decreases as the number of components in series increase.

FIGURE B-61 Reliability Equations for Components in Series

On the other hand, when the components of a product are in a parallel arrangement, as shown in Figure B-62, the system reliability, *Rs*, is defined by a more complicated function of those of the individual components. In parallel arrangement, the system reliability is greater than that of any individual component.

System with Components in Parallel

In this case, the system fails only if all its components fail.
The probability of system failure, F_s, is:

$$F_s = (1 - R_1) \times (1 - R_2) \times ... (1 - R_n)$$

$$= \prod_{i=1}^{n}(1 - R_i)$$

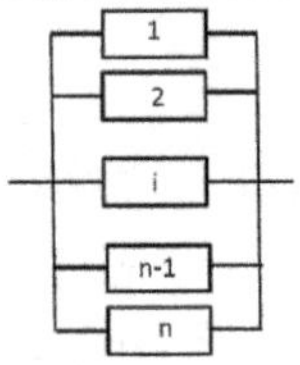

Where, R_i is the reliability of component i.

Reliability of the system is the complement of F_s:

$$R_s = 1 - F_s = 1 - \prod_{i=1}^{n}(1 - R_i)$$

FIGURE B-62 Reliability Equations for Components in Parallel

Companies with established product lines use their own warranty and field data to establish the failure rates of their systems and components. Companies that do not have this data can use generic estimates to help guide their design decisions although generic failure rates can lead to large errors depending on the specific design circumstances. For generic electronic devices, *Military Handbook 217* in its most recent revision is a good source. Failure rate estimates for mechanical components can be found in the Reliability Analytics Toolkit webpage.[1]

---

1 Reliability Analytics Corporation. 2020. *Reliability Analytics Toolkit*. Accessed July 29, 2020. https://reliabilityanalyticstoolkit.appspot.com/mechanical_reliability_data.

Table B-38 provides a list of generic mechanical component failure rates that the authors have found useful during conceptual and preliminary design phases. This data assumes that the system is operating in ground-based applications where temperature and vibration conditions are not extreme.

Table B-38 Generic Failure Rates for Common Mechanical Components	
DESCRIPTION	Failures/Million Hrs
Bearing Assembly	2.5
Bolt, Hex, Flanged	1.0
Bushing	0.2
Flange	0.2
Housing	0.4
Nut, Hex, Self-lock	0.1
Nut, Rotor	1.5
Retainer	0.1
Screw, Jam	1.2
Screw, Socket, Cap	1.2
Seal	4.0
Sensor	2.0
Shaft	1.0
Spring	1.5
Spring, Bellville	0.2
Structural Member	0.2
Washer	0.02
Washer, tab	0.2

The components in a system can be arranged in more complex reliability arrangements, such as two different components in parallel or a component in standby mode. In these cases, the mathematics becomes more complex, such as that shown in Figure B-62, and above and beyond the scope of this handbook. However, the situation where two identical components are placed in parallel is often used to increase the reliability of a system and is presented in Figure B-63.

Two Identical Components in Parallel

In this case, the system fails only if both components fail.

The probability of system failure, F_s, is:

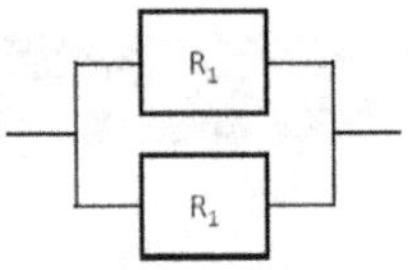

$$F_s = \prod_{i=1}^{2} (1 - R_i) = (1 - R_1)(1 - R_1)$$

$$F_s = (1 + R_1^2 - 2R_1),$$

Where, R_1 is the reliability of each of the two components.

The reliability of the system is the complement of F_s.

$$R_s = 1 - F_s = 1 - (1 + R_1^2 - 2R_1) = 2R_1 - R_1^2$$

For the exponential distribution model with $R_i(t) = e^{-\lambda_i t}$, the system reliability is:

$$R_s = 2R_1 - R_1^2 = 2e^{-\lambda t} - e^{-2\lambda t}$$

FIGURE B-63 Reliability of Two Identical Components in Parallel

B.3.16.4 Example Illustrating How Reliability Can Be Designed into the Product

To illustrate the benefits of adding a second identical component in parallel to the first (redundancy), the problem of designing a shaft spinning in a housing filled with oil is used. The shaft is supported on bearings, and outboard of each bearing is a sliding shaft seal to prevent oil leakage. The system is required to operate for one year with a reliability of at least 0.920. The upper portion of Figure B-64 shows the initial design for this application. The failure rates of the components are listed and used to find a system failure rate, which is then used to calculate the reliability for operating without failure for one year (8,760 hours). This configuration has only a reliability of 0.881 and does not meet the reliability requirement.

The design team notes that the seals have the highest failure rate of any of the components and are a relatively inexpensive component in the system. Based on these observations, the team decides to redesign the system and make the seals redundant, as shown in the lower portion of Figure B-64. In this case, the system's reliability is now 0.945, which exceeds the engineering requirement.

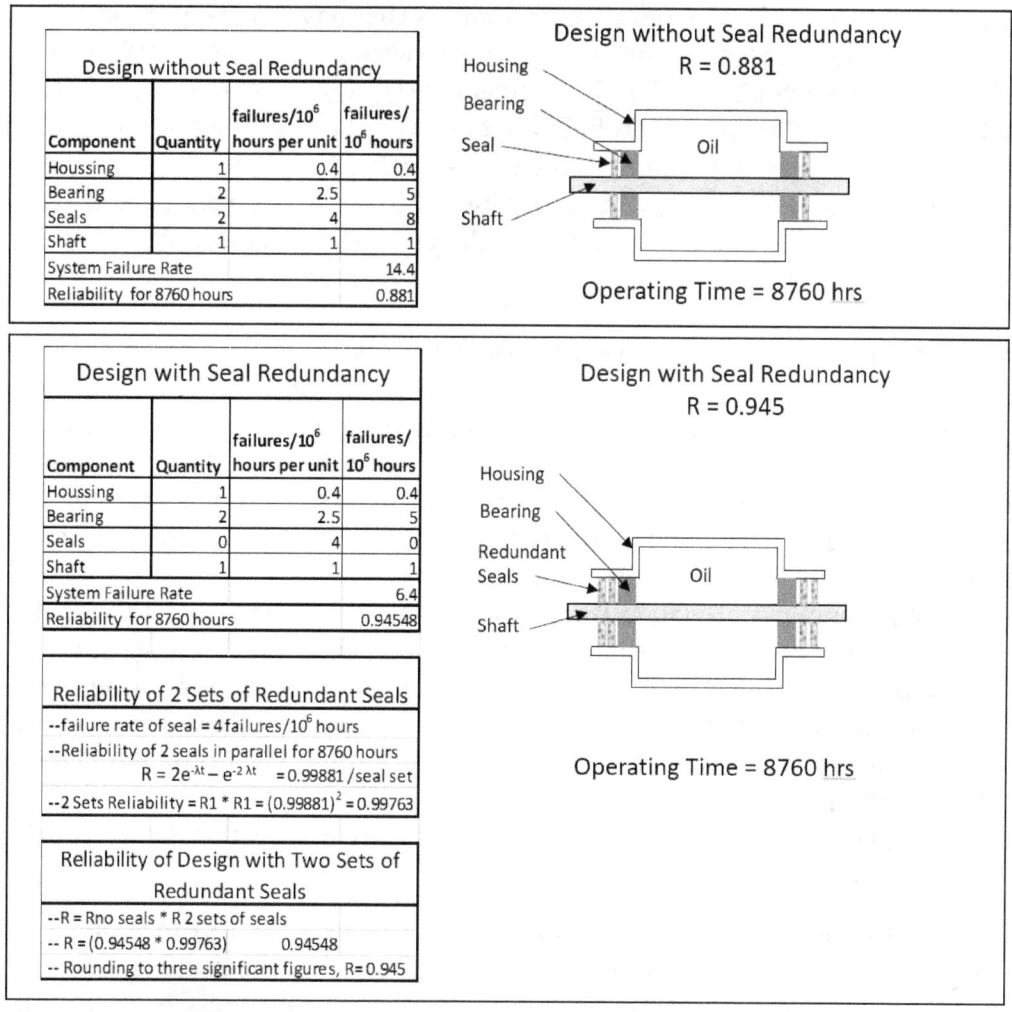

FIGURE B-64 Example of the Benefits of Redundancy in Meeting a Reliability Requirement

A number of key concepts in designing reliability into a system are illustrated in the above example. The team did not have their own database of component reliability based on their firm's prior testing and field data and instead had to use generic data. Hence, the reliability could be in error by a substantial amount. In this case, the team must include reliability testing of the system during development to gain a better understanding of the design's inherent reliability. Even though the team was limited to the use of generic failure rate data, they were able to identify the use of redundant seals as a relatively low-cost means of gaining a higher probability that the final design will meet the system reliability requirement.

Based on the team's understanding of the bathtub curve, they need to be sure that there are adequate inspections during the manufacturing process and that each unit undergoes burn-in testing prior to shipping. The bathtub curve also emphasizes the need for the deign team to make sure that none of their components have wear-out or aging modes of failure that could occur during the unit's useful operating life. Finally, the team should be sure that the probability of random failures is reduced by using components that are designed for the environment and operating conditions expected in actual use of the product.

Sometimes a product will incorporate components that have wear-out modes that occur during the product's useful life. In this case, there are two design choices: (1) Provide a means of replacing the component before it fails or (2) replace the component when it fails. The first choice will result in a higher product reliability; however, the costs involved in changing the component to avoid a potential failure may be substantial. The designer must take this into consideration. For example, many products include the use of electrochemical batteries. In many cases, the battery life is much shorter than the product life. This topic will be discussed in more detail in the next section.

B.3.17 Maintainability, Availability, and Warranties

For some products, such as disposable razors, maintenance is not an issue. However, for many products, such as automobiles, maintenance is an important factor. Automobile customers want their cars to be affordable to repair. That means the replacement parts need to be low-cost and the time required to diagnose and fix the problem must be kept to a minimum. These desires influence both component selection and packaging.

There are two types of maintenance: (1) scheduled maintenance and (2) unscheduled maintenance. For a car, scheduled maintenance includes tasks such as changing the engine oil at regular intervals. Unscheduled maintenance is needed when a component on the car fails. Sometimes a component is replaced before it fails. This is called preventive maintenance. Other times, the user waits until the component fails, and then it is repaired. This is called on-condition maintenance.

When preparing the requirements for a product, systems engineers consider the following relative to maintenance:

- How important is it that the product is always ready to fulfill all its functions?
- How important is it that any repairs can be accomplished by the user?
- How important is it that user maintenance does not require any special tools or equipment?
- How important is it that the user knows when a component is about to fail?

To illustrate these considerations, a jet aircraft engine will be used as an example.

- If the aircraft has only one engine, then it is highly important that the engine does not fail while it is operating in flight. On the other hand, a multiengine aircraft is usually designed to operate with one engine not working so that the flight can be safely ended.
- An aircraft jet engine may have turbine blades that have an operating life less than the operating life of the entire engine. In this case, these blades can be replaced before they enter the timeline for their wear-out modes. This type of major scheduled repair during which a major part of the product is repaired is called an overhaul.
- Jet aircraft engine turbine blades usually fail due to the excessive growth of a fatigue crack. For many aircraft, instead of replacing the turbine blades at regular intervals, a borescope is used at regular intervals to identify small blade cracks before they grow to failure. If no cracks are found, then the blades do not have to be replaced.

To illustrate the levels of maintenance, an aircraft will be used as an example.

- If the aircraft's tire fails, then it can be repaired on the flight line. This is called on-line maintenance, and the tire is referred to as a line replaceable unit (LRU).

- If the aircraft's radio fails, it is replaced as an LRU and repaired in the airport's shop. This is called intermediate maintenance.
- If the aircraft engine needs new turbine blades, then the engine must be sent to a major repair center called a depot, where it undergoes depot-level maintenance. Overhauls, for example, are a depot maintenance procedure. If the engine can be replaced on the line, then it is an LRU. If the aircraft needs to go somewhere other than the line to have the engine replaced, then it is not an LRU.

There are several variables associated with maintenance, as listed below:

- Mean time to repair (MTTR)—the average time it takes to repair the product
- Mean cost to repair (MCTR)—the average cost to repair the product
- Mean time between overhauls (MTBO)—the time intervals between scheduled major repair

MTTR is the total maintenance hours divided by the total number of repairs. Repair maintenance time includes:

- Time to troubleshoot and diagnose the problem
- Time to acquire repair parts
- Repair time
- Testing time
- Time to assemble and start up the asset

Operational availability in systems engineering is a measurement of how long a system has been available to use when compared with how long it should have been available to be used. It can be determined by the following equations:

$$\text{Availability} = \frac{\text{Up time}}{\text{Total time}} = \frac{\text{Mean up time}}{\text{Mean up time + mean down time}} = \frac{\text{MTBF}}{\text{MTBF + MTTR}}$$

In this definition, it is assumed that:

- The failure was noticed as soon as it occurred
- It was decided that the equipment be repaired as soon as the failure occurred
- The repaired equipment was immediately put back in service

In the case of an aircraft, for example, availability can be significantly reduced if replacement parts are not immediately available.

The designer must clearly understand the customer's maintenance philosophy and resources before decisions are made relative to product reliability, maintenance, and availability.

A warranty is a written guarantee issued to the purchaser of an article by its manufacturer that promises to repair or replace it if necessary, within a specified period of time. The warranty can often be an important feature of the product. In general, the better the warranty is, the higher the cost to produce the product is, and the more the customer is willing to pay for the product. This is not always true. For many consumer goods, the cost of sending the item for warranty repair and the inconvenience of not having the item in the interim leads the customer to just throw away the failed item and buy a new one. It is also important to understand the length of time that the warranty applies and whether it includes the labor and material costs to repair the item.

B.3.18 Logistics Support

Logistics support is related to maintenance. This systems engineering task considers what materials and training are needed to support the product. Some maintenance tasks can be done by the user, while others must be done by a qualified technician. Still other maintenance items may require that the product be sent back to the factory. Another consideration is the type of warranty provided with the product. Logistics support considers the following:

- Spare parts distribution
- Operating manuals
- Maintenance manuals
- Special tools and test equipment
- Warranties
- Where maintenance will occur

B.3.19 System Safety

The first duty of an engineer is to protect public health, safety, and welfare. The professional societies such as the American Society of Mechanical Engineers (ASME), the National Society of Professional Engineers (NSPE), and Accreditation Board for Engineering and Technology (ABET) all have statements similar to this one in their codes of ethics. To meet this duty, the engineering project team must perform a number of tasks, as listed in Table B-39, to ensure that the product being designed is safe for all concerned such as the customer, user, maintainer, and manufacturer.

Table B-39 Product Safety Design Tasks

1. Identify and comply with applicable safety standards and regulations.
2. Research and identify applicable technologies for safe hardware design.
3. Consider how the user might misuse the product.
4. Provide allowances for safety issues that might occur due to wear, etc.
5. Consider hazards to those manufacturing and repairing the product.
6. Provide proper warnings on the product and in product documentation.
7. Prepare and document hazards analysis (HA) and FMEA.
8. Implement hazard mitigation actions identified in the HA and FMEA.
9. Design the development test plan to address potential safety concerns.
10. Document all safety-related activities in the project notebook.

Two effective safety analysis tools are (1) an HA and (2) an FMEA. The HA will be discussed in this section, and the FMEA will be presented in the following section.

The steps to perform an HA are given in Table B-40. A sample form for the HA is provided in Table B-41. A good reference for hazards analysis is *Standard Practice for System Safety MIL-STD-882D* 10 February 2000.

Table B-40 Steps to Perform a Hazards Analysis

1. Identify known hazards.
2. Determine the cause(s) of the hazards.
3. Determine the effects of the hazards.
4. Determine the probability that an accident will be caused by a hazard.
5. Establish initial design and procedural requirements to eliminate or control hazards.

Table B-41 Hazard Analysis Form

Hazard	Cause	Effect	Probability of Accident due to Hazard	Corrective or Preventive Measures

Each team must complete a hazard and safety analysis on their design. This should take into account the perspective of all stakeholders in the device (e.g., customer, user, maintainer, manufacturer, etc.). Any potential hazards must be mitigated, and the steps taken during this activity must be documented. The final deliverable is a completed hazard analysis form. To aid in hazards identification, the checklist given in Table B-42 may be used.

Table B-42 Hazard Identification Checklist[2]

Temperature Effects
- [ ] Hot surfaces
- [ ] Cold surfaces
- [ ] Freezing
- [ ] Melting
- [ ] Changing material properties
- [ ] Pressure changes
- [ ] Humidity
- [ ] Moisture

Pressure
- [ ] Overpressure
- [ ] Rupture
- [ ] Implosion
- [ ] Explosion
- [ ] Unexpected release
- [ ] Unexpected flows
- [ ] Noise

Hazardous Materials or Biological Agents
- [ ] Skin
- [ ] Inhalation
- [ ] Swallowing

Ionizing Radiation
- [ ] Alpha
- [ ] Beta
- [ ] Gamma
- [ ] X-ray
- [ ] Neutron
- [ ] Electron beam

Mechanical
- [ ] Pinch points
- [ ] Crushing
- [ ] Ejected material
- [ ] Sharp edges
- [ ] Moving parts
- [ ] Unstable parts

Electrical
- [ ] Arcing
- [ ] Explosion
- [ ] Unwanted feedback
- [ ] Shock
- [ ] Burns
- [ ] Ignition
- [ ] Loss of power
- [ ] Spurious signals
- [ ] Loss of signal
- [ ] Electrostatic discharge

Radiation
- [ ] Microwave
- [ ] Ultraviolet
- [ ] Infrared
- [ ] Laser

Accelerations
- [ ] Unexpected movement
- [ ] Impacts
- [ ] Binding

Credits

Fig. B-56: Source: NASA Systems Engineering Handbook, SP-6105.

---

[2] Checklist was adapted from "Preliminary Hazard Analysis (Lecture Presentation)," R.R. Mohr, Sverdup Technology, Inc., June 1993 (Fourth Edition).

MODULE 23

FMEA, DFMA, and DTC
Part 1

OVERVIEW

Modules 23, 24, and 25 address FMEA, DFMA analysis, and design-to-cost (DTC) analysis. This module is focused on the FMEA. It will describe the FMEA process and have the team apply it to their product design. The FMEA process requires work to be done by individual team members and the team as a whole. In this module, the team will perform the initial FMEA cycle and arrive at an understanding of the areas of the design that need to be improved. In the next module, the team will make those improvements and quantify the benefits of these changes by using risk priority numbers (RPN).

LEARNING OBJECTIVES

- Understand the importance of the FMEA in the design process.
- Comprehend how to conduct a FMEA.
- Gain FMEA experience by conducting this form of analysis on the product's preliminary design.

PRE-LECTURE ASSIGNMENT

- Read and study ssections B.3.20 and B.3.21 and Appendix BA5.
- Review the information in the project notebook describing the current state of the product's preliminary design.
- Select one product component and use the form in the handbook to conduct an FMEA of this component. Bring the completed form to the class meeting.
- Review the project's management metrics and determine what the team needs to do to keep the project on schedule and within the resources budget.

POST-LECTURE ASSIGNMENT

- Conduct a team meeting to 1) review the project management metrics and make adjustments in the project's plan that will keep the project on schedule and within the resources budget, 2) discuss how the FMEA process will be applied to the product, 3) review each member's example component FMEA, and 4) assign portions of the FMEA to each team member.
- Each team member completes their FMEA assignment and sends the results to the other team members prior to the start of Module 23.
- Conduct a second team meeting to compile the members' input into a complete initial product FMEA and decide on what design changes must be made to lower the system's RPNs.

TEAM DELIVERABLES

- Results of the initial team FMEA analysis
- List of areas of the design that need to be improved by improving the RPNs
- Minutes of the team meetings
- Team activity as documented in the team notebook

B.3.20 Failure Mode and Effects Analysis (FMEA)

The FMEA is a systemized group of activities designed to:

- Recognize and evaluate the potential modes of failure of a product and their effects
- Identify actions that could eliminate or reduce the chance of potential failures occurring
- Document the process

Ideally, FMEAs are conducted during the product design or process development stage. A FMEA supports the design process in reducing risk of failure by the following:

1. aiding in the objective evaluation of design requirements and design alternatives;

2. increasing the probability that potential failure modes and their effects on systems and product operation have been considered in the design and development process;

3. providing additional information to aid in the planning of efficient design testing and product development programs;

4. developing a list of potential failure modes ranked according to their effect on the "customer," thus establishing a priority system for design improvements and development testing;

5. providing an open issue format for recommending and tracking risk-reducing actions; and

6. providing future reference to aid in analyzing field concerns, evaluating design changes, and developing advanced designs.

The FMEA is initiated early in the design process and continually updated as changes occur or additional information is obtained throughout the phases of product development. The FMEA addresses the design intent and assumes that the design will be manufactured and assembled to this intent. It does not rely on process controls to overcome potential weaknesses in the design, but it does take the technical and physical limitations of manufacturing and/or assembly process into consideration.

The FMEA must be accomplished by a team that possesses the background experience and necessary skills to successfully manage the process. The FMEA process is summarized in Table B-43. Procedures for conducting a FMEA are provided by several organizations, such as the US military and the SAE International. A detailed description on how to prepare an FMEA is provided in Appendix BA5.

Table B-43 FMEA Process
1. Assemble a team with the necessary skills and product information.
2. Define the system through schematics, drawings, and specifications.
3. Assemble additional information, such as customer input and performance of similar systems.
4. List the functions the system must provide.
5. Utilize a worksheet to document the analysis
6. For each component, identify the ways it could fail (i.e., failure modes).
7. For each failure mode, list the effects on the component and on the system.
8. Establish the criticality of each failure mode by determining its risk priority number (RPN), which is the product of the severity (S), probability of occurrence (O), and ability to detect (D).
9. For each failure mode, list the action items that need to be accomplished to reduce the RPN by identifying actions to be taken to mitigate the cause of failure.
10. Assign a person to be responsible for each action item.
11. Establish a date for the action item to be completed.
12. Once the worksheets have been completed, list the failure modes by descending risk number.
13. Identify the critical failure modes by selecting those with a high RPN.
14. Designate a team leader to be responsible for assuring that all actions recommended have been implemented or adequately addressed before the detailed design review.
15. Make the FMEA a living document that always reflects the latest design level as well as the latest relevant actions, including those occurring after the start of production operations.
16. Identify and implement any necessary design changes in the product as a result of the actions taken to mitigate the causes of failure during FMEA.

B.3.21 Example Travel Iron Team Report for Week 12

Week 12 proved to be a challenging week for the team as they learned to use the system engineering tools of RMS analyses and FMEA to improve the travel iron product design.

Subteam 1 (team members 1 and 2) completed the following reliability tasks during the first part of the week:

- Made a list of manufacturing inspections and testing prior to product shipment to ensure that all infant mortality failures will be removed from the lot of units being shipped
- Reviewed all component specifications and analyses to make sure that each component would not have any failure modes during its useful product life
- Prepared a reliability block diagram of the product components
- Estimated the MTBF of the product using generic-component failure rate data
- Made a list of failure modes for each component and estimated the failure rates for use in the FMEA

Subteam 2 (team members 3 and 4) completed the following safety tasks during the first part of the week:

- Reviewed the product safety design tasks in the handbook
- Identified the following Underwriters Laboratories (UL) standard and tried to acquire it from the university library:
  - UL 60335-2-3: Standard for Safety of Household and Similar Electrical Appliances, Part 2: Particular Requirements for Electric Irons

- *Note:* The library had to order this standard, and it would not be available until the beginning of the next semester. This was reported to the whole team in an email with the recommendation that the team revisit the standard during their Phase 4 activities next semester. The team agreed by return emails, and this was noted in the team notebook.
- Conducted a safety hazard analysis (HA)
- Considered that (1) the operator may not switch to the right voltage setting, (2) the operator may try to open the unit and repair it, and (3) the operator may leave the unit on with the sole plate on a surface rather than in an upright position
- Made a note in the project notebook that a safety test during both engineering prototype and production unit prototype development should be conducted
- Made a list of warning labels to be attached to the product and in the instruction booklet
- Made a list of warning notes that should be included on the assembly drawing

Subteam 3 (team members 5 and 6) incorporated the product changes into the CAD model Rev 2.

The team spent their remaining time this week conducting the first 60% of the FMEA on the production unit design. This included the determination of the initial product RPNs and a decision about the improvements that must be made to lower the overall RPN.

It should be noted that the initial FMEA required more time by the team than budgeted. The team agreed to spend the extra time this week to have an effective initial FMEA.

MODULE 24

FMEA, DFMA, and DTC
Part 2

OVERVIEW

This module will continue work on the FMEA and start the DFMA and DTC analyses. DFMA analyses seek to reduce the design's UPC. It is addressed initially in this module and will be addressed again during Phase 4: Detailed Design. This module will also address DTC studies. The team will complete Worksheet 3 on DFMA and DTC in this module.

LEARNING OBJECTIVES

- Continue to understand how to use RPNs in the FMEA to improve the design
- Understand the process of materials selection
- Understand the process of designing a product for manufacturing and assembly (DFMA)
- Understand the different types of product costs and how to conduct DTC and life cycle cost (LCC) analyses.
- Gain experience by applying the above tools to the product's design
- Gain experience in managing a team through a time of many tasks to be completed in a short amount of calendar time

PRE-LECTURE ASSIGNMENT

- Read and study sections B.3.22–B.3.24.

POST-LECTURE ASSIGNMENT

- Conduct a team meeting to 1) review team progress and agree to mid-course corrections to keep the project on schedule and within resource budgets and 2) form the following subteams:
  - Subteam 1: Redesign selected aspects of the product's design to lower the system FMEA RPN.
  - Subteam 2: Complete 50 percent of the process of redesigning selected aspects of the product's design relative to materials and DFMA.
  - Subteam 3: Complete 50 percent of the process of redesigning selected aspects of the product's design to meet the desired DTC goals.
- Conduct a second team meeting to review the results of each subteam's work, complete Worksheet 3, and decide on a go-forward plan for the next module.

TEAM DELIVERABLES

- Update of the FMEA
- Worksheet 3 for DFMA and DTC
- Minutes of the team meetings
- Team activity as documented in the team notebook

B.3.22 Design for Manufacturing and Assembly (DFMA)

In the case of both the engineering prototype and the production unit, once detailed design is completed, the next task is to manufacture the parts and components and then assemble them into a final product. Traditionally, the designer gives the designs to the manufacturing engineers, who were not involved in the design effort, to deal with the various manufacturing and assembly problems arising during the actual manufacturing process. Many of these problems can be avoided by including the manufacturing and production engineers in the project's design phases. This concurrent engineering approach is being used extensively in industry.

It ensures that the issues related to manufacturing, cost, quality, assembly, and serviceability are addressed at the design stage. Design for manufacture (DFM) deals with ensuring that the design achieves ease of manufacture of the collection of parts that will form the product after assembly. Design for assembly (DFA) deals with ensuring that the design achieves ease of product assembly. The importance of using DFM in product design is highlighted by the fact that about 70% of manufacturing costs of a product are determined by design decisions.

Specialized books, such as the ones listed as follows, present detailed coverage of DFMA and related topics, including their principles and guidelines. They also include examples and applications of DFMA in product design.

- Boothroyd, Geoffrey, Peter Dewhurst, and Winston Knight. 2010. *Product Design for Manufacture and Assembly.* Third ed. Boca Raton, FL: CRC Press.
- Dieter, George D. and Linda C. Schmit. 2021. *Engineering Design.* New York: McGraw-Hill Education.
- Stoll, Henry W. 2020. *Design for Manufacture: Principles and Practices.* Independently published.
- Ulrich, Karl T., Steven D. Eppinger, and Maria C. Young. 2020. *Product Design and Development.* New York: McGraw-Hill Education.

In this section, each capstone design team is required to apply DFMA to the engineering prototype design CAD model Rev 2. To achieve this goal, the design teams apply the principles and guidelines of DFM and DFA that are listed in the following subsections. The schematic in Figure B-65 shows how to implement the DFMA process during product design. This figure shows that DFM and DFA are applied concurrently during the product design process to arrive at a product design for a problem or concept.

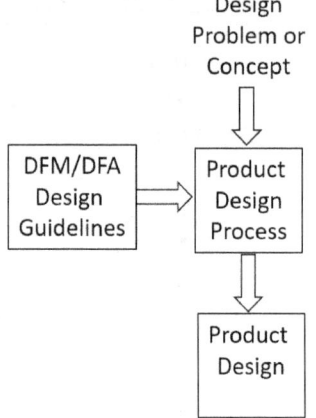

FIGURE B-65 Implementing the DFMA Process in Product Design

B.3.22.1 DFM Guidelines

DFM is a development practice, emphasizing manufacturing issues throughout the product development process. Its purpose is to ensure ease of manufacturing of each of the collection of parts and components that will form the product after assembly. In industry, DFM is a concurrent engineering tool employed by a cross-functional team that usually includes from the start design, materials, manufacturing, and production engineers. The main goal of the team is to design a product that is easily and economically manufactured but not alter the product's concept or functionality. The team members' combined efforts also lead to reducing the product development cycle time.

In capstone design projects, the design team has limited resources and does not have the means of appropriately representing the manufacturing, materials, and production areas. Furthermore, these team members have relatively limited experience in these fields. The team is advised to seek the help of their instructors or professors who teach courses in these areas as subject-area experts. At Arizona State University (ASU), capstone design teams seek and receive advise from the course instructor, who has extensive experience in manufacturing, materials, and design, and from the student machine shop director.

To properly implement an effective DFM process during design, the design team applies widely known guidelines, which are a collection of good design practices that are compiled through years of experience. A list of DFM guidelines that are considered during component design are included below.

1. **Minimize total number of parts in the product:** This guideline leads to savings in the cost and time of making the eliminated parts without altering the product's concept or functionality. Minimize parts count by optimizing the number of parts during preliminary design and incorporating multiple functions into single parts. Avoid making the remaining parts, which also serve the functions of the eliminated components, too complex.

2. **Use standard components:** This guideline recommends using both COTS and in-house standard components that are developed by the design firm. These parts are expected to be less costly and possess consistent specifications that allow parts interchangeability.

3. **Reduce the number of operations to produce a component:** This guideline will reduce the number and variety of machine tools, tooling, and fixturing used in making the component. Carefully study the component's functional features from its drawings and identify the opportunities that lead to reducing the number of operations.

4. **Select economical manufacturing processes:** Many alternative manufacturing processes can be used to make the component. Select the processes that are technically and economically capable of making the component. Technical capability includes the ability to produce the component features, specified surface finish, and specified tolerances.

5. **Reduce material waste or scrap:** This guideline aims at selecting materials whose properties do not deteriorate or fail during processing. For example, components made from high-carbon steels using welding have poor weldability, which may lead to cracks and eventual discard of the welded component. Likewise, avoid selecting machining processes, which produce chips, when there is an alternative technically capable and economically viable process.

6. **Avoid specifying unnecessary and excessively tight tolerances for the component:** Tolerances are specified to satisfy certain levels of fit between assembled components. Processes that can produce tight tolerances, such as grinding operations, are usually more expense. Specifying tolerances that are tighter than functionally needed results in increased cost. Select the manufacturing processes that can produce the specified tolerance and surface finish.

7. **Standardize the components' design features:** Design features include holes, screw threads, chamfers, bend radii, and splines. Using when possible the standard or the same type pf features will reduce the number and types of tools used in making them.

8. **Design components for ease of fabrication:** This guideline includes selecting for the component a material that has better manufacturability (e.g., formability, castability, machinability, weldability, etc.). Also consider reducing errors and worker efforts by decreasing the number of times that are required to reorient a part during its making. It is recommended to use generous fillets and radii on molded, cast, and machined components.

9. **Avoid using unnecessary secondary operations:** Examples of secondary operations include heat treatment for welded and cast parts, machining for cast and formed parts, grinding and polishing, etc. Use secondary operations only when there is a specified functional requirement to save time and cost.

10. **Reduce production cycle time:** Use manufacturing processes, process parameters, and supporting tooling and resources to achieve higher productivity.

Capstone design teams will apply the above DFM guidelines for their product design and introduce any modifications to the CAD model Rev 2 that arise from the DFM process. The team should list the guidelines used to improve the product design and the impact of each on the CAD model Rev 2.

B.3.22.2 DFA Guidelines

The manufactured and COTS components are assembled into subassemblies and products using the assembly operations, namely (a) handling and (b) insertion and fastening. As mentioned above, DFA is a tool used to assist the design teams in designing the product for ease of assembly by using a set of compiled best practice guidelines. The main goal of DFA is to address costly assembly issues in the design and prototyping stage of the product. In industry, DFA is another concurrent engineering tool employed by a cross-functional team of product design and development. They apply DFA to reduce product assembly cost and shorten the product development cycle time.

In capstone design projects, the design team has limited resources and does not have the means of assembling a concurrent engineering product design team to apply DFA during the product/prototype design process. However, each team is also required to employ concurrent engineering on their capstone design project by applying DFA guidelines to the engineering prototype design CAD model Rev 2.

The DFA guidelines that are considered during product design are included in the list below. These guidelines belong to the two types of assembly operations: part handling and part insertion and fastening. Insertion includes grasping, orienting, and positioning of the part.

1. **Minimize the number of parts of the product:** Fewer parts reduce the product assembly time. Eliminate parts by combining several parts into a single component that performs both its function and those of the eliminated parts. When possible, exchange hardware to ease assembly (e.g., replace screws and washers with snap or press fits).

2. **Design parts with self-locating features:** Design the mating features for easy insertion of parts (e.g., add tapers at the end of bolts).

3. **Provide parts with adequate alignment features:** This guideline will ease the insertion of a component in a mating one (e.g., use dowels on one part to insert the part in into holes or slots of the mating part).

4. **Minimize assembly directions and reorientation of parts:** Enable assembling the product from one direction (e.g., emphasize z-axis top-down assemblies). Avoid rotation of an assembly, which requires extra time, motion, and additional transfer station and fixtures. Consider using base parts to locate other components.

5. **Design parts for ease of retrieval, handling, and insertion:** If possible, design the parts so that they cannot be tangled or nested into each other. For example, design for component symmetry for ease of insertion and provide chamfers or recesses for part alignment. Also provide generous clearance between parts to ease orientation and insertion during assembly.

6. **Standardize and minimize use of fasteners:** This guideline helps save the cost of the labor involved in handling fasteners during assembly, which can be 75 percent of the assembly costs. Possible actions include using fits, whenever possible, to reduce the cost associated with fasteners by standardizing and using a few types and sizes of fasteners and associated tools for assembly.

7. **Unobstructed access for parts and tools:** This guideline aims to improve productivity and product quality by allowing workers to easily perform assembly. Design for ease of locating and fitting parts in their respective locations and provide room for the operator's arm and fastening and joining tools.

8. **Design components with minimum handling during assembly:** Make the required position of insertion or joining clear and easy to achieve. Include design features that help guide and locate parts in the proper position.

Capstone design teams must apply the above DFA guidelines for their product design and should introduce any modifications to the CAD model Rev 2 that arise from the DFA process. The team should list each of the guidelines used to improve the product design and the parts that have been eliminated.

B.3.23 Design to Cost (DTC) and Design to Value

B.3.23.1 DTC Thinking

When the authors of this handbook started their engineering careers in the 1970s, many companies did not permit their design engineers to consider cost; they were only to consider performance. Cost—that is, production cost—was determined by other departments after the design was completed. Overhead and profit margin markups were then added to arrive at the sell price to the customer. If this price was too high to be competitive, then an expensive cost reduction redesign project ensued.

EM-(i)

EM-(n)

Today's engineering design practice differs greatly from the one described above. Now most firms recognize that about 70 percent of a product's production cost is determined before detailed design. Cost has become an integral part of the engineering design process. In fact, it is often the driving factor in the product's design. This means that the engineering design team must work with the firm's marketing and business areas to establish cost targets for the product during conceptual design.

By the time the product is in preliminary design, the design team should have a clear understanding of what the product's production cost has to be in order to be competitive. The approach is to find the least costly design that still has the functionality to meet all the other engineering requirements.

The example travel iron project is a good example of how to use the DTC approach early in the design process. In this case, the team performed their first UPC estimate near the end of Phase 2 (refer back to Figure B-32).

B.3.23.2 Life Cycle Costs (LCC)

Figure B-66 shows the stages of the life cycle of a product from both the producer's and customers' point of view. There are costs associated with each of these stages. Systems engineers are interested in working with the business and marketing departments in their firm to determine what these costs must be in order to result in a customer LCC that is attractive to the customer and a producer LCC that is profitable for the producer. In addition, there may be constraints on certain types of costs, such as what the customer is willing to initially pay for the product, that will influence the overall product cost targets.

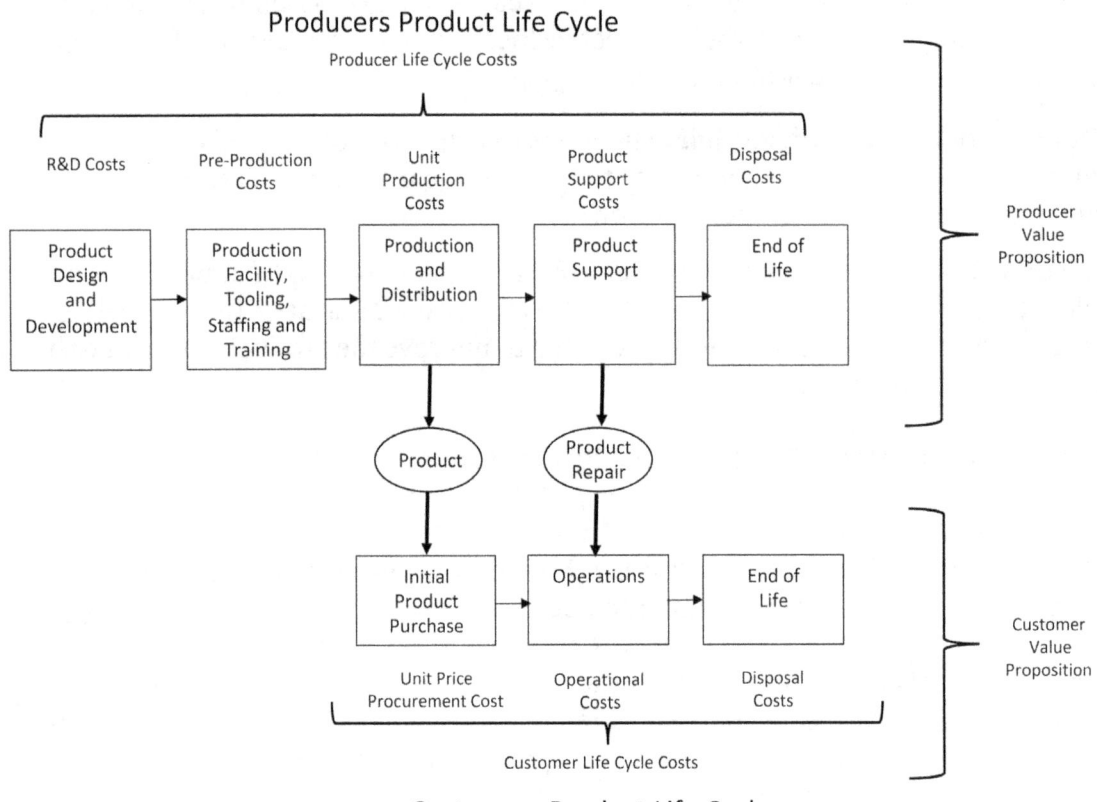

FIGURE B-66 Producer's and Customer's Product Life Cycles

The design team must work with the marketing and business experts in their company to determine which type of costs will drive the design. For the capstone design, the selection of cost targets is especially difficult because the design team is not associated with a company that makes similar products. Such a company would usually have a database of both customer cost preferences and production cost estimates that can be used to effectively determine producer profitability.

Many producer costs, especially UPCs, are dependent on the number of units being produced. As more units are produced, the cost of production decreases due to learning curve effects. Costs also come down with increased number of units produced because fixed costs, such as facilities and tooling, can be amortized over a greater number of units.

With so many types of costs, it is important for the team to set cost targets for each of the cost categories in the referenced Figure B-66. It is important to remember that labor is also a cost. For example, it takes the cost of engineering labor to design and develop the product.

B.3.23.3 Example of Life Cycle Costing and Design to Value

To illustrate these concepts, the two-semester capstone design course at ASU usually has six team members. The university provides only $100 per team member for material costs to develop the product. In addition, most students plan to spend about 10 hours out of class each week working on the project. The semesters are 15 weeks long. Therefore, the team has about 1,800 hours available to develop the product to the point at which an engineering prototype has been built and tested and the actual detailed design package of the product has been created. These material and labor costs all fit in the first producer box labeled "Product Design and Development" in Figure B-66. So one important task for ASU capstone design teams is to define and execute a scope of product development that produces these deliverables within the resource constraints.

To set targets for the other types of costs shown in Figure B-66, the team needs to start with an understanding of what the customer is willing to pay for a product that will meet their need. That cost includes the product's initial purchase as well as the operational costs of using it and the disposal costs when they are finished using it. A good example is the customer who needs a car for a long commute to work. First cost is important, but so is the cost of fuel and repairs. The team may use a cost measure of dollars per commuter mile. However, a car that yields the minimum dollars per commuter mile may have a first cost that is more than the customer can afford. In this case, the design team has to select only the fuel-saving features that keep the first cost of the car affordable. If we add maintenance costs as well as fuel costs, this analysis becomes even more complex. To effectively set these cost targets, the company's marketing and business aspects must be integrated with the engineering product development effort as illustrated in the Goldsmith Commercialization Model discussed in previous parts of this handbook.

The car example given above becomes more complicated if the following is assumed: (1) The minimum cost per commuter mile occurs when the car's reliability is moderate at a MTBF of 500 hours, but (2) most buyers want a very reliable car with a MTBF of at least 1,000 hours. In other words, the buyers value reliability more than driving a car with the least cost per commuter mile. Now we have a different design parameter: design to value.

There are many other design features that customers value, but they are difficult to put in economic terms. Another feature example is being environmentally friendly. Some buyers of the hypothetical car above may highly value the fact that the car was assembled in a plant that uses only renewable energy even if this energy is procured at a higher cost than that of using conventional fossil fuels.

A design-to-value figure of merit is harder to define than one for DTC. However, it is an important design consideration that must be factored into the design selection process. For capstone students, it is important to understand this concept and include it in their design activities. In industry, the understanding of how to effectively use this figure of merit is still being developed by many companies. These companies start from the same place from which the capstone students begin, with the understanding that designing for value is even more important than designing for cost.

The capstone design team must quickly research similar products and gain insight into the LCCs involved. The team should work with the course instructor to select appropriate cost targets in terms of type of cost and the target value in dollars. These decisions and their rationale must be fully documented first in the project notebook and eventually in the project final report. These concepts will be developed more in the next module.

B.3.24 DFMA and DTC Worksheet

Below is Worksheet 3: DFMA and DTC. The team needs to complete this worksheet and submit it to the instructor at the end of Module 24.

Worksheet 3 DFMA and DTC

A. DFMA

Review Section B.3.22 and conduct a design review to your product design to improve the design using the DFM and DFA guidelines. Summarize the team's DFM and DFA review and consequent improvement decisions in the following table. Based on the results of applying DFM and DFA to the product design, identify design changes that are needed in the CAD solid model Rev 2. Include the identified design changes in the parts, subassemblies, and assembly drawings of the product. Include neat sketches and drawings to support your results.

DFM/DFA Guideline	Results of Applying the Guideline on Product Design

B. DTC

Review Figure B-66 in the handbook and then provide an estimate for each of the following costs for the team's project.

Producer's LCCs

1. research and development costs: _____

2. Preproduction costs: _____

3. UPC: _____

4. Product support costs: _____

5. Disposal costs: _____

Customers' LCCs

1. Unit price and procurement cost: _____

2. Operational costs: _____

3. Disposal costs: _____

MODULE 25

FMEA, DFMA, and DTC
Part 3

OVERVIEW

In this module, the team completes the FMEA, DFMA, and DTC studies by completing Worksheet 4 and makes its final updates to the product's preliminary design. These actions result in the CAD model Rev 3.

LEARNING OBJECTIVES

- Understand the process of materials selection.
- Understand the process of designing a product for manufacturing and assembly (DFMA)
- Understand the different types of product costs and how to conduct DTC and LCC analyses.
- Gain experience by applying the above tools to the product's design.
- Gain experience in managing a team through a time of many tasks to be completed in a short amount of calendar time.

PRE-LECTURE ASSIGNMENT

- Reread applicable handbook sections.

POST-LECTURE ASSIGNMENT

- Conduct a team meeting to review the subteams' work, complete Worksheet 4, and finalize the preliminary design in preparation of assembling a slide presentation for the preliminary design review.
- Subteams make any changes as a result of the team meeting and document all work in the project notebook.

TEAM DELIVERABLES

- Worksheet 4: Design Changes due to FMEA, DFMA, and DTC Analyses
- Updated production unit's CAD model Rev 2 to Rev 3
- Minutes of the team meetings
- Team activity as documented in the team notebook

B.3.25 Worksheet 4: FMEA, DFMA, and DTC

To complete the work on the subjects of FMEA, DFMA, and DTC, the team needs to complete Worksheet 4, which describes how each of these studies were applied to the team's production design. Worksheet 4 is provided below.

Worksheet 4   Design Changes due to FMEA, DFMA, and DTC Analyses

A. What changes were made in the production unit design as a result of the FMEA?

B. What changes were made in the production unit design as a result of the DFMA?

C. What changes were made in the production unit design as a result of the DTC analyses?

B.3.26 Example Travel Iron Team Report for Week 13

The main efforts by the team during Week 13 were to complete the FMEA, DFMA, and DTC tasks and use the results to update the CAD model to Rev 3. The outside, interior, and exploded views are given in figures B67, B-68, and B-69, respectively.

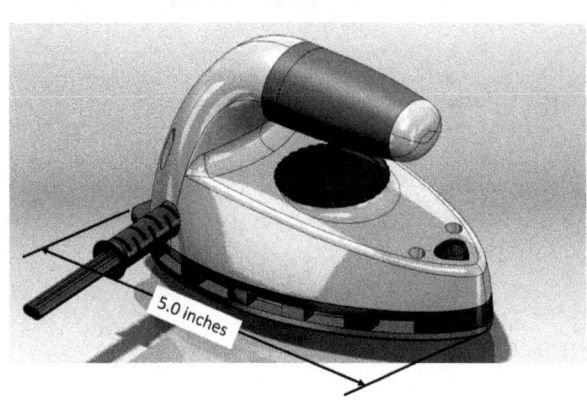

FIGURE B-67 Travel Iron CAD Model Rev 3 External View

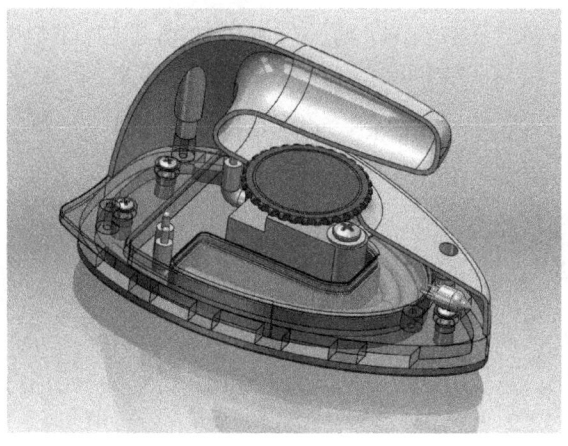

FIGURE B-68 Travel Iron CAD Model Rev 3 Internal View

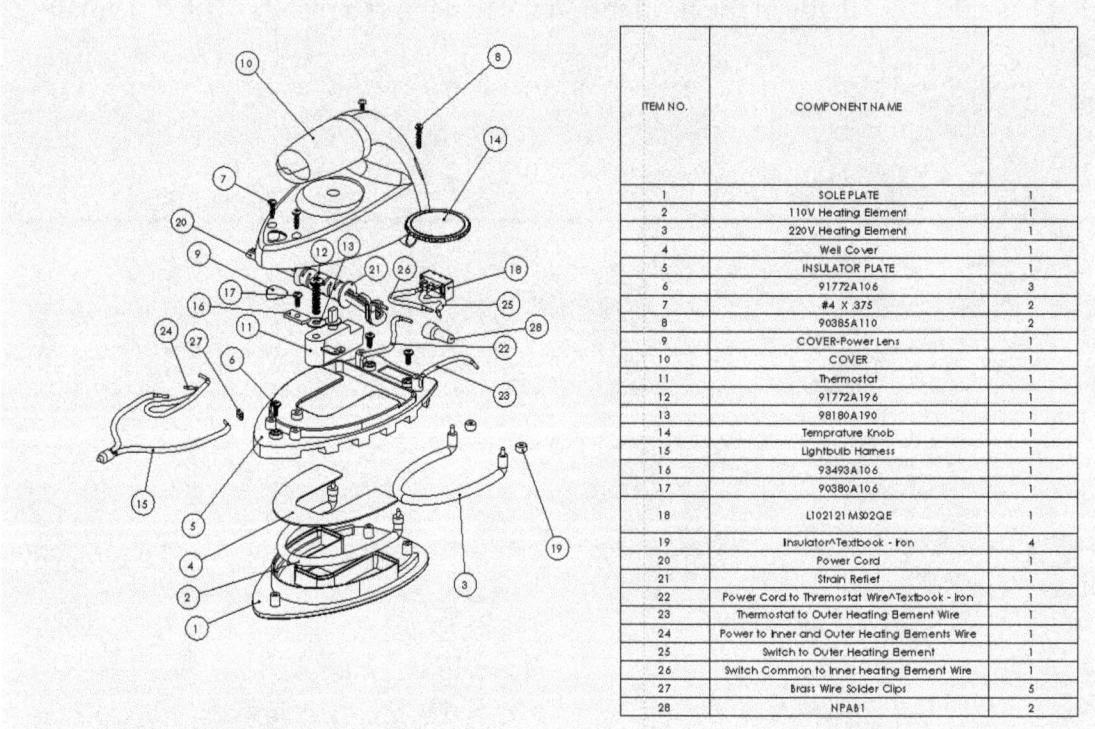

ITEM NO.	COMPONENT NAME	
1	SOLE PLATE	1
2	110V Heating Element	1
3	220V Heating Element	1
4	Well Cover	1
5	INSULATOR PLATE	1
6	91772A106	3
7	#4 X .375	2
8	90385A110	2
9	COVER-Power Lens	1
10	COVER	1
11	Thermostat	1
12	91772A196	1
13	98180A190	1
14	Temprature Knob	1
15	Lightbulb Harness	1
16	93493A106	1
17	90380A106	1
18	L102121MS02QE	1
19	Insulator^Textbook - Iron	4
20	Power Cord	1
21	Strain Relief	1
22	Power Cord to Thremostat Wire^Textbook - Iron	1
23	Thermostat to Outer Heating Element Wire	1
24	Power to Inner and Outer Heating Elements Wire	1
25	Switch to Outer Heating Element	1
26	Switch Common to Inner heating Element Wire	1
27	Brass Wire Solder Clips	5
28	NPA61	2

FIGURE B-69 Travel Iron CAD Model Rev 3 Exploded View with Component Identifiers

MODULE 26

Final Preliminary Design and Commercialization

OVERVIEW

This module covers the items that are needed in addition to the CAD model Rev 3 to define the final preliminary design. The task of designing, building, and testing the engineering prototype is also covered. It is important to obtain early feedback from a fabrication expert on the engineering prototype scope. This module also discusses the further product commercialization.

LEARNING OBJECTIVES

- Demonstrate the ability to define the preliminary design.
- Demonstrate the ability to conceptualize the goals and basic design of the engineering prototype to be accomplished in Phase 4.
- Demonstrate the ability to gain feedback from a fabrication expert on the conceptual design of the engineering prototype.
- Demonstrate the ability to integrate commercialization into the Phase 3 design.

PRE-LECTURE ASSIGNMENT

- Read sections B.3.27–B.3.34.

POST-LECTURE ASSIGNMENT

- Conduct a team meeting to 1) list the items needed to define the preliminary design and assign tasks to prepare these items, 2) establish the goals of the engineering prototype and prepare a conceptual sketch of this unit, 3) prepare Worksheet 5: Goldsmith Commercialization Stage 3 Report, 4) assign a subteam to have the conceptual engineering prototype design reviewed by a fabrication expert by the end of Module 27, and 5) schedule a second team meeting to review the work done by team members in addressing item 1.
- Each team member completes the assigned preliminary design definition tasks and brings the results to the second team meeting.
- Conduct a second team meeting to review, edit, and approve the team member write-ups for defining the final preliminary design.

TEAM DELIVERABLES

- Set of items that define the final preliminary design (see Table B-44)
- List of goals for the engineering prototype to be accomplished in Phase 4
- Conceptual design of the engineering prototype
- Worksheet 5: Goldsmith Commercialization Stage 3 Report
- Minutes of the team meetings
- Team activity as documented in the team notebook

B.3.27 Defining the Final Preliminary Design

Once the final preliminary design has been chosen, it must be clearly defined with the items listed in Table B-44.

Table B-44 Items to Define the Final Preliminary Design
• Narrative description • Table of features and benefits • Labeled sketch with overall dimensions • Configuration block diagram • Performance tables (weight, power output, UPC, etc.) • Requirements validation matrix (include results from analyses and POC testing) • Drawing package of CAD Rev 3

B.3.28 Initial Engineering Prototype Planning

According to the phased product development (PPD) process, the engineering prototype is not started until the project enters Phase 4 detailed design. However, many students have trouble with scoping down their production unit design into something that can be built in the limited time available in the capstone course. To address this issue, the student shop manager at ASU arranges time to work with students on their conceptual engineering prototype designs near the end of Phase 3. This approach of having a manufacturing expert work with the students near the end of Phase 3 is recommended for other capstone courses.

The goal of the engineering prototype is to retire risk in the production unit design before Phase 5. This means that the engineering prototype does not have to address all the engineering requirements for the production unit. Instead, it should address those requirements that are still a concern to the project team.

In the case of the travel iron example, the team's concern is making sure that the internal components are able to be calibrated so as to enable the unit's sole plate to be heated to three temperatures: 275, 350, and 400 °F. In addition, the iron must be able to hold these temperatures in the range of ±10 °F.

The project team's conceptual engineering prototype design should be included in the preliminary design review.

B.3.29 Commercialization

EM-(a)

The sociopolitical landscape is not static. As a product goes through its timeline of commercialization, this landscape can change. It is important to understand the trends of these changes. One major trend is that more people are becoming concerned about sustainability of the ecosystem. One measure of this is the concern people express about climate change. For example, a recent Pew Research Center survey found that "two-thirds of US adults say the federal

EM-(e)

EM-(f)

EM-(k)

EM-(o)

government is doing too little to reduce the effects of global climate change."[1] It is reasonable to assume that many of these adults are also willing to do something themselves to reduce the effects of global climate change. However, additional market research is needed to confirm this assumption.

As shown previously in Figure A-11, the PPD process phases correlate with the stages of the Goldsmith Commercialization Model. Specifically, PPD Phase 3: Preliminary Design correlates with the first part of the Goldsmith model's Stage 3 Development. In Stage 3, the commercialization team prepares a strategic marketing plan and a strategic business plan.

The capstone project team has limited resources to prepare the above plans in depth. However, they can invest some of their resources in preparing top-level strategic plans in the form of Worksheet 5. These plans will be updated near the end of Phase 4: Detailed Design.

A simple internal rate of return (IRR) analysis is listed in Worksheet 5. IRR is the rate of return on investment potential investors can expect from the commercialization of the product. The more risky the commercialization of the product is, the higher IRR the investor needs to expect if no risks develop. Investors can be grouped according to the level of risk versus reward they are willing to take. Banks, for example, tend to make investments when these loans are backed by collateral such as ownership of the equipment if the project is unable to repay the loan. Bank loan officers are risk-averse, so they are mostly interested in potential investments that have less risk and therefore for which the bank can expect a lower IRR. On the other hand, joint venture capitalists are investors who are interested in more risky projects because they are interested in earning higher rates of return.

EM-(j)

EM-(l)

---

[1] Pew Research Center. 2020, April 21. *How Americans See Climate Change and the Environment in 7 Charts*. Accessed March 13, 2021. https://www.pewresearch.org/fact-tank/2020/04/21/how-americans-see-climate-change-and-the-environment-in-7-charts/.

Worksheet 5 Goldsmith Commercialization Stage 3 Report

A. What is the name of the product?

B. Engineering Status:

In Phase 3, the goal is to show by analysis and/or POC testing that the design of the product has a high probability of meeting all of the customer-based engineering requirements. Are there still some engineering requirements that need to be addressed in Phase 4? Describe what needs to be done.

<u>Strategic Marketing Plan</u>

C. State the customer value proposition:

D. Market characteristics:
Describe the customer market in terms of size, segmentation, competition, price points, and sales volume.

E. Type of marketing:

There are six types of marketing described in the handbook: influencer, relationship, viral, green, keyword, and guerilla. What type of marketing is planned for this product?

F. Marketing mix:

Provide a description for each of the following 4P's:

1. <u>Product</u> is a group of functions and benefits in the form of a good and/or service that is capable of exchange or use.

2. <u>Price</u> is the quantity of money needed to acquire a given quantity of goods and/or services.

3. <u>Place</u> or distribution is the process of marketing and transferring products from producers to customers.

4. <u>Promotion</u> is providing information to potential and existing customers about the product, the brand, and the company.

G. Describe the customer's DTC considerations:

H. Market strategy based on the five competitive forces:

   1. Competition rivalry:

   2. Threat of substitution:

   3. Buyer power:

4. Supplier power:

5. Threat of new entry:

Strategic Business Plan

I. Producer's value proposition:

J. Business initiation and growth plans:

K. Producer DTC considerations:
    o   The investment required prior to start of production is estimated to be the following:

Cost Category	Basis for Estimate	Estimate
Research and development (R&D) through Phase 4		
R&D Phase 5		
Marketing and Business Setup		
Facilities and Equipment		
Total Investment		

- Profit/unit:

Cost Category	Basis of Estimate	Estimate
Sell Price		
UPC		( )
Markups		( )
Profit/unit	Sell Price-UPC-Markups	

- Product support
- Disposal

L. Simple IRR analysis:

M. Financing plan:

N. Business growth plan:

MODULE 27

End-of-Phase 3 Deliverables

OVERVIEW

This module covers the following Phase 3 documentation: 1) preliminary design portion of the final report, 2) Phase 3 exit checklist, 3) project notebook, and 4) preliminary design review presentation.

LEARNING OBJECTIVES

- Demonstrate the ability to properly document the design process.

PRE-LECTURE ASSIGNMENT

- Read sections B.3.30–B.3.34.

POST-LECTURE ASSIGNMENT

- Conduct a team meeting to 1) make final team assignments for completing the following documents by the second team meeting scheduled for this module: preliminary design portions of the final report, initial Phase 3 exit checklist review, project notebook and preliminary design review (DR3) presentation slides; and 2) schedule a second team meeting prior to the start of Module 28.
- Team members complete their assigned documentation tasks prior to the second team meeting.
- Conduct a second team meeting to 1) review, edit, and approve team member write-ups for the Phase 3 documentation; 2) practice the design review presentation; and 3) assign a team member to schedule the team to meet with the capstone instructor to do a green run of DR3 prior to the beginning of Module 28.

TEAM DELIVERABLES

- Assignment sheet for completing Phase 3 documentation
- Sheet reporting status of all Phase 3 documentation assignments after the second team meeting
- Phase 3 preliminary design review presentation slides
- Minutes of the team meetings
- Team activity as documented in the team notebook

B.3.30 Draft of Preliminary Design Section of the Final Report

One of the deliverables at the end of Phase 3 is the preliminary design section of the final report. The team should use the final report outline given in Section C of the handbook to prepare this draft. Once individual team members prepare their parts of the draft, the team as a whole must review and edit the entire section until everyone on the team approves of everything in the draft.

B.3.31 Phase 3 Exit Criteria Checklist

The team officially prepares the Phase 3 exit checklist in the next module. However, it is important for the team to be proactive and make sure that all the required tasks are either completed or scheduled for completion before the end of Phase 3. Any incomplete checklist items should be assigned to a team member during the first team meeting of this module.

B.3.32 Team Project Notebook Preparations

The project notebook is due at the end of Module 28. During this module, the team needs to review and edit the notebook so that it conforms to the outline provided in Table A-1. Special attention needs to be taken to be sure that there is documentation in the notebook for each entrepreneurial mindset checklist item for phases 1, 2, and 3.

B.3.33 Phase 3 Preliminary Design Review (DR3)

The DR3 presentation is a key deliverable for Phase 3. The team needs to take adequate time to prepare the slides. This includes the team as a whole editing each slide until all team members approve of its content. Practicing the presentation to make sure the team stays within the allotted time is vital. Figure B-70 presents an outline of slides for DR3.

The design review needs to be prepared for the audience. In this case, the sponsor is the key member of the audience. It is important to remember that the sponsor will decide whether to continue funding the project based on the DR3 presentation. A crisp, well-prepared, and well-presented preliminary design review will build confidence in the mind of the sponsor that this is the right team to take the project into Phase 4: Detailed Design.

EM-(h)

EM-(m)

During the presentation preparation process, the team should schedule a meeting with the capstone instructor to review the presentation once the team has had an opportunity to practice giving it. Instructor feedback should be incorporated into the final presentation slides.

In industry, most managers schedule a practice presentation with them before the team presents to the manager's superiors. This practice presentation is often called a *green run*. It is important to remember that managers don't like surprises, especially if it is a negative one in front of their manager.

> **Outline of Slides for DR3: Preliminary Design Review**
>
> 1. **Title Page**—call it "DR3: Preliminary Design Review and include date, name of project, picture of project, team picture, team names and sponsor name.
> 2. **Problem Statement**—succinct one or two sentences.
> 3. **Voice of Customer**—describe customers and their segmentation and provide a weighted list of their needs.
> 4. **Engineering Requirements**—don't have too much on the slide. List the key requirements and their target values.
> 5. **Phase 2 Final Conceptual Design**—sketches, performance tables, etc.
> 6. **Phase 3 Design Drivers**—list drivers and describe how they were analyzed and the design impacts.
> 7. **Goal Function Chart**—show the changes in the goal function versus time.
> 8. **CAD Model Rev 3 Outside and Internal Views**-- labeled with dimensions.
> 9. **CAD Model Rev 3 Exploded View**—components labeled.
> 10. **Performance Table**—weight, power, etc.
> 11. **How It Works**—use words and drawings to show how device works.
> 12. **Preliminary Design Improvements**—table with change, benefit, analysis type. For analysis type use design driver, requirement analysis, requirement testing, RMS analysis, FMEA, DFMA or DTC.
> 13. **Analysis or Testing Example**—one slide, not cluttered, include key points
> 14. **Commercialization**—summarize in bullets the plan to commercialize the product
> 15. **Prototype Scope**—requirements and initial drawing of unit.
> 16. **Key Issues for Detailed Design**—list and briefly describe.
> 17. **Phase 3 Schedule Status**—must be readable, include one sentence summary
> 18. **Phase 3 Labor Status Char**t—include reporting date, budget line, actual line, ETC line and one sentence summary.
> 19. **Lessons Learned**—bulleted list of top five things the team has learned so far about the PPD process as applied to their project.

FIGURE B-70 Outline of Slides for DR3: Preliminary Design Review

B.3.34 Example Travel Iron Team Report for Week 14

B.3.34.1 Defining the Final Preliminary Design

In addition to the CAD model Rev 3, the team assembled the following information:

- Narrative description — Figure B-71
- Table of features and benefits — Table B-45
- Configuration block diagram — Figure B-72
- Travel iron key characteristics — Table B-46
- Requirements validation matrix — Figure B-73

Narrative Description:

The travel iron is designed to be a reliable, compact, easy to use and affordable method of removing clothing wrinkles during travel. It will operate on both 120 volts and 240-volt electrical systems. This product is targeted for customers in the United States that want to travel world-wide. The differentiating factor for this travel iron is that it is locally made and environmentally friendly. The team plans to stay with the project after the capstone course and complete Phase 5 Production Unit Prototype Development and Validation. When this milestone is reached, the team intends to sell the technology package to a producer for $60,000. The producer would then invest another $140,000 prior to start of production. The unit is expected to have a sell price of $30 with a profit margin of $6.50/unit. The production life is estimated to be three years at a production rate of 15,000 units per year. Based on these assumptions, the IRR is over 23 percent. The team would like to find a producer interested in using employees with major disabilities to produce and market this product. This would be another differentiator in a market that is becoming more interested in helping workers with major disabilities find meaningful employment.

FIGURE B-71 Narrative Description of Travel Iron Product

Table B-45 Travel Iron Features and Benefits

Feature	Benefit
Compact and light	Fits easily in travel suitcase
Three temperature settings	Safely removes wrinkles from all fabric types
Dual voltage	Can be used worldwide
Sell price less than or equal to $30	Affordable
High-quality components	Reliable operating life
Fabricated from recycled plastic	Environmentally friendly
Locally produced components and assembly	Supports local economy

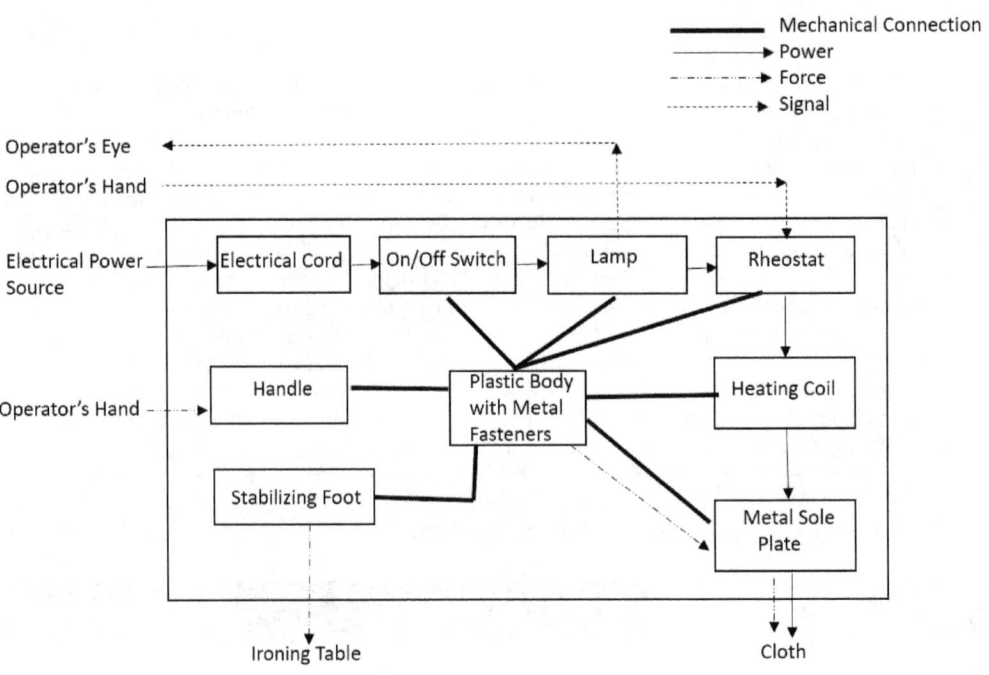

FIGURE B-72 Configuration Block Diagram

Table B-46 Travel Iron Key Characteristics	
Characteristic	**Value**
Compact	Less than 60 cubic inches
Lightweight	Less than 1 pound
Three temperature settings	Removes wrinkles from all types of fabric
Powerful heating elements	400 watts
Dual voltage	120 volts, 60 Hz and 240 volts, 50 Hz
Sell price	Less than or equal to $30
Maintenance-free	All components have high useful life reliability

Requirement	Target Value	Status	Confidence of Meeting Target
Weight	<1 lbs	Analysis predicts 0.XX lbs	High
Volume	<100 cubic inches	Analysis predicts approx 60 cubic inches	High
Length	<7.0 inches	Designer selected 5 inches	High
Input Power	60 Hz 110-120 vac; 50 Hz 220-240 vac	Analysis shows two heating elements with a switch will meet requirement	High
Press Time	<2.0 minutes for pillow case	Heat transfer analysis and POC testing estimate of approx 1.5 min	High
Iron Operating Temp. Range	Three settings; 275 de F, 375 deg F, 400 deg F	Thermostat is used in other irons, but testing needed to verify	Medium
Visual On Signal	Red on-light	This feature in CAD Model	High
On/Off Switch	On/Off Switch	Part of thermostat	High
Operating Life Reliability	MCBF = 2000 cycles	Generic estimate in excess 1000 cycles	High
Drop Test	4 feet onto concrete floor	POC test suggests shell and handle will survive drop. Needs to be checked in development	Medium
Operating Life	> 200 cycles	Estimate in excess of 1000 cycles	High
Shelf Life	5 years (0-150 deg F ambient temperature range)	Material analysis indicates shelf life greater than 5 years	High
Unit Production Cost	< $4.00 (2020 dollars)	UPC analysis estimates $10/unit. This requirement changed to that value.	High
Time to Production	< 2 years	Estimate is less than 1 year after capstone course.	High
Development/Tooling/Facility Cost	< $100k (2020 dollars)	Rough estimate is a conservative $200k. Requirement changed to this value.	High
Initial Prototype Project Materials Cost	< $600 (2020 dollars)	Total cost of components, fabrication, assembly tools, test rig and test equipment estimated <$600	High

FIGURE B-73 Requirements Validation Matrix for Travel Iron

B.3.34.2 Engineering Prototype Planning

Due to resource constraints, the team decided that the engineering prototype did not need to include the outer plastic shell and handle. Instead, the team will focus on the internal components and their ability to remove wrinkles from a variety of fabrics. The iron's housing and handle will be simulated with wood shapes that may vary from the exact design of the plastic shell housing and handle.

A critical part of the engineering prototype testing program will be to calibrate the thermostat to enable the unit's sole plate to be heated to three temperatures: 275, , 350, and 400 In addition, the iron needs to be able to hold these temperatures in the range of ±10 °F.

B.3.34.3 Worksheet 5: First Draft of Goldsmith Commercialization Stage 3 Report

During the team's first meeting of the week, Worksheet 5 on commercialization was completed. This completed worksheet is shown below.

Worksheet 5   Travel Iron First Draft of Goldsmith Commercialization Stage 3 Report

A. What is the name of the product?

Travel iron

B. Engineering status:

In Phase 3, the goal is to show by analysis and/or POC testing that the design of the product has a high probability of meeting all the customer-based engineering requirements. Are there still some engineering requirements that need to be addressed in Phase 4? Describe what needs to be done.

1. Need more supplier quotes to verify that a UPC of $10 per unit for a production rate of 15,000 units per year

2. Need to find multiple made-local component suppliers

3. Need to find multiple suppliers who will use recycled plastic

4. Need to verify that unit will operate and hold at three temperatures: 275, 350, and 400 °F.

Strategic Marketing Plan

C. State the customer value proposition:

Affordable, compact, lightweight, dual voltage, environmentally friendly, locally made product for the traveler to easily remove wrinkles from packed clothing

D. Market characteristics:

Describe the customer market in terms of size, segmentation, competition, price points, and sales volume.

- There is a worldwide market for clothes irons.
- The market is segmented into home, travel, commercial, and industrial users.
- Most irons feature steam, but dry irons are still popular due to their simplicity.
- In the United States, there are home-use and traveler-use markets.
    - There are numerous brands and offerings; most if not all are imported.
    - Emphasis is on ease of use.
    - Retail price ranges from $20 to $90 per unit.
    - Research did not find any offerings emphasizing locally made and/or environmentally friendly.
    - Revenue in the irons segment amounts to $691 million in U.S. dollars in 2021. The market is expected to grow annually by 3.05% (compound annual growth rate 2021–2025).[1]

---

1 Statista. 2021. *Irons, United States*. https://www.statista.com/outlook/cmo/household-appliances/small-appliances/irons/united-states.

- Increasing demand among the millennials to look well-dressed is expected to promote the utility of these products.[2]
- Sales volume is estimated to be greater than 15,000 units per year based on the team's professional judgment.

E. **Type of marketing:**

There are six types of marketing described in the handbook: influencer, relationship, viral, green, keyword, and guerilla. What type of marketing is planned for this product?

The major marketing effort will be green (i.e., emphasizing the product is produced in an environmentally friendly manner).

F. **Marketing mix:**

Provide a description for each of the 4P's.

1. **Product is a group of functions and benefits in the form of a good and/or service that is capable of exchange or use.**

See customer value proposition.

2. **Price is the quantity of money needed to acquire a given quantity of goods and/or services.**

The price will be in the range of $20 to $30 per unit. Currently, the lowest-priced travel iron is $20. It is expected that customers will pay a premium for a travel iron that is locally made and environmentally friendly.

3. **Place or distribution is the process of marketing and transferring products from producers to customers.**

There is a niche market of being produced and marketed by a firm that uses employees who are disabled. This will appeal to many customers who are looking for an environmentally friendly device due to their interest in a sustainable society.

4. **Promotion is providing information to potential and existing customers about the product, the brand, and the company.**

News stories, websites, and catalogs will be used to promote the product.

G. **Describe the customer's DTC considerations:**

- Initial cost is the overriding cost concern for the customer. Travelers generally are not charged for the electricity they use in their hotel room, so operational costs are not a concern. There are no scheduled maintenance costs. If the unit fails, the customer will generally just dispose of it and buy another iron. However, the failure may cause the customer to not buy the same iron again. Disposal cost is generally not a concern. Most users currently dispose of an unwanted travel iron by putting it in garbage can. This practice may not be acceptable in the future.

---

2   Grand View Research. 2019, July. *Electric iron market size, share & trends analysis report by function (automatic, non-automatic), by product (dry, steam), by application, by distribution channel, and segment forecasts, 2019–2025.* https://www.grandviewresearch.com/industry-analysis/electric-iron-market.

- Sell price can be higher than that of the lower-priced competitor because this unit has the additional features of being environmentally friendly and locally made. The low-cost competitor price is about $20. A price of up to $30 is considered appropriate for this niche market.
- Warranty coverage (i.e., parts and labor) and time period are not important because the cost and inconvenience of sending a failed unit back to the producer is higher than just buying a similar unit.

H. Market strategy based on the five competitive forces:

1. **Competition rivalry:**

Differentiators are locally made and environmentally friendly.

2. **Threat of substitution:**

Possible substitutions are wrinkle-free clothing and hotel ironing services. These are not considered major threats for the niche market being targeted.

3. **Buyer power**

The buyer has many travel iron options; however, all these existing options are imported units. In addition, no current units feature themselves as environmentally friendly.

4. **Supplier power**

There are a large number of off-shore suppliers; however, local suppliers may be difficult to find for some components. It is important to have more than one supplier for each component. Finding at least two injection molding suppliers that use recycled plastic may also be difficult to find.

5. **Threat of new entry**

The niche market being pursued is too small for more than one producer. Once this travel iron becomes established in this niche market, it will be difficult for another competitor to enter unless there is a substantially lower cost, which is considered unlikely.

Strategic Business Plan

I. Producer's value proposition:

Design that meets the market need for an environmentally friendly and locally made travel iron that is easy to produce, provides a profitable revenue stream, and yields an IRR of greater than 20 percent.

J. Business initiation and growth plans:

The capstone team wants to continue the project beyond the capstone course that stops at the end of Phase 4: Detailed Design. The team wants to complete Phase 5 production unit development and validation and then sell the technical package to a firm interested in producing and marketing this product. In addition to the technical package, the team plans to prepare a strategic marketing plan and strategic business plan with the goal of convincing an existing firm of adding this product to their production portfolio.

K. Producer DTC considerations:

- The investment required prior to start of production is estimated to be $200,000 based on the following:

Cost Category	Basis for Estimate	Estimate
R&D through Phase 4	Capstone project	$ 0
R&D Phase 5	1 year time frame; 900 hours (hr) x $100/hr = $90,000	$ 90,000
Marketing and Business Setup	Concurrent with Phase 5; 500 hrs x $100/hr = $50,000	$ 50,000
Facilities and Equipment	Last 3 months of Phase 5; engineering judgment.	$ 60,000
Total Investment		**$200,000**

- Profit/unit:

Cost Category	Basis of Estimate	Estimate
Sell price	Market study	$30.00
UPC	See Figure B-32	($10.00)
Markups	Engineering judgement	($13.50)
Profit/unit	**Sell Price-UPC-Markups**	**$ 6.50**

- Product support is included in markups.
- Disposal after three years of production provides a net salvage value of $10,000.

L. Simple IRR analysis:

From Figure B-41, the estimated IRR using the costs listed above is 23.2 percent.

M. Financing plan:

- Capstone project team continues to do Phase 5 that includes further commercialization and then sells the technical package, marketing plan, and business plan to producer for $90,000.
- Producer finances the costs prior to production.

N. Business growth plan:

- The producer may want to grow this project by extending the production beyond three years. The producer may also want to expand beyond its market niche, but this will probably be too costly a growth project.

MODULE 28

Preliminary Design Review and Exiting Phase 3

OVERVIEW

In this module, the team conducts their DR3, completes the Phase 3 documentation tasks, and has the Phase 3 exit checklist approved by the sponsor.

LEARNING OBJECTIVES

- Demonstrate the ability to properly conduct a DR3 and complete the Phase 3 exit checklist.

PRE-LECTURE ASSIGNMENT

- Read sections B.3.35 and B.3.36.
- Team members complete all their assigned work to complete Phase 3.

POST-LECTURE ASSIGNMENT

- Conduct a team meeting after the DR3 to 1) review and document the DR3 feedback, 2) submit Section 6 of the final report, 3) submit the project notebook, 4) complete and make sure that the sponsor has approved the Phase 3 exit checklist, and 5) plan a team celebration for successfully completing Phase 3.

TEAM DELIVERABLES

- Phase 3 DR3 presentation
- Section 6: Phase 3 preliminary design portion of the final report (see Appendix C)
- Project notebook
- Phase 3 preliminary design exit checklist for approval
- Minutes of the team meetings
- Team activity as documented in the team notebook

B.3.35 Example Travel Iron Team Report for Week 14.5

The day before DR3, the team met and reviewed the DR3 slides. They decided to delete some of the material to stay within the time limit given by the instructor. The team had to run through their presentation three times before they were able to complete it within the given time constraint. The team also made a list of questions they thought might be asked and they agreed on what the team would say if asked each question. The team also completed the Phase 3 exit checklist and planned a team celebration for the following weekend.

B.3.36 Phase 3 Exit Criteria Checklist

The checklist for this phase is provided on the next page.

Exit Criteria Checklist: Phase 3: Preliminary Design

Team Name: _____ Team Leader: _____ Members: _____

_____ _____

Date Started: Date Completed:

Done	Exit Criteria	Comments
	1. An updated functional block diagram and component block diagram are complete.	
	2. FMEA is complete.	
	3. Component trade studies are complete.	
	4. Analyses to support the trade studies and requirements validation were planned and are now complete.	
	5. POC testing to support trade studies and requirements validation were planned and are now complete.	
	6. The system optimization is complete.	
	7. Solid model of product is complete.	
	8. Preliminary design review was completed and approved.	
	9. Final report chapters 1, 3, 4, 5, and 6 were completed and approved.	
	10. Notebook is checked and approved.	

Yes or No	Entrepreneurial Mindset Indicators Exit Criteria	Questions Related to Criteria
	a. Critically observes surroundings to recognize opportunity	Is there a list of new design opportunities identified during Phase 3 in the project notebook?
	b. Explores multiple solution paths	Is there a description in the project notebook of how the team evaluated multiple potential suppliers for key COTS items?
	c. Gathers data to support and refute ideas	Is there a discussion in the project notebook on what new design ideas were generated during Phase 3 and how data was gathered to explore these new ideas?
	d. Suspends initial judgment on new ideas	When a new idea was presented to the team by a team member, did the team suspend judgment until it was explored? Explain in the project notebook how this was done.
	e. Observes trends about the changing world with a future-focused orientation/perspective	Is there a discussion in the project notebook on how the team integrated trends of change in the world into the Phase 3 design? How were these trends integrated?
	f. Collects feedback and data from many customers and customer segments	During Phase 3, what customer feedback and data were collected and integrated into the design? Is this documented in the project notebook?
	g. Applies technical skills/knowledge to the development of a technology/product	How did the team use the skill of project management to keep the project on schedule and budget?
	h. Modifies an idea/product based on feedback	Did the team meet with the instructor to have a green run of DR3? Did the team improve their presentation based on this feedback?
	i. Focuses on understanding the value proposition of a discovery	In Phase 3, how did the team use the concept of design to value?
	j. Describes how a discovery could be scaled and/or sustained, using elements such as revenue streams, key partners, costs, and key resources	How did the team's IRR analysis influence the design of the product?

(continued)

	k. Defines a market and market opportunities	Did the team fill out all the marketing portions of Worksheet 5?
	l. Engages in actions with the understanding that they have the potential to lead to both gains and losses	How did the team determine if the calculated IRR of the project is high enough for the level of risk in commercializing the product?
	m. Articulates the idea to diverse audiences	Did the team focus DR3 on convincing the sponsor to fund the team moving into Phase 4?
	n. Persuades why a discovery adds value from multiple perspectives (technological, societal, financial, environmental, etc.)	Were the multiple value streams included in the customer's and producer's value propositions?
	o. Understands how elements of an ecosystem are connected	Does the project design address the sustainability of the ecosystem?
	p. Identifies and works with individuals with complementary skill sets, expertise, etc.	Did the team demonstrate that they were a high-performing team? Did they conduct team member assessments?
	q. Integrates/synthesizes different kinds of knowledge	Did the team improve their marketing and business acumen to properly integrate these considerations into the technical design of the product?
Approved by	Name and Title	Completion Date

Section B

Phase 4: Detailed Design

MODULE 29

Phase 4 Plan, Prototype Scope, and Requirements

OVERVIEW

Module 29 focuses on detailed planning of the activities for Phase 4. The design team will first review feedback from the preliminary design review and then start the detailed planning of Phase 4. To investigate certain aspects of the production unit, the capstone design team develops in Phase 4 the engineering prototype, performs detailed design analyses, and prepares the drawing package and bill of materials (BOM) for the final production unit. In addition to Phase 4 planning, this module covers the selection of the engineering prototype scope and requirements.

LEARNING OBJECTIVES

- Understand the detailed design phase goals and processes.
- Apply project management principles to create a detailed Phase 4 plan.
- Demonstrate the ability to define the engineering prototype scope and requirements so that it will fit within the Phase 4 schedule and budgets.

PRE-LECTURE ASSIGNMENT

- Read sections B.4.1–B.4.4.
- Study the second semester capstone course syllabus.
- Each team member prepares:
  - The engineering prototype scope description and requirements list
  - A list of the Phase 4 tasks, identifying which ones they would like to take the lead in planning

POST-LECTURE ASSIGNMENT

- Conduct and document in the project notebook a team meeting to 1) review the feedback from the Phase 3 preliminary design review, 2) identify the prototype scope and requirements, and 3) create an outline for the Phase 4 detailed plan and assign tasks to each member that must be initially completed prior to the beginning of Module 30.

TEAM DELIVERABLES

- Engineering prototype scope and requirements
- Outline for the Phase 4 detailed plan and list of tasks assigned to each team member
- Minutes of team meetings
- Team activity as documented in the team notebook

B.4 Detailed Design (Phase 4)

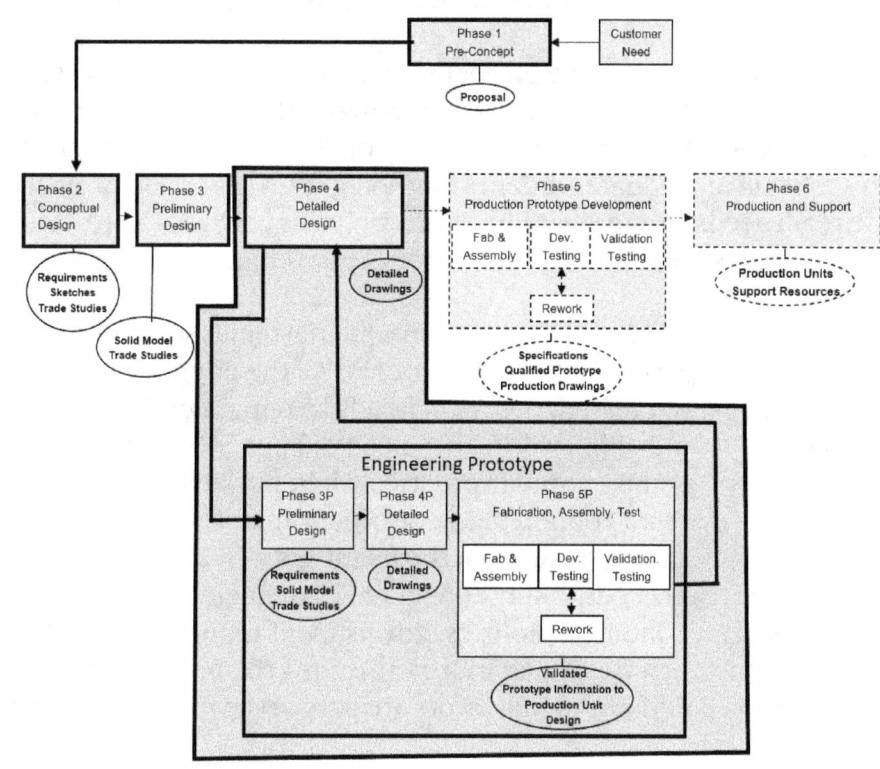

B.4.1 Overview

Once the preliminary design of the production unit is completed, the team exits Phase 3 and starts Phase 4, where any final detailed analyses are completed and a complete package of production drawings is prepared. At the end of Phase 4, the production design goes into Phase 5, where it is developed and validated prior to the beginning of production in Phase 6. It should be noted that, due to its limited budget and time resources, the capstone course goes only through Phase 4.

Engineering prototypes are usually developed during either Phase 3 or 4 to investigate certain aspects of the production unit design. The process of designing, building, developing, and validating an engineering prototype is its own subproject. It has its own phases: Phase 3P: Preliminary Design, Phase 4P: Detailed Design, and Phase 5P: Development and Validation. Results of the engineering prototype subproject are integrated into the production unit design.

For the capstone course, the engineering prototype subproject is conducted at the beginning of Phase 4: Detailed Design of the production unit. In parallel with the engineering prototype subproject, the team may conduct final analyses of the production unit. When the engineering prototype subproject is completed and the other detailed design analyses are finished, the capstone design team prepares the drawing package and bill of materials (BOM) for the final production unit. The capstone project team also prepares an updated top-level commercialization plan to help the next team to take over the product development process (Phase 5) and take the product into production (Phase 6).

To help illustrate the various design tasks for the engineering prototype and production final design, the travel iron example presented in prior sections of this handbook will be used again.

By the end of the second capstone project semester, the team has invested a great deal of creativity, analysis and testing in their project. It is highly desirable that this project continues after the capstone course is completed. This can be done by the capstone team itself or by others, such as a team of graduate students for their graduate applied project. In either case, the capstone team needs to end their Phase 4 by preparing a go-forward commercialization plan. This plan includes both production prototype development (Phase 5) and production (Phase 6).

Entrepreneurial Mindset:
Curiosity, Connections, Creating value

The use of the entrepreneurial mindset continues to be an integral part of the design process. As the engineering prototype is built and tested, the team needs to remain curious about what worked and didn't work. They need to know when to be contrarian in going against popular practice when possible innovative and entrepreneurial opportunities are identified. They need to stay connected to subject experts on the faculty as well as to their customer and suppliers. In addition, the team needs to stay connected to their environment, which is changing in terms of technology, economics, politics, government regulations, and climate. Finally, the team must be constantly designing more value into the product. As mentioned in the previous phases, the 17 EM@FSE 2.0 indicators are part of the Phase 4 exit criteria checklist.

B.4.2 Detailed Phase 4 Planning

The design team has gained experience with project management as they completed the first three product development phases (i.e., Preconcept Design, Conceptual Design, and Preliminary Design). This new learning needs to be factored into the detailed planning of Phase 4. The scope of work must match with the resources in terms of money, team labor, facilities, equipment, supplier capabilities, and the availability of experts. The team must identify all tasks necessary to design, build, test, and validate an engineering prototype and create the production unit drawings package and BOM. For many of the tasks to be performed, it is helpful to divide the team into subteams at various points in the conduct of this phase to enable certain tasks to be accomplished in parallel.

During phases 1, 2, and 3, two modules were covered each week. In Phase 4, there is only one module per week. This results in a larger scope of work for the Phase 4 modules. Teamwork is very important in being able to accomplish the quick pace of the Phase 4 activities.

EM-(d)

The development of an engineering prototype will have unplanned events. For example, a fabricated part is out of print, a supplier part is back-ordered, the unit fails a development test, etc. The team needs to have a plan for how they will manage these events to stay on schedule and within budget.

Quality control is a key planning item. The team must decide how they will make sure that all individual and team assignments will be submitted complete, on time, and in a professional format. There will be times when the assignment is unclear. It is the team's responsibility to obtain clarification from the sponsor. The team, not the sponsor, is in charge of the project.

The planning process should be similar to that conducted in Phase 3. The team should start by studying Section B.4 of this handbook and making a list of all of the tasks that need to be accomplished. As a guide, team members should plan on spending at least 10 hours outside of classroom time on the project each week. The plan should anticipate interruptions in the flow of the program due to special events, holidays, and scheduled work in other courses.

Figure B-74 shows the top-level schedule for Phase 4. It is important to note that each learning module covers an entire week. This is different from the first semester plan, where two modules were covered each week.

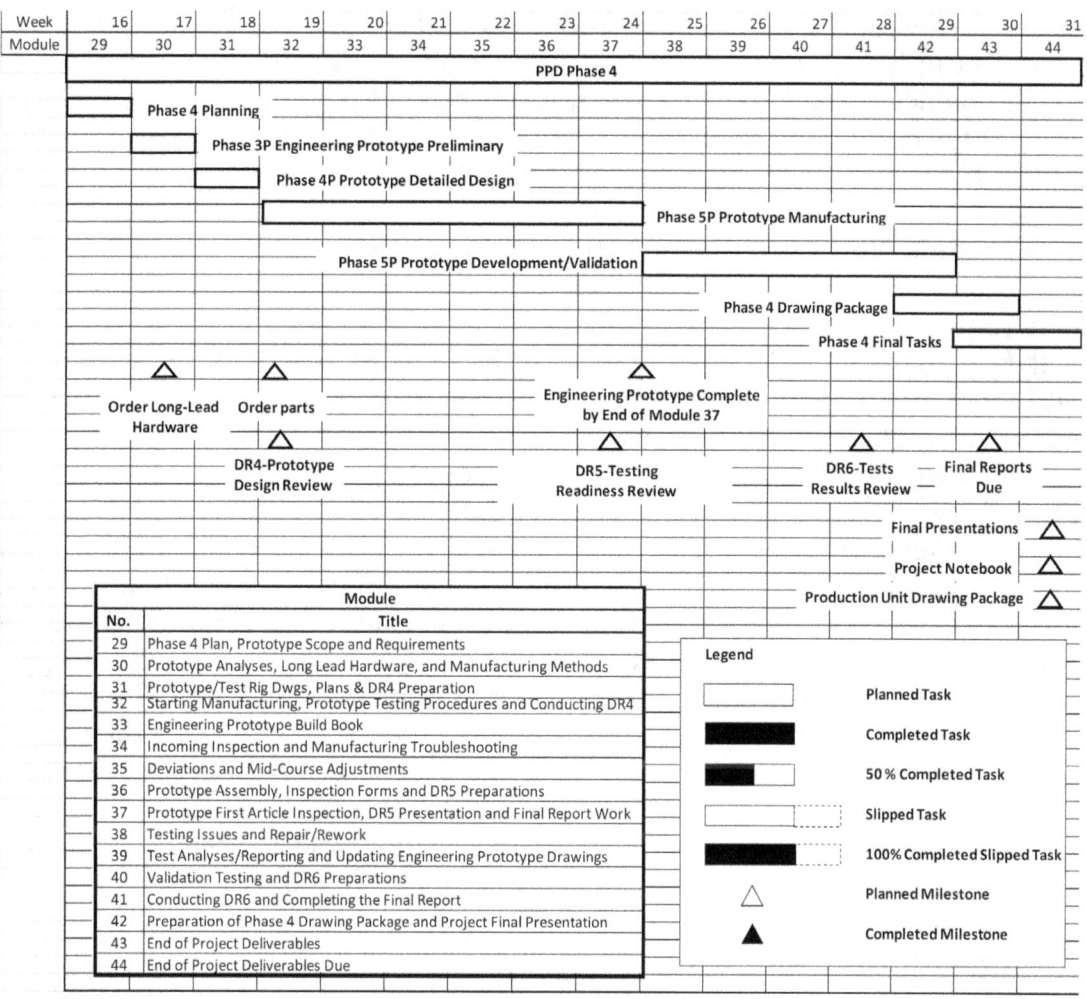

FIGURE B-74 Phase 4 Detailed Design Schedule

A key milestone is that the engineering prototype must be built by the end of Week 24. This is a hard date. That means that the team must continually be evaluating how well the engineering prototype design and manufacturing tasks are going so that the engineering prototype hardware is designed, manufactured, and assembled by the end of Week 24. To accomplish this, the team may have to reduce the scope of the engineering prototype during phases 3P or 4P to stay on schedule. In this case, the team must secure the approval of the sponsor (i.e., instructor) first by presenting a sound justification for reducing the scope.

B.4.3 Prototype Scope and Requirements

A key task in Phase 4 is deciding what production unit requirements will be addressed in the engineering prototype subproject. Often there are not enough resources for the engineering prototype to address all of the production unit requirements. The team needs to select the production requirements based on obtaining the highest value for the resources used. This means identifying what requirements may be difficult to achieve and prioritizing them. Then the engineering prototype is designed to best use the available resources to learn the most on how the production unit will meet these difficult requirements. Sometimes the engineering prototype can be scaled down from the production unit configuration to meet the available resources. The prototype requirements may also include items unique to the prototype unit due to safety concerns, certain parts that are not available, test limitations, etc. In summary, the scope of the engineering prototype subproject depends on the resources available and the team's core competencies in the areas being addressed.

B.4.4 Example Travel Iron Team Report for Week 16

The first item on the team meeting agenda was to prepare a detailed plan for Phase 4. Each member came to the meeting with a list of tasks they wanted to do. Each member also committed to spending at least 10 hours outside of class each week on the project. The team agreed to use the Phase 4 schedule presented in Figure B-75. The team then identified what tasks needed to be accomplished during each module (i.e., each week during Phase 4). Table B-47 shows the tasks scheduled for each module. This table also shows how many hours each team member will spend on each task during the module week. The work was divided among the six team members so that each member was scheduled to spend a total of 10 hours outside of class on the project each module week. The team realized that the scope of each task had to be limited to what could be done within the team member hours allocated. These limitations helped determine the scope of the engineering prototype requirements.

> EM-(b)
> EM-(l)

The team then made a list of unexpected events that could impact their plan. The team agreed that they would manage these unexpected events by (1) bringing the unexpected event immediately to the team's attention, (2) identifying a number of potential solutions, (3) selecting the solution that results in the project staying on schedule and budget, and (4) obtaining the instructor's approval for any needed changes in the project scope and plan.

The team also discussed how they would maintain a high level of quality in their assignments. It was decided that team member 1 would serve as the quality control (QC) engineer. It is the QC engineer's responsibility for clearly understanding the level of detail expected by the instructor for each team assignment and conveying that information to the team. Further, the QC engineer is responsible for reviewing each team assignment before it is submitted with enough time for the team to make any needed changes.

The team decided to (1) continue using a revolving action item list during their team meetings, (2) prepare team minutes for all meetings, and (3) document all work in the project notebook. The QC team member agreed to be responsible for keeping the project notebook up-to-date.

The team also prepared a detailed Gantt chart schedule for Phase 4. A portion of this chart is shown in Figure B-75. The chart shows that at the end of Week 16, the team completed the

Table B-47 Travel Iron Project Detailed Plan for Phase 4 (Page 1 of 2)

Module	Task	Team Member 1	2	3	4	5	6
29	Detailed Phase 4 Plan	**2**	2	2	2	2	2
29	List of Prototype Requirements	1	**1**	1	1	1	1
29	Start Requirements Analyses	6	7	**7**			
29	Start RMS Analyses				**7**		
29	Start FMEA					**7**	7
29	QC	1					
30	Requirements Analyses	5	9	**8**			
30	RMS Analyses				**9**		
30	FMEA					**9**	9
30	Requirements Matrix with Status			**1**			
30	Mtg 2–Prepare for DR-1	**4**	1	1	1	1	1
30	Order and approve long lead hrdwr						
30	QC	1					
31	CAD Model and Detail Dwgs					5	**9**
31	Manufacturing Plan	**2**	2	2	2	2	
31	Hardware Fabrication	**6**	7	7	7	2	
31	Team Meeting 2	**1**	1	1	1	1	1
31	QC	1					
32	Test Plans and Procedures	**4**	2	2	2	2	2
32	Remaining Parts Orders	3					
32	Hardware Fabrication	2	8	8	8	8	8
32	QC	1					
33	Start Build Book	2					
33	Manufacturing	**6**	10	10	10	10	10
33	Manufacturing Issues Report	2					
	QC	1					
34	Manufacturing	**7**	8	8	8	8	8
34	Start Manufacturing Report	**2**	2	2	2	2	2
34	QC	1					
35	Peer Assessments	**2**	2	2	2	2	2
35	Mid-Course Changes	**2**	2	2	2	2	2
35	Manufacturing	5	6	6	6	6	6
35	QC	1					

(continued)

Note: The lead team member for a task has a black border and number of hours in bold—for example: **1**

Table B-47 Travel Iron Project Detailed Plan for Phase 4 (Page 2 of 2)

Module	Task	Team Member					
		1	2	3	4	5	6
36	Detailed Test Procedures	2	2	2	2	2	2
36	Prepare for DR-2	1	1	1	1	1	1
36	Assign Final Report Sections	2	2	2	2	2	2
36	Mtg 2-Prepare for DR-2	1	1	1	1	1	1
36	Finish Manufacturing	3	4	4	4	4	4
36	QC	1					
37	Test Article Inspection	1	1	1	1	1	1
37	Minor Test Article Rework	1	1	1	1	1	1
37	Start Testing	4	5	5	5	5	5
37	Sections of Final Report & Editing	3	3	3	3	3	3
37	QC	1					
38	Testing	10	10	10	10	10	10
39	Testing and Test Reports	10	10	10	10	10	10
40	DR-3 Assignments	2	2	2	2	2	2
40	Final Report Asssignments	5	6	6	6	6	6
40	Mtg 2-Practice DR-3	2	2	2	2	2	2
40	QC	1					
41	Final Report Asssignments	5	5	5	5	5	
41	Production Dwg Pkg Assignments	4	5	5	5	5	10
41	QC	1					
42	Final Presentation Assignments	2	2	2	2	2	
42	Final Report Editing and Approval	4	4	4	4	4	
42	Notebook Review Assignments	1	2	2	2	2	
42	Continue Prod Dwg Assignments	2	2	2	2	2	10
42	QC	1					
43	Approve Dwg Package	1	1	1	1	1	1
43	Practice Final Presentation	2	2	2	2	2	2
43	Checklists	6	7	7	7	7	7
43	QC	1					
44	Team Celebration	2	2	2	2	2	2

Phase 4 plan, including a review by the QC team member. However, the team slipped schedule on some of the tasks to start the reliability, maintainability, and safety (RMS) and failure mode effects analysis (FMEA) activities. Specifically, the percentages complete for the start of the requirements analysis, RMS analysis, and FMEA were 50 percent, 50 percent, and 25 percent, respectively.

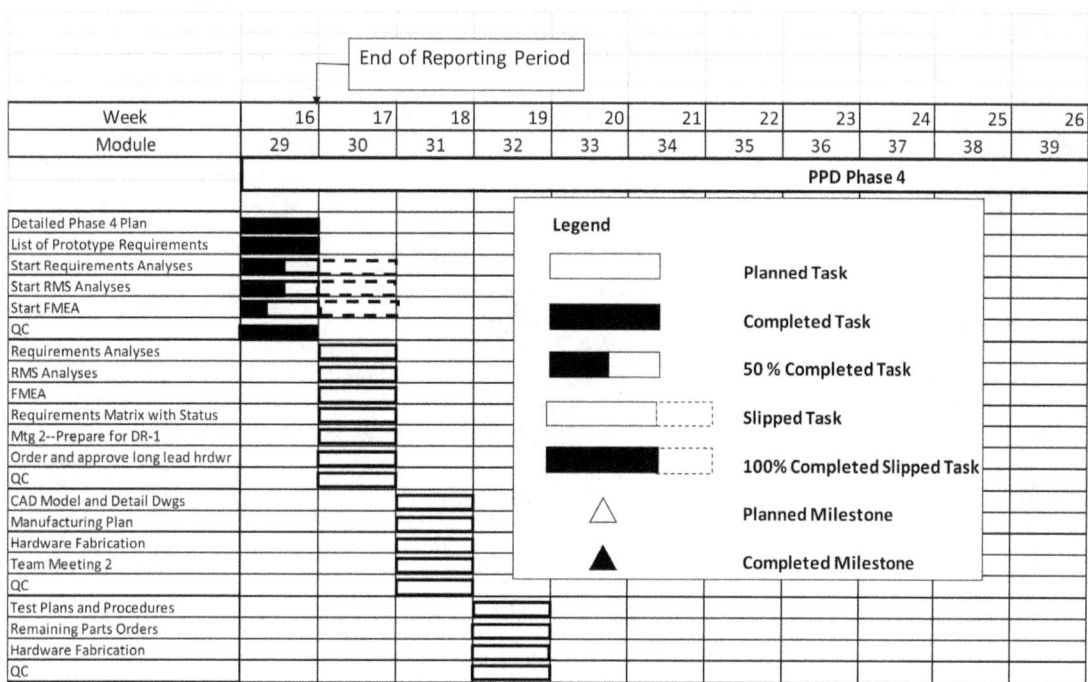

FIGURE B-75 Partial Phase 4 Schedule for Week 16 Showing Slips

As shown in Table B-48, the engineering prototype requirements were selected from the production unit requirements. The selected requirements are in grey. The criteria for selecting the engineering prototype requirements are as follows: (1) The requirement cannot be easily validated by analysis, and (2) the prototype and test rigs will not cost more than $600. In addition, the team agreed to review the Underwriters Laboratories (UL) requirements for clothes irons and add any requirements that impact safety.

During the team meeting, the Phase 3P preliminary design tasks were assigned and started. The team did not complete all planned tasks for Week 16. This means Week 17 will be very busy to make sure that all tasks are on schedule by the beginning of Week 18.

Table B-48 Engineering Prototype Requirements Shaded in Grey

Engineering Requirement	Target Value
Weight	<1 lbs
Volume	< 100 cubic inches
Length	< 7.0 inches
Input Power	60 Hz 110-120 vac; 50 Hz 220-240 vac
Press Time	< 2.0 minutes for pillow case
Iron Operating Temp. Range	Three settings: 275 deg F, 375 deg F, 400 deg F
Visual On Signal	Red on-light
On/Off Switch	On/Off Switch
Operating Life Reliability	MCBF = 2000 cycles
Drop Test	4 feet onto concrete floor
Operating Life	> 200 cycles
Shelf Life	5 years (0-150 deg F ambient temperature range)
Unit Production Cost	< $4.00 (2020 dollars)
Time to Production	< 2 years
Development/Tooling/Facility Cost	< $100k (2020 dollars)
Initial Prototype Project Materials Cost	< $600 (2020 dollars)

Note: MCBF is mean cycles between failures

MODULE 30

Prototype Analyses, Long Lead Hardware, and Manufacturing Methods

OVERVIEW

Module 30 focuses on the engineering prototype preliminary design (Phase 3P). The design team uses a process like the one used in Phase 3: Preliminary Design to design the prototype and analyze that design to make sure it meets the engineering prototype requirements. The team also conducts additional analyses, such as reliability, maintainability, safety, FMEA, and design for manufacturing and assembly (DFMA), during Phase 3P. Outcome of this phase are creating a computer-aided design (CAD) model of the engineering prototype, identifying and ordering items with a long lead time, and designing any needed engineering prototype test rigs.

LEARNING OBJECTIVES

- Gain knowledge and demonstrate the ability to complete all of the tasks necessary to conduct the engineering prototype preliminary design (Phase 3P).
- Demonstrate the ability to efficiently modify existing Phase 3P deliverables to quickly complete the preliminary design of the engineering prototype.

PRE-LECTURE ASSIGNMENT

- Read sections B.4.5–B.4.12.
- Each team member prepares a one-page summary of their work on their assigned Phase 3P tasks and how they plan to complete their tasks by the beginning of Week 18.

POST-LECTURE ASSIGNMENT

- Conduct a team meeting to 1) review each team member's one-page progress report and 2) modify any plans for the week to make sure that all work is done by the second team meeting.
- All team members complete their Phase 3P tasks.
- Conduct a second team meeting to (1) review all the Phase 3P work, (2) finalize the Phase 3P design, and (3) enter all results into the team notebook.

TEAM DELIVERABLES

- CAD model of the engineering prototype based on the analysis performed to ensure that the design meets the engineering prototype requirements
- Order of the long lead time items of the engineering prototype
- Phase 3P exit checklist
- Design review 4 slides that relate to Phase 3P
- Written sections of the final report associated with Phase 3P as presented in Section C of the handbook
- Minutes of team meetings
- Team activity as documented in the team notebook

B.4.5 Phase 3P: Engineering Prototype Preliminary Design

A process similar to the one used in Phase 3 preliminary design is used to design the engineering prototype and analyze that design to make sure it meets the engineering prototype requirements. A requirements validation matrix is used to communicate the results of the analyses. If the engineering prototype is substantially similar to the production unit, then the analysis task is not very large. However, if the engineering prototype must be substantially different from the production design due to resource limitations, then more analyses are required. This also applies to analyses beyond those directed at the prototype requirements and may include other analyses areas, such as reliability, maintainability, safety, FMEA, DFMA, and design to cost (DTC). In many cases, the methods of fabrication may differ between the high-volume production unit and the single manufacture of the engineering prototype. For example, the production unit housing may be cast, while the engineering prototype housing would usually be machined.

An important task in this phase is the creation of a CAD model of the engineering prototype. Again, if the engineering prototype is similar to the production unit, then modifying the production CAD model for the engineering prototype design is relatively simple. However, if the engineering prototype differs substantially from the production unit, then the creation of the CAD model and subsequent detailed drawings can become a sizeable task.

When the preliminary design of the prototype is completed, the team will move into Phase 4P: Prototype Detailed Design. At the end of Phase 4P, the team will hold a design review with the sponsor and other experts if available to review both the design and the drawings. All work needs to be documented in the team's project notebook. Due to the tight prototype development schedule, the prototype long lead time items should be identified near the beginning of the design process so that these items can be ordered. The preliminary design must also include the design of any test rigs needed during prototype development.

B.4.6 Nonlinearity of the Product Development Process

EM-(q) The block flow diagrams of the product development process imply a linear flow from one phase to the other. This type of diagram clarifies the overall flow of the product development effort. However, it is a simplification of most product development projects. The team keeps gaining new insights into the design as it progresses. This often requires the team to step back and revisit an assumption or decision made in a previous phase. The team needs to be aware of this phenomenon and welcome the opportunity to integrate new information into the design process. An example of this nonlinear design is presented in the example travel iron project described in Section B.4.12.

B.4.7 Human Factors Engineering and Industrial Design

Engineers focus on the product's functionality. Human factors engineering focuses on human characteristics, capabilities, and limitations in relation to machines, tasks, and environments. This includes such functional attributes as ease of use, effectiveness, and safety. Industrial designers are concerned with human factors but also aesthetics. One way of characterizing the difference between industrial design and engineering is that the industrial designer is more concerned with the outer characteristics of the product, while the engineer is more concerned with the inner characteristics of the product. Clearly, there is an overlap between these two design disciplines.

The example travel iron project requires all of the above concepts. The iron must press cloth by moving a heated, flat, and smooth plate in specific motions in both normal and lateral directions relative to the cloth/plate interface. The plate is heated by an outside electrical energy source. The normal force is supplied partially by the weight of the iron. Additional normal force and lateral force is supplied by the operator. The operator controls the specific motions of the iron by seeing what needs to be pressed and then moving their arm and hand to remove the crease in the material. The hand physically interfaces with the iron's handle. This human/iron process of pressing the cloth must be done in a safe way and, ideally, efficiently.

Most customers buy products not only for their functionality but also for their aesthetics. Are they pleasurable to the senses? Good product design requires both engineering and industrial design. Some companies have industrial designers on their staff. Other companies expect their engineers to provide both functionality and aesthetics. In the travel iron example presented below, the project team is focused on providing both of these characteristics.

EM-(q)

B.4.8 Long Lead Hardware

Supplier lead times are often the critical path on a development project schedule. This is especially true if some of the items are coming from offshore places such as China. The team needs to review the prototype preliminary design and identify any long lead items. This includes raw material that the team will use to manufacture parts or have them made by a supplier. Before any long lead hardware is ordered, it must be approved by the sponsor.

B.4.9 Manufacturing Methods

In parallel with the team completing the engineering prototype preliminary design, each team member should prepare themselves for the task of deciding during Phase 4: Detailed Design how production unit and prototype components will be manufactured. This section provides a summary of the most common manufacturing methods used.

Specialized books, such as those listed below, present detailed coverage of manufacturing processes and related topics such as their capabilities and limitations.

- DeGarmo, E. Paul, Black, J.T., and Kosher, Ronald A. 2003. *Materials and Processes in Manufacturing*. Ninth ed. Hoboken, NJ: John Wiley & Sons, Inc.
- Groover, Mikell P. 2019. *Fundamentals of Modern Manufacturing: Materials, Processes, and Systems*. Seventh ed. Hoboken, NJ: John Wiley & Sons, Inc.
- Kalpakjian, Serope and Steven Schmid. 2017. *Manufacturing Processes for Engineering Materials*. Sixth ed. London: Pearson.

This section provides a top-level summary. Selecting a manufacturing process for a part requires adequate knowledge of shapes or features incorporated into the part and their corresponding methods of manufacturing. The most important process selection criterion is the process capability with respect to tolerance and surface finishes specified on the detailed drawing. Industrial standards tabulate these values in handbooks. Decisions for selecting a manufacturing process is based on evaluating the relative technical capability, productivity, and economy among alternatives.

The most common manufacturing process classes are summarized below.

1. Casting and Molding Processes

In casting processes, parts are produced by pouring molten metal into a mold cavity, allowing it to solidify and cool before removing it from the mold. The cavity is made in the mold using a pattern, which is a replica of the part plus any required allowances. These allowances include shrinkage allowance, post casting machining allowance, warping allowance, and draft allowance.

Metal casting processes are classified by type of mold, type of pattern, or force used to fill the mold. Molds can be either expendable/nonpermanent (breakable and used once) or nonexpendable/permanent (nonbreakable and used several times). Depending on the casting process, patterns can be made from wood, expandable polystyrene, plaster, metal, wax, or rubber. The force used to fill the mold can be atmospheric, low-pressure, or high-pressure.

Factors affecting casting process selection include the following:

- Quantity of castings required
- Design of the casting
- Tolerances required
- Complexity
- Metal specification
- Surface finish required
- Tooling costs
- Economics of machining versus casting costs
- Financial limits on capital costs
- Delivery requirements

Some common casting/molding production processes are plastic thermal injection molding and metal die casting. Less well-known is powder metallurgy, in which parts are formed by pressing powder metal and alloys in molds to form a green compact, which is further sintered to acquire the desired strength and toughness.

Casting/molding processes are rarely used in prototyping because of their high capital cost. Instead, material removal processes are generally used.

2. Forming and Shaping Processes

Forming and shaping processes are applied on solid metals and alloys and are classified into two major classes: a) bulk deformation processes and b) sheet metal forming processes.

Bulk deformation processes are further classified as hot forming, when performed above the metal recrystallization temperature, or cold forming, when performed below the recrystallization temperature. Materials are shaped under high-force equipment and a special types of dies. These processes, which are not used for prototyping, include forging, rolling, extrusion, drawing, and swaging.

Sheet metal forming processes are applied to metal sheets. The most common processes are shearing operations, in which the sheet is cut to form a blank by shearing and blanking operations. The sheet blank can be formed into shape by bending, rolling, deep drawing, ironing, rubber forming, spinning, superplastic forming, and several other specialized forming processes.

Sheet metal processes are used in prototyping to produce shapes and enclosures by bending and drawing. During sheet metal forming, care must be taken to compensate for spring-back arising from elastic recovery of the formed part after removing the forming load. Another important consideration in sheet metal forming is including bending allowances when calculating the blank size of the sheet. Sheets can also be cut and put together to form a shape by one of the joining processes such as spot welding, riveting, or mechanical fasteners. The wide variety of sheet metal forming equipment include stamping presses, press breaks, rollers, and box shapers.

3. Material Removal (or Machining) Processes

This is the conventional group of processes that are used for prototyping before the advent of additive manufacturing era and is still a main means of prototyping. It has the advantage of being suitable for economically making one or a few parts. In this class of manufacturing processes, the shape of a part is produced from a solid shape by selective removal of material in chip form from a solid piece. This class of processes can be applied to metals, polymers, ceramics, and composite materials. Machining processes are used to make complex shapes and parts that require good surface finish and relatively more precise levels of tolerances. The conventional material removal processes include (1) lathe work (straight and taper turning, straight and taper boring, thread cutting, knurling, drilling, parting or cut off, etc.), (2) drilling or hole-making operations (drilling, core drilling, step drilling, counterboring, countersinking, reaming, center drilling, and gun drilling), and (3) milling operations. Milling operations use milling machines equipped with work holding devices and cutting tools in making horizonal, vertical, or inclined flat or curved surfaces.

Machining processes use machine tools (such as lathes, milling machines, drill presses, and machining centers) for mounting the workpiece, provide relative motion between the part and cutting tool, and adjust the feed rate and cutting speed of the operation. They use single-point tools as in turning and multipoint tools, such as drills and milling cutters. A user of these operations consults the appropriate handbooks and machine tool and cutting tool manufacturers' catalogs to select the right tool material and process parameters for machining a part.

Grinding class of machining includes surface and cylindrical grinding operations, which produce high-quality surface finish and more precise tolerances. Grinding operations are more classified as finishing operations than material removal ones.

4. Heat Treatment Processes

Whereas the preceding classes of manufacturing are used to create or change the shape of the workpiece, heat treatment processes change the properties of the workpiece by changing the microstructure. Heat treatment processes that change the properties of the bulk of the workpiece include quenching (in a fluid such as water, oil, or brine), annealing (heating to the desired temperature and then letting the part to cool in the furnace), and normalizing (heating the part to the required temperature and removing it to cool in air). Hardening can also be performed on the part surface by processes like carburizing, carbonitriding, nitriding, cyaniding, boronizing, flame hardening, and induction hardening. Parts that are surface-hardened include gears, cutting tools, and dies. Stress relieving and tempering are also heat treatment processes. Several technical sources exist on the heat treatment processes.

5. <u>Joining or Assembly Processes</u>

Joining or assembly processes connect various parts to form a working product. This is a necessary process in prototyping. There are three main classes of assembly operations: adhesive bonding, mechanical fastening, and welding.

Adhesive bonding joins parts using glue, epoxy, or one of a wide range types of adhesives. Adhesive bonding is not suitable for carrying large loads. Adhesive bonding is a handy process in prototyping.

Mechanical fastening is also an essential prototyping group of processes. These processes use mechanical fasteners (bolts, nuts, crews, rivets) to join the parts. They are more suitable for assemblies that require disassembling during their service lives for servicing or maintenance.

Welding processes are also used for prototyping. Welding has three major classes of processes: liquid-liquid (fusion welding), solid-liquid (brazing and soldering), and solid-solid (such as friction welding, roll bonding, ultrasonic bonding, and diffusion bonding). Welding processes are used to create permanent joints.

In *liquid-liquid welding processes*, the two parts to be joined are molten at their interfaces and allowed to fuse into each other to solidify and form a very strong bond. These include nonconsumable and consumable electric arc, electron beam, laser beam, and oxy-fuel gas welding processes.

In solid-liquid welding processes (such as brazing and soldering), the parts to be joined are not molten, but a filler alloy is molten in a gap between the two parts to make the bond. For each combination of parts, there is a special corresponding filler alloy for the joint that has to be used. The difference between brazing and soldering is that the former is performed above 800 °F, whereas the latter is performed below this temperature. Both brazing and soldering can be used in prototyping, but their weld strengths do not support high loads.

In solid-solid welding processes, the joining of the two parts is performed while they are solid. They may be subjected to heat and pressure during welding. These processes include cold, ultrasonic, friction, resistance, explosion, and diffusion welding processes. These processes are not used in prototyping.

6. <u>Surface Treatment Processes</u>

Surface treatment processes are applied to the part surface to either protect or decorate it. Surface protection can be done by coating using various techniques, including electroplating and painting. These will protect the surface from the surrounding environment, including corrosion or wear. Surface decoration can be done by polishing or coating processes. These processes are generally not used in prototyping.

7. <u>Quality Assurance, Testing, and Inspection Processes</u>

To ensure that the manufacturing processes are performed at the desired level of quality, the design team has to inspect, measure, and validate the quality of the process. Visual as well as appropriate metrological instruments and techniques are used for this purpose. This class of manufacturing is essential in prototyping.

B.4.10 Preparing for Phase 3P: Engineering Prototype Preliminary Design Review

For the engineering prototype the preliminary design review is combined with the engineering prototype detailed design review and occurs in Module 32. The outline for this design review (referred to as *DR4*) is provided in Table B-49. The team should prepare the presentation slides for the Phase 3P portion of the design review during this module.

Table B-49 Outline for the Engineering Prototype Phases 3P and 4P Design Review (DR4)

- Phase 3P: Engineering Prototype Preliminary Design
    - Summary of Phase 3 production unit design (CAD model and requirements)
    - Rationale for the scope of the engineering prototype
    - Engineering prototype requirements validation matrix (with updated status)
    - Examples of key Phase 3P engineering prototype preliminary design analyses
    - CAD model of the engineering prototype
    - List of long lead items ordered
- Phase 4P: Engineering Prototype Detailed Design
    - Changes to the Phase 3P CAD model (if any and why)
    - Indented drawing list and BOM
    - Top engineering prototype assembly drawing
    - Example of fabricated part drawing
    - Example test rig drawing
    - Phase 5P engineering prototype manufacturing plan
    - Phase 5P engineering prototype test plan
- Top-level (easy-to-read) project schedule and status
- Team labor hours chart
- Reasons why team will be successful in conducting Phase 5P

B.4.11 Phase 3P Exit Criteria Checklist

The final tasks for Phase 3P are to complete the (1) phase-exit checklist and (2) sections of the final report associated with Phase 3P as outlined in Section C of the handbook.

Phase 3P Exit Criteria Checklist

Team Name: _____ Team Leader: _____

Members: _____   _____
         _____   _____
         _____   _____

Date Started: _____ Date Completed: _____

Done	Exit Criteria		Comments
	1.	Functions and configuration block diagrams of the engineering prototype are complete.	
	2.	FMEA is complete.	
	3.	Component trade studies are complete.	
	4.	Analyses to support the trade studies and requirements validation were planned and are now complete.	
	5.	Proof-of-concept testing to support trade studies and requirements validation were planned and are now complete.	
	6.	The goal function analysis is complete.	
	7.	Solid model of prototype is complete.	
	8.	Preliminary design review portion of DR4 is completed and approved.	
	9.	Final report chapters 1, 3, 4, 5, and 6 are completed and approved.	
	10.	Notebook is checked and approved.	
Approved by	Name and Title		Completion Date

B.4.12 Example Travel Iron Team Report for Week 17

B.4.12.1 Integrating New Information into the Design

The team started their in-class meeting by recognizing that they were continuing to learn new engineering requirements for both the production unit and the engineering prototype unit. Even though the simplified phased product development process indicates that the production unit requirements are developed only during Phase 2: Conceptual Design, it was recognized that the actual design process is not linear. The requirements need to be updated as the design effort progresses. The team identified four major areas where new information has been gained: (1) The UL standard for clothes irons finally arrived, (2) a safety study was recommended in the preliminary design safety analysis report, (3) to the team must decide how electrical wires will be connected, and (4) the initial research into using recycled plastic for the shell indicated this effort is beyond the scope of the planned engineering prototype.

The team reviewed this information and realized that there was not enough calendar time in the capstone course to complete Phase 4 with a final detailed design drawing package that will lead immediately to Phase 5: Production Unit Development. Therefore, the team decided that an additional engineering prototype would be needed after the current one before a final drawing package would be ready for Phase 5. The scope of the current engineering prototype effort is discussed in the following sections of the team's weekly report.

B.4.12.2 UL Requirements Study

The UL requirements for clothes irons that was ordered during Phase 2 was finally provided by the university's library. The specific document is *UL 60335-2-3 Standard for Safety: Standard for Safety of Household and Similar Electrical Appliances, Part 2: Particular Requirements for Electric Irons*. Unfortunately, this document states that it only covers changes to the more general requirements listed in UL 60335-1. The team realized that there is not enough time to obtain this additional standard. It was decided that only the changes in Part 2 would be considered and the final report would note that Part 1 was not used and should be considered by the team that moves forward on the design at the conclusion of the capstone project.

The key items in the UL document are summarized in Table B-50. The UL requirements to be addressed with the engineering prototype are in **bold**. The UL requirement to have the automatic temperature control hold the temperature settings with a tolerance of ±10 °C is of special concern. The team has been unable to find documentation regarding the temperature control tolerances of the iron thermostats advertised for sale on the internet. The team decided to purchase one of these thermostats and determine its tolerances during development testing.

Table B-50 Summary of UL Requirements[1]

1.	**An automatic iron has automatic temperature control.**
2.	The stand is the heel of the iron on which the iron is placed when at rest.
3.	**Appliances must be marked with their rated power input.**
4.	Specific wording is required in the instructions booklet included with the production unit.
5.	A probe is used to verify that a finger cannot come into contact with a hot surface other than the sole plate or an electrical shock hazard.
6.	**There is a heating test where the sole plate is horizontal and supported by three metal pins. A thermocouple is used to measure the temperature of the sole plate. The sole plate on its highest setting cannot be higher than 662 °F.**
7.	**In the above heating test, the following temperatures above ambient cannot be exceeded:**
	a. **Insulation of supply chord and wiring is 60 kelvin.**
	b. **Electrical components when power is 1.15 times rated power.**
	c. **Surface iron rests upon is 65 kelvin.**
	d. **Nonmetallic iron handle is 50 kelvin.**
	e. **Any nonmetallic knob, grip, or other surface held for only a short period of time is 60 Kelvin, including surfaces below handle.**
8.	Leakage current and electrical strength are covered in Part 1 of the UL document.
9.	There is a separate heating test in which the iron is placed on a stack of cloth and does not ignite the cloth.
10.	**Stability test on an inclined plane.**
11.	Iron drop test and handle drop test.
12.	**The supply chord shall be type HPD, HPN, HS, HSJ, HSJO, or HSO or an equivalent heater cord.**
13.	Cord flex test.
14.	Plastic ignition test.

---

1 Requirements in bold will be included for the engineering prototype.

B.4.12.3 Safety Study

The team used the Phase 3 preliminary design safety report and FMEA along with the UL document to identify the safety decisions and actions listed in Table B-51.

Table B-51 Safety Decisions and Actions for Production and Prototype Units

1. A thermal fuse is to be placed in the circuit to limit the sole plate temperature to less than 662 °F.
2. To reduce unit production cost and because it is not a required safety item by the UL document, the travel iron will not incorporate circuit-off features for a tipped-over iron or an iron left on for a long time.

B.4.12.4 Connection of Electrical Wires Study

The method of connecting the wiring in the iron is a significant unit production cost factor. The team considered three methods: (1) screws, (2) wire nuts, and (3) soldering. The first two methods make disassembly easy, while desoldering is more difficult. The use of screws means that there must be a backing plate for each screw and also connectors at the end of each wire. With the number of connections needed, this approach can be expensive. The maintainability strategy is to not repair the production iron. It is a relatively inexpensive item, and it would be difficult for the manufacturer to justify having a distribution system for spare parts. The potential of the customer becoming injured while attempting a repair is also a consideration. The only potential replacement part is the electrical cord, which can be obtained at a local hardware store. The team decided to use wire nuts for this connection. The rest of the connections will be soldered.

One of the team members pointed out that the connection to the heating element is a wire that must operate above 400 °F. This is higher than the melting point of solder. The heating element wire operates at a high temperature due to the high resistivity of the wire (i.e., there is high I^2R thermal power loss into the wire). The heater wire needs to be connected to a low resistance wire that has low I^2R thermal power and adequate surface area to let heat transfer from the wire to the ambient conditions so that the end of the wire opposite the heater wire connection is at a temperature lower than the melting point of the solder. There are three options: (1) Braze or weld a terminal pin to the heating element, (2) provide a screw connection with a solder tab distant enough to ensure a relatively cold condition, or (3) use an aluminum sleeve over the Nichrome wire of the heating element and a copper pin that can provide the soldering connection at opposite end to the aluminum sleeve. The team selected the third option as shown in Figure B-76.

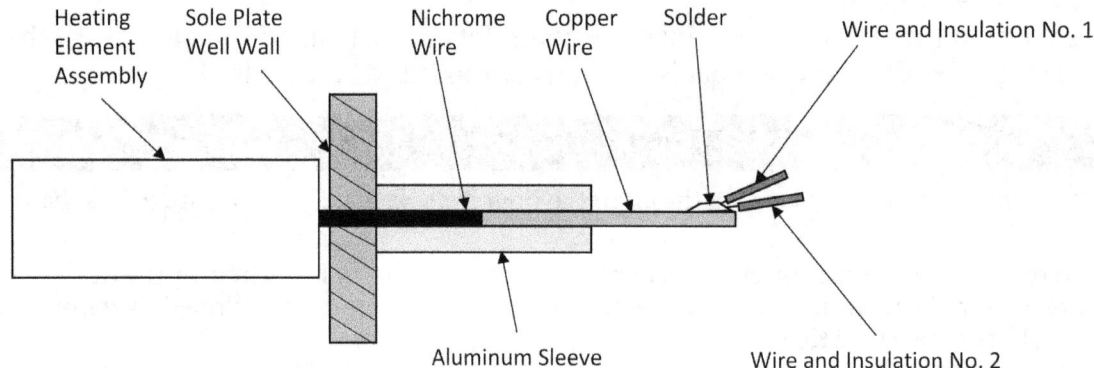

FIGURE B-76 Copper Soldering Wire Connection to Heating Element Wire

B.4.12.5 Engineering Prototype Requirements Validation Matrix

Table B-52 shows the requirements validation matrix developed by the team for the engineering prototype. The size and the weight of the engineering prototype were not included in the requirements because the shell and handle were not of the production configuration; however, the sole plate was the same as the production configuration.

Table B-52 Engineering Prototype Requirements Matrix

Engineering Requirement	Target Value	Validation Method	Status
Weight	<1 lhs	Demonstration with scale	CAD estimate = X.XX lbs
Volume	<100 cubic inches	Demonstration with scale	CAD estimate = X.XX cu inch
Length	<7.0 inches	Demonstration with scale	CAD estimate = 5.0 inches
Input Power	60 Hz 110-120 vac; 50 Hz 220-240 vac	Test by press cycle with power input and time	Wiring diagrams, analysis and selection of components complete
Press Time	<2.0 minutes far pillow case	Test and measure time	Analysis complete
Iron Operating Temp. Range	Three settings: 275 deg F, 375 deg F, 400 deg F	Test rig A with IR thermometer	Heating analysis and thermostat selection complete.
Visual On Signal	Red on-light	Inspection	LED harness designed
On/Off Switch	on/off switch	Inspection	Part of temperature selection dial
Operating Life	>200 cycles	Test by cycling test rig	Material and stress analyses complete
Initial Prototype Project Materials Cost	<$600(2020 dollars)	Demonstration by tracking all manufacturing and assembly material and supplier costs	initial budget less than $600.
UL Reqt	automatic temperature control—iron holds temperature in each of three settings ±10 deg F	Test rig A and IR thermometer	Not addressed beyond selecting a thermostat used in other irons.
UL Reqt	rated power marking of 400 watts	Inspection	Incorporated in drawings
UL Reqt	The sole plate on its highest setting cannot be higher than 662 deg F.	IR thermometer and test rig A where the sole plate is Horizontal and supported by three metal pins.	Not addressed beyond selecting a thermostat used in other irons.
UL Reqt	In the above heating test the following temperatures above ambient cannot be exceeded: a) insulation of supply cord and wiring is 60 deg K, b) non-use iron resting position surface is 65 deg K, c) non-metallic iron handle is 50 deg K, d) any non-metallic knob, grip, or other surface held for only a short period of time Is 60 deg K including surfaces below handle.	See validation method above.	Analyses indicate this requirement will be met in testing.
UL Reqt	stable when in the non-use sitting position on a TBD inclined surface and the iron is rotated about its major axis every 45 degrees.	Inspection using an inclined plane surface.	Estimated CG analyses indicates the iron will be stable under these conditions.
UL Reqt	The supply chord shall be Type HPD, HPN, HS, HSJ, HSJO, HSO, or equivalent heater cord.	Inspection	Supply cord ordered is Type TBD.

B.4.12.6 Reliability Analyses

The key reliability concern for the engineering prototype is avoiding manufacturing defects (infant mortality). The team agreed that each supplier part will have a team member monitoring its progress and ensuring that the supplier is following good quality control practices.

B.4.12.7 Maintainability Analyses

As opposed to the production unit, the engineering prototype will require multiple disassembly and assembly operations as the unit goes through its development testing. The team concluded that the production unit design can be assembled and disassembled multiple times. The soldered connections are not considered difficult to disassemble. The team did decide that some spare parts should be ordered as listed in Table B-53.

Table B-53 Spare Parts and Tools for Engineering Prototype Unit Maintenance

Spare parts	Maintenance tools
• Thermostat • 120- to 240-volt switch • Extra fasteners • Extra wire • Solder and flux	• Volt-ohm-meter (VOM) • Soldering iron • Phillips screwdriver • Slotted screwdriver • Wire cutters • Wire strippers • Infrared thermometer

B.4.12.8 Safety Analyses and FMEA

The team decided that instead of using a thermocouple to measure the component temperatures during the development and validation testing, an infrared (IR) thermometer will be used. This device uses a laser to point to the center of the area being measured. The laser light is a potential eye damage hazard. Mitigation steps will include safety warnings in the test procedures and diagrams to show how to use the IR thermometer in conjunction with the sole plate test stand. Soldering and desoldering the wire connections is a safety hazard due to the soldering iron's temperature. A warning note will be included in both the assembly and testing procedures.

B.4.12.9 Design for Manufacturing and Assembly (DFMA)

The team decided to eliminate the shell and handle in the prototype after conducting a DFMA analysis.

B.4.12.10 Design to Cost (DTC)

The team made sure that the cost of the prototype remained under the $600 budget.

B.4.12.11 CAD Model

The team did not have time to modify the production CAD model for the engineering prototype configuration. Instead, the team created the cross-section shown in Figure B-77. Some team members expressed concern regarding the ability of the nylon 6 plastic insulation plate to operate in the temperature environment created by the sole plate operating at a maximum of 400 °F. The team decided that the sole plate/plastic plate interface should be substantially less than 400 °F and nylon 6 will be an acceptable material. Temperature readings will be taken during engineering prototype testing to verify that this assumption is correct.

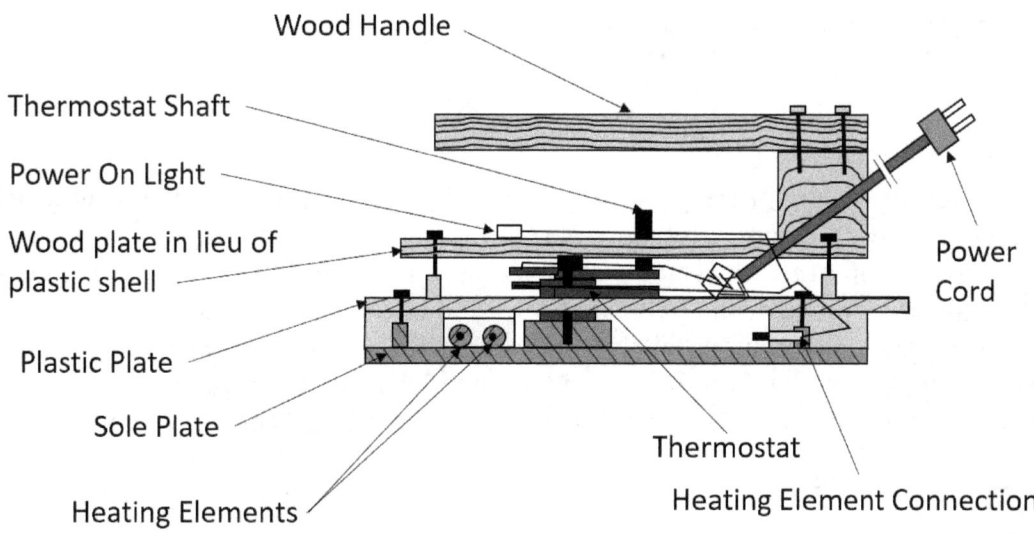

FIGURE B-77 Cross-Section of Travel Iron Engineering Prototype

B.4.12.12 Test Rig Design

The sole plate test rig (see Figure B-78) will be based on the sketch included in the UL document; however, a thermocouple and its support rod will be replaced using the IR thermometer.

B.4.12.13 Final Deliverables for Phase 3P

The team held a second meeting this week. The Phase 3P analyses were reviewed, the engineering prototype requirements validation matrix updates were approved, the Phase 3P exit checklist was completed, the final report sections applicable to Phase 3P were edited, and the Phase 3P portion of the DR4 was prepared based on the outline given in Table B-49.

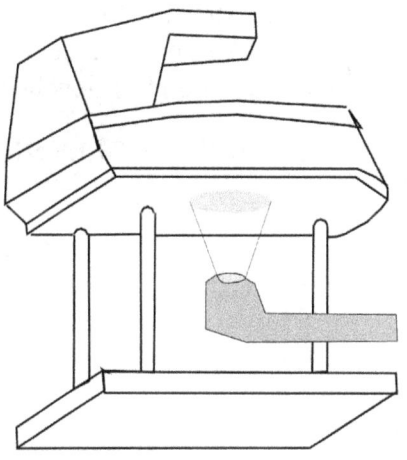

FIGURE B-78 Sole Plate Test Stand with IR Thermometer

Module 30 Prototype Analyses, Long Lead Hardware, and Manufacturing Methods | **255**

MODULE 31

Prototype and Test Rig Drawings, Manufacturing and Test Plans, and DR4 Preparation

OVERVIEW
Module 31 focuses on engineering prototype detailed design (Phase 4P), during which the prototype preliminary design moves into detailed design. This includes any remaining analyses and the creation of the prototype drawing package. A similar drawing package is also created for any test rigs to be used during prototype development. The team also creates a manufacturing plan that includes fabrication instruction sheets and a Phase 5P testing plan. The team also prepares for DR4 to be held in Module 32. DR4 presents phases 3P and 4P to the sponsor and subject experts prior to starting the manufacturing of the engineering prototype.

LEARNING OBJECTIVES
- Gain knowledge and demonstrate the ability to complete all the tasks necessary to conduct the detailed design of the engineering prototype (Phase 4P).

PRE-LECTURE ASSIGNMENT
- Read sections B.4.13–B.4.29.
- Each team member selects a part to be fabricated and develops for it a manufacturing plan.
- Each team member develops a manufacturing plan for the test rigs.

POST-LECTURE ASSIGNMENT
- Conduct a team meeting to review the team individual pre-lecture work and then assign subteams tasks to create: 1) the prototype drawing package, 2) the drawing package for any test rigs to be used during prototype development, 3) the manufacturing plan, and 4) the prototype testing plan.
- Conduct a second team meeting to (1) review the drawings and testing plans and (2) prepare/practice the DR4 to be given to the sponsor and others in Module 32.
- The team should hold a design review with the sponsor and subject experts prior to starting the prototype development.

TEAM DELIVERABLES
- Engineering prototype drawing package
- Drawing package for any test rigs to be used during prototype development
- Engineering prototype manufacturing plan
- Engineering prototype testing plan
- DR4 presentation slides
- Minutes of team meetings
- Team activity as documented in the team notebook

B.4.13 Phase 4P: Engineering Prototype Detailed Design Overview

In this phase, the engineering prototype preliminary design in the form of a CAD solid model is converted into a detailed design drawing package. As an example, Figure B-79 shows the CAD solid model for a trolley wheel product.

While the CAD solid model has nominal dimensions, the detailed drawings provide dimensions with tolerances so that they can be manufactured. Other key information, such as materials, manufacturing processes, and quality control measures, are included as notes on the drawings. In addition to the engineering prototype's drawings, the test rigs needed for the prototype's development and validation must be defined with detailed drawings. Other important deliverables of the prototype detailed design phase are the manufacturing plans and the test plans that will be used in Phase 5P: Engineering Prototype Development and Validation.

Phase 4P concludes with a design review, the completion of applicable final report sections, and the Phase 4P checklist. The outline for the design review was provided in Table B-49.

FIGURE B-79 Trolley Wheel CAD Solid Model Created by a Team During Preliminary Design

Sections B.4.13 through B.4.29 cover topics relative to Phase 4P. Information concerning how to create the detailed drawing package are provided in sections B.4.14 through B.4.21. A presentation on tolerances is covered in Section B.4.22. Information on how to create a manufacturing plan and a Phase 5P testing plan are discussed in sections B.4.23 and B.4.24, respectively. The end-of-phase deliverables of final report draft sections, DR4, and phase-exit checklist are given in sections B.4.26 through B.4.28. Section B.4.29 provides the example travel iron team report for Week 18.

B.4.14 Detailed Drawing Package Contents

The first step in learning how to prepare an engineering drawing package is to define some important terms. A product is the complete device. It is made up of components. A component can be a part, a standard component, or a subassembly. A part is an object fabricated from a single piece of material. A subassembly is the integration of two or more parts into a component that serves one or more functions. A standard component is an unaltered component for which no detail drawing is included because the part is to be procured from a source which fabricates that component to that source's specifications. An assembly (or top assembly) is the integration of all the parts, standard components, and subassemblies into the final product. The drawing package must have a detail drawing for each part in the product. It must also have drawings for each subassembly and for the top assembly.

The major deliverable for Phase 4P is a detailed drawing package that includes all of the information necessary to manufacture the engineering prototype. The drawing package includes the following:

- Outline drawing that gives the overall shape and dimensions of the product, identifies interfaces, and lists the product's weight. (Here, the product is the engineering prototype.)
- Assembly drawing that labels each component of the product and includes notes on how these components are to be joined to make the final assembly. For relatively simple

products the assembly drawing can use cross-sections to show all the parts. For complex products, an exploded-parts drawing is generally used.
- Subassembly drawings.
- Detail drawings, which are provided for all parts that are fabricated either in house or by a supplier.
- Control drawings or supplier specification sheets, which are provided for all standard components.
- BOM, which is a centralized source of information used to manufacture a product. Items included in a BOM are the part number, part name, quantity, unit of measurement, assembly references, method of parts construction, and additional notes.
- Process sheets, which are also included if there are any specialized processes required to fabricate or assemble the product.

B.4.15 Engineering Drawing Format

Drawings are an important method for engineers to communicate design intent to others. In the United States, engineering drawings generally follow the American National Standard Engineering Drawing and Related Documentation Practices (ASME Y14/ANSI Y14) for constructing working drawings and subassembly and assembly drawings. The example CAD models and engineering drawings presented in this handbook are based on the SolidWorks CAD software. There are a variety of other CAD software available, such as CATIA and Pro/E.

A detailed discussion of engineering drawing procedures is beyond the scope of this handbook. There are several good engineering graphics textbooks available for the team to use in preparing their drawings. This handbook covers the following key considerations for the detailed design drawing package:

- Key parts of a detailed drawing
- Key parts of a subassembly or assembly drawing
- How to draw third angle projections
- Detailed and cross-sectional views
- How to prepare a BOM

Figure B-80 shows the following major parts of an engineering part drawing:

- Graphical representation(s) of the item
- Border
- Title block
- Drawing number
- General tolerancing instructions
- Scale
- Drawing approvals
- Material
- Drawing revision information
- Manufacturing and inspection notes

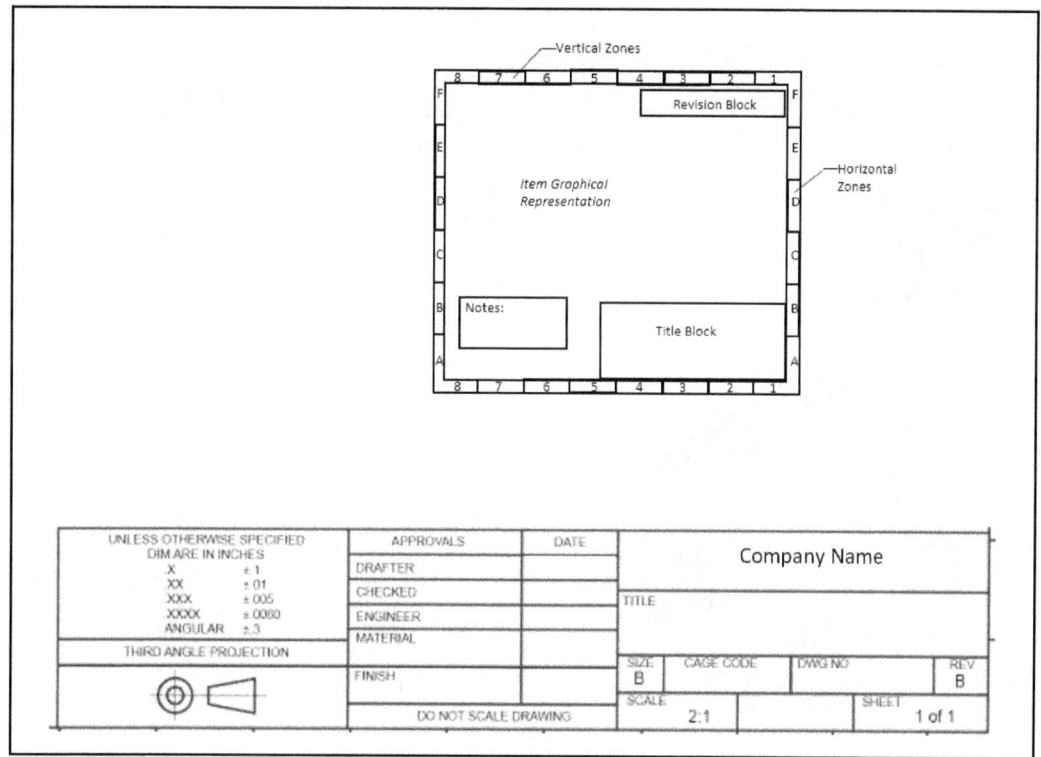

FIGURE B-80 Parts of an Engineering Part Drawing

A typical part drawing is the trolley wheel drawing shown in Figure B-81. This drawing uses a slightly different title block than the one provided in the previous figure. It should be noted that many of the dimensions do not have tolerances. This means that there needs to be a note somewhere on the drawing to state what the tolerances are as a function of the number of figures to the right of the decimal point. A suggested tolerance note is as follows:

**Unless Otherwise Specified,
Dimensions Are in Inches**

.X	± .1
.XX	± .01
.XXX	± .005
.XXXX	± .0005
ANGULAR	± 3 degrees

Note that the center hole in the trolley wheel has an ambiguous tolerance of 0.200 to 0.200 inches. There is no tolerance as written.

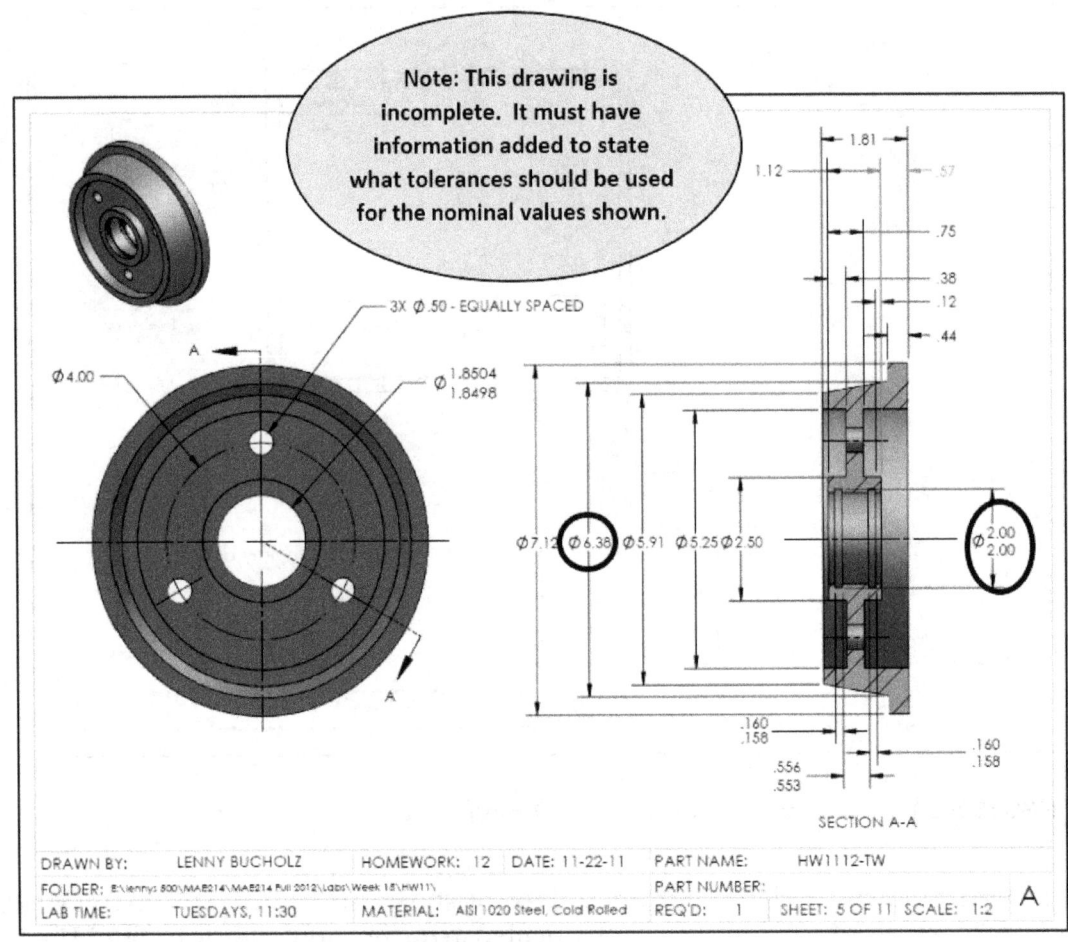

FIGURE B-81 Example of an Incomplete Detailed Part Drawing

B.4.16 Third Angle Projection

Figure B-82(a) shows a three-dimensional (3-D) object similar to the 3-D objects included in the CAD model created in Phase 3P. To convert this 3-D object into a detailed drawing, viewing planes are used, as shown in Figure B-82(b). The vertical plane is perpendicular to the horizontal plane dividing the space into four quadrants. Each quadrant is given an angle label. To convert the 3-D object into a detailed drawing, the object needs to be viewed from one of these quadrants. As shown in Figure B-83, the object is placed in the third angle quadrant and its top and side views are projected onto the horizontal plane and vertical planes, respectively.

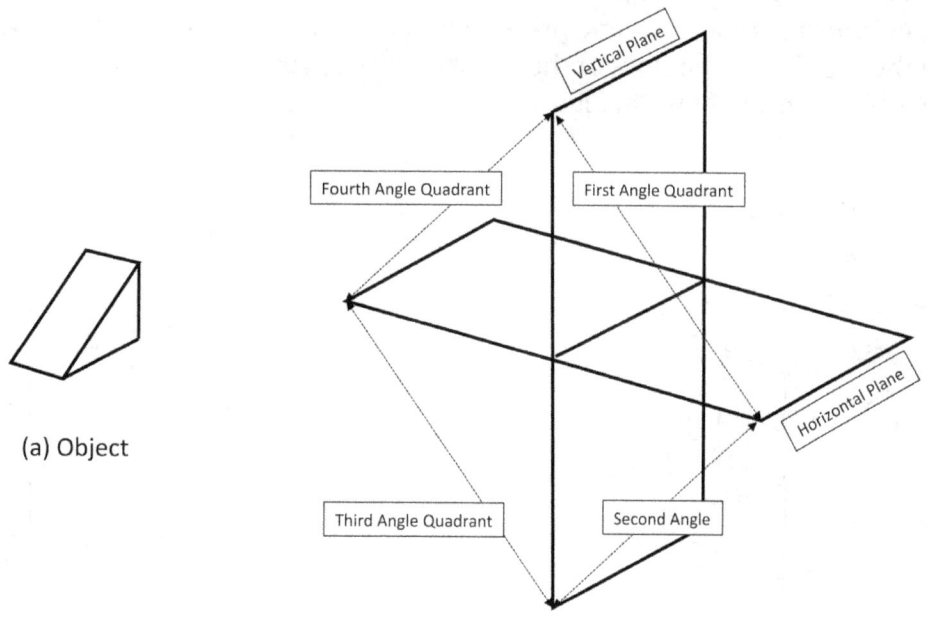

(a) Object

(b) Vertical and Horizontal Viewing Planes

FIGURE B-82 Possible Viewing Quadrants for a 3-D Object

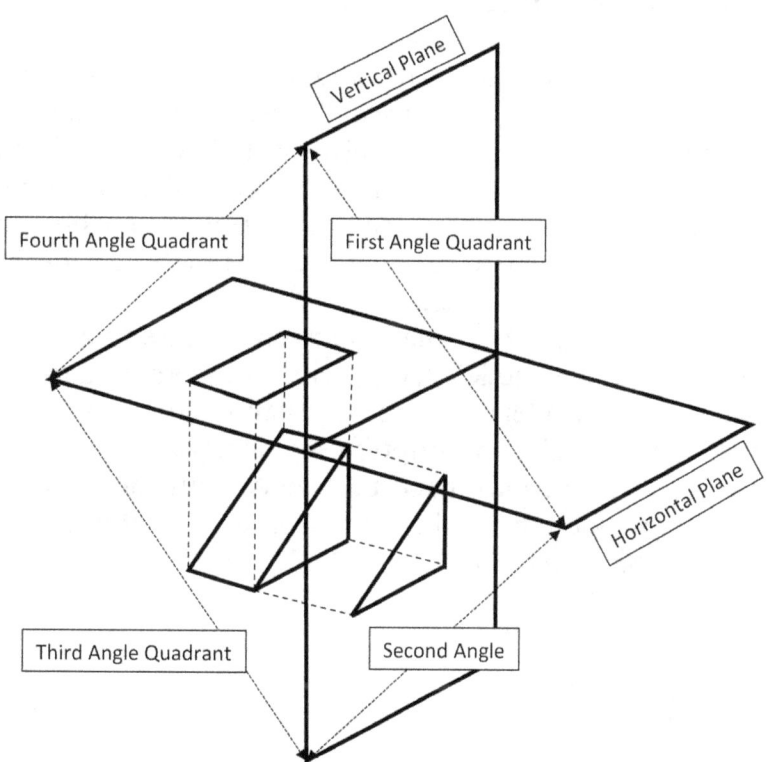

FIGURE B-83 Projecting a 3-D Object from the Third Angle Quadrant

Module 31 Prototype and Test Rig Drawings, Manufacturing and Test Plans, and DR4 Preparation

Figure B-84 shows how a 3-D object can be converted into a third angle projection by placing the object in an imaginary box and projecting the front, top, and sides of the object onto the sides of the box. The three sides of the box with object views can be unfolded to form the third angle projection, as shown in Figure B-85.

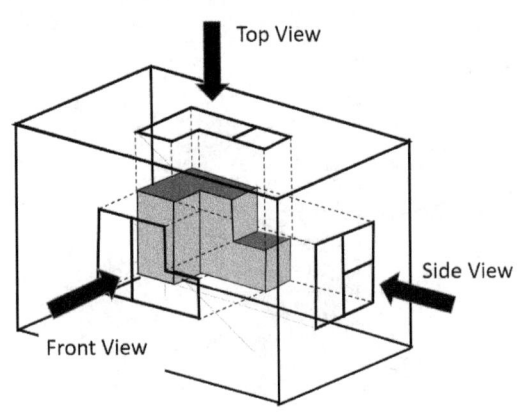

FIGURE B-84 3-D Object's Front, Top, and Side Views Projected onto an Imaginary Third Angle Projection Box

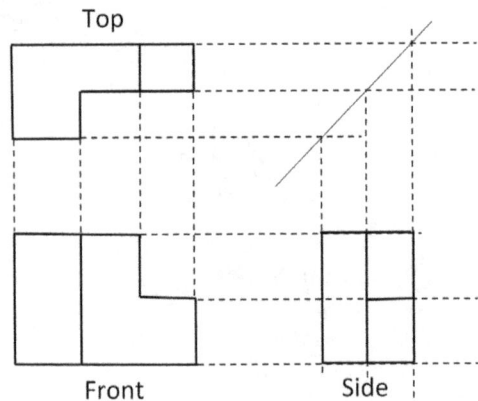

FIGURE B-85 How the Third Angle Projection of the Part in Figure B-84 Will Display in a Detailed Drawing

B.4.17 Subassembly and Assembly Drawings

The product or prototype are made up of parts and standard components. These items are connected to form the top assembly. Often some of the items are arranged into subassemblies, which are then incorporated into the top assembly. The designer often forms a subassembly when the items that make up the assembly are assembled at a place different from where the final assembly is assembled. Subassemblies are often formed to enable subsystem tests to be performed prior to testing the entire assembly. A top assembly can be made up of subassemblies, parts, and standard components.

A subassembly or assembly drawing usually shows all the items in an assembled state. Sometimes cross-sections are needed to reveal some of the items. Each item is identified by a number in a balloon and a leader going from the balloon to the item. A list of the items, including item number, part number, description, and material, is provided above the title block. There may be notes to describe how certain items should be assembled.

Figure B-86 shows the graphic and item list taken from an assembly drawing for the trolley wheel. A cross-section is included to show items 11, 12, and 13.

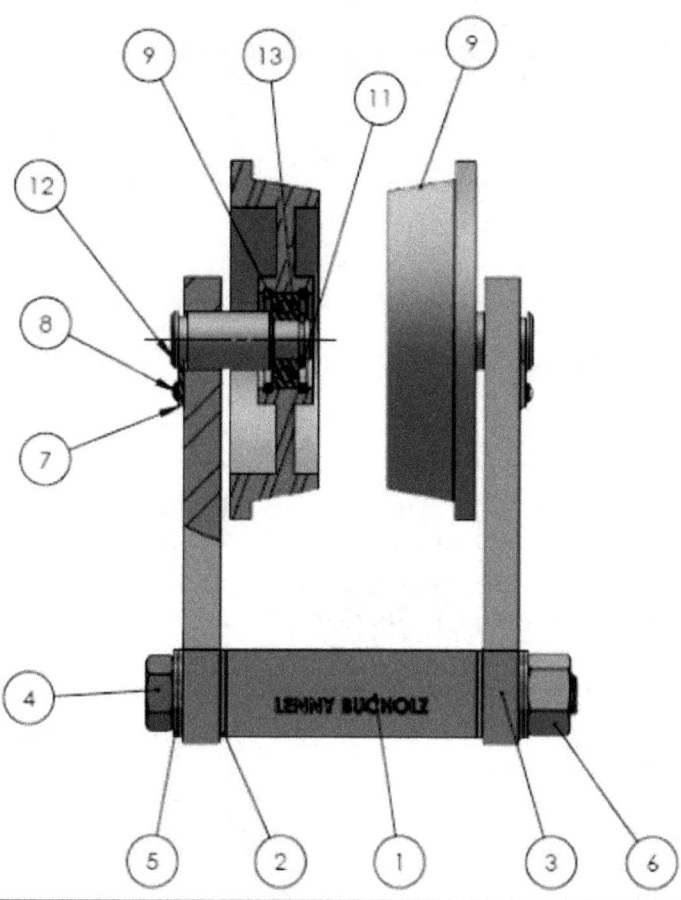

ITEM NO.	PART NUMBER	DESCRIPTION	Material	QTY.
1	HW1112-CH	CROSS HEAD	AISI 1020 Steel, Cold Rolled	1
2	HW1112-AS	AJUSTING SHIM	AISI 1020 Steel, Cold Rolled	2
3	HW1112-SP	SIDE PLATE	AISI 1010 Steel, hot rolled bar	2
4	HW1112-HHB	7\8-9 X 8.5 hex head bolt	GRADE 5 STEEL	2
5	HW1112-PWTA	7\8 FLAT WASHER	Plain Carbon Steel	8
6	HW1112-HHN	7\8-9 HEX NUT	GRADE 5 STEEL	2
7	HW1112-KP	KEEPER PLATE	AISI 1020 Steel, Cold Rolled	2
8	HW1112-MS1024	PAN HEAD 10-24 MACHINE SCREW	ZINC PLATED STEEL	4
9	HW1112-TW	TROLLEY WHEEL	AISI 1020 Steel, Cold Rolled	2
10	HW1112-WSR	WHEEL SNAP RING	SPRING STEEL	4
11	HW1112-ASR	AXLE SNAP RING	SPRING STEEL	2
12	HW1112-AX	AXLE	GRADE 5 STEEL	2
13	HW1112-BBD	BALL BEARING, DEPARTURE	Steel	2

FIGURE B-86 Trolley Wheel Assembly Drawing Graphic and Item List

B.4.18 Detail Views

As shown in Figure B-87, a detail view is a separate larger-scale view of a small section of another view. It is used to show features that are difficult to see in the main drawing. A detail view is designated by a letter, and its scale is indicted.

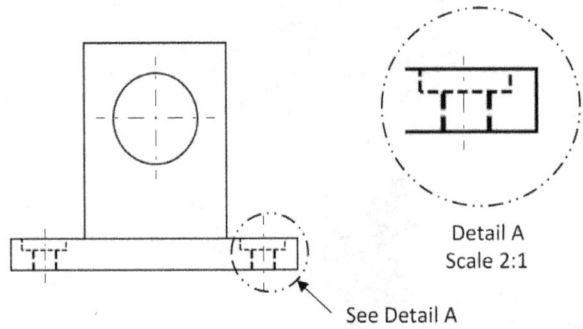

FIGURE B-87 Example of a Detail View

B.4.19 Section Views

A sectional view is used to show parts that are not visible when using just the front, top, and side views of the device. An example of this is given in Figure B-88. This device consists of a cylindrical pipe (Part 3) with end caps (Part1). Within Part 3 is a second pipe (Part 2). The wall thickness of the two pipes are shown in Section AA. The lengths of the pipes are shown in Section BB.

A section line is a line with arrows on each end perpendicular to the line. Each arrow has the same letter that designates the section. The line cuts through the object and the arrows point to the view that is used.

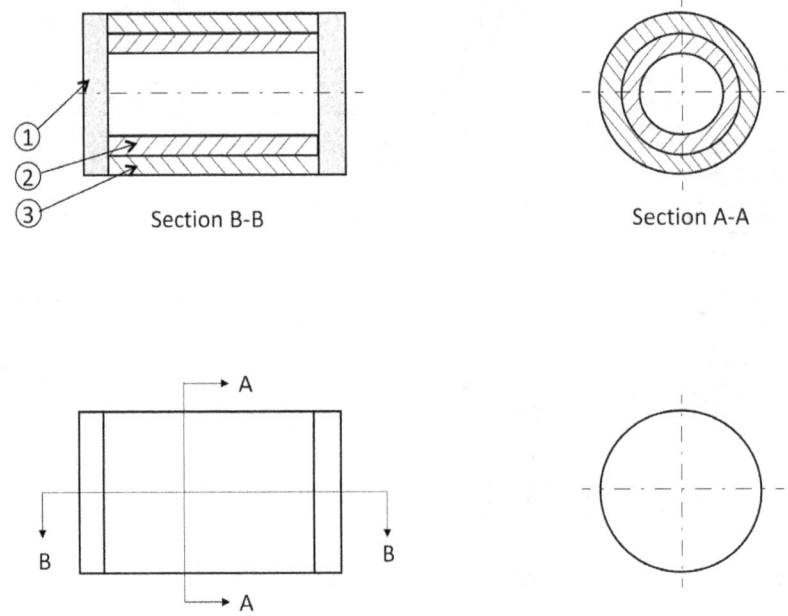

FIGURE B-88 Example of Using Section Views

B.4.20 Questions to Ask When Making a Working Drawing

When starting a detailed drawing, the following questions should be addressed:

- What material should be used for the part so that it will satisfy the design requirements?
- Is the part properly designed for the manufacturing methods to be used?
- Have standardized sizes and parts been used when possible?
- What are the critical dimensions of the part?
- To help reduce manufacturing costs, are the tolerances as large as possible for the part to still meet the design requirements?

B.4.21 Configuration Management

When a turbine blade fails in an aircraft engine, there is an investigation to find the cause. Suppose that it is determined that there was a problem with the material used to fabricate the turbine blade. There needs to be a system of serial numbers, part or drawing numbers, a database of revisions, and a manufacturing database that allow the team to trace the failure all the way back to the specific lot of material used in the blade's fabrication. The team then needs to determine what other aircraft engines have turbine blades that were fabricated from that same lot of material. To effectively accomplish this task, a process is needed to manage all the data that relate to the configuration of the turbine blade. This process is called configuration management (CM).

All engineering organizations need a configuration management system. For the aerospace industry, the configuration management systems are extensive and provide a great deal of traceability. On the other hand, consumer products tend to have less involved configuration systems that do not require as much traceability.

A good definition of CM is provided by MIL-STD 641A as follows:

> Configuration management (CM) is a systems engineering process for establishing and maintaining consistency of a product's performance and functional and physical attributes with its requirements, design, and operational information throughout its life.

For teams in an existing company, it is important to follow the CM procedures for that company. For teams in a startup company, it is the team's responsibility to develop their own CM system. In general, there is one CM process for products under development and another one for products that are in production.

Most of the aspects of CM are beyond the scope of this handbook. The discussion of CM in this section will be limited to those aspects that pertain to the capstone team. In most cases, the team is functioning like a startup company, so they need to develop a configuration management system that, as a minimum, covers the items listed in Table B-54.

Table B-54 Topics for a Minimal Configuration Management Plan

- Engineering prototype at the end of Phase 5P: Engineering Prototype Development
  - Specification for the engineering prototype that covers what level of completion each engineering prototype requirement achieved
  - Engineering prototype hardware build book
  - Drawing numbers for all parts
  - Drawing numbers for all subassemblies
  - Drawing number for the top assembly
  - Drawing number for the outline drawing
  - Database for all standard components used
  - Process sheets for special processes identified on the drawings
  - Method of marking all fabricated parts, including serial numbers
  - Name plate for the assembled engineering prototype, including part number
  - Drawing numbers for all test rig parts, subassemblies, top assembly, and outline drawing
  - Inspection plan for all incoming hardware
  - Plan and procedures for drawing changes
- Production unit at the end of Phase 4: Detailed Design
  - Specification for the product that covers what level of completion each production requirement achieved
  - Detail design drawing package:
    - Drawing numbers for all parts
    - Drawing numbers for all subassemblies
    - Drawing number for the top assembly
    - Drawing number for the product outline drawing
    - Database for all standard components used
    - Process sheets for special processes identified on the drawings

This section on configuration management will conclude with an example of a drawing numbering system and its method of display as either a drawing tree or an indented parts list. More aspects of configuration management will be presented in later sections of this handbook.

Figure B-89 provides an example drawing number system for an engineering prototype. Figure B-90 shows how this numbering system can be displayed as a drawing tree. Figure B-91 shows how this same numbering system can be displayed as an indented parts list.

Hypothetical Drawing Numbering System

Outline Drawing	1000A
Top-Assembly Drawing	1000B
Subassembly 1 Drawing	1101
Subassembly 2 Drawing	1102
Part 1	1001
Part 2	1002
Part 3	1003
Part 4	1004
Standard Components	
Component 1	Mfg #1
Component 2	Mfg #2
Component 3	Mfg #3

FIGURE B-89 Example Drawing Numbering System for a Hypothetical Engineering Prototype

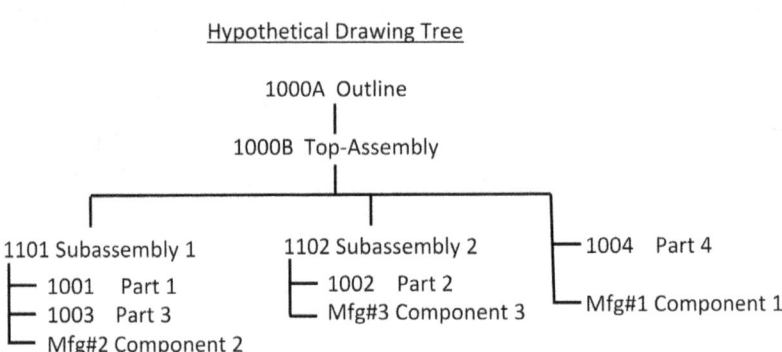

FIGURE B-90 Example Drawing Tree for a Hypothetical Engineering Prototype

```
Hypothetical Indented Parts List
1000A Outline
1000B Top-Assembly
          1101 Subassembly 1
                    1001 Part 1
                    1003 Part 3
                    Mfg #2 Component 2
          1102 Subassembly 2
                    1002 Part 2
                    Mfg#3 Component 3
          1004 Part 4
          Mfg#1 Component 1
```

FIGURE B-91 Example Indented Parts List for a Hypothetical Engineering Prototype

B.4.22 Tolerances

As discussed in the previous sections, engineers communicate through drawings. To define an object on a drawing, the engineer starts with the object's shape. The size of the shape is accomplished by dimensioning. In preliminary design, the object's dimensions were just one value. This is called the nominal dimension. When the object is manufactured, there will be some variation in the object's actual dimensions. Engineers must account for this variation by assigning tolerances to the nominal dimension. A measurement with a zero tolerance is impossible to manufacture in the real world. A part has been successfully manufactured if each dimension value is within the acceptable tolerance for that dimension.

B.4.22.1 Linear Tolerances

Tolerances limit the error a machinist is allowed on all dimensions, unless otherwise specified. There are two types of tolerances. Linear tolerances provide specific error limits for a particular linear measurement, such as length or angle. Geometric tolerances are error limits not on the size but on the shape of a feature, such as roundness or parallelism. Geometric tolerancing is beyond the scope of this handbook. This discussion of tolerances will be limited to linear tolerances.

As shown in Figure B-92, there are three major ways to define part tolerances: (1) unilateral, (2) bilateral, and (3) limit.

If no tolerances are specified at the dimension level, then general tolerances may be applied by deliberately controlling the number of values past the decimal point on each dimension in a note on the part drawing. An example note is given in Figure B-93.

The total tolerance is the value that describes the maximum amount of allowable variation in the dimension. A measuring device should be able to accurately measure within one-tenth of the total tolerance identified on the drawing.

In general, as the total tolerance of an object's dimension decreases, the cost of manufacturing that object increases. Conversely, as the total tolerance of the object's dimension increases, it is generally harder for the designer to ensure that the parts will function properly when mated to each other. Table B-55 provides typical tolerances that can be achieved with different manufacturing processes. Greater tolerances should be used when appropriate to reduce manufacturing costs.

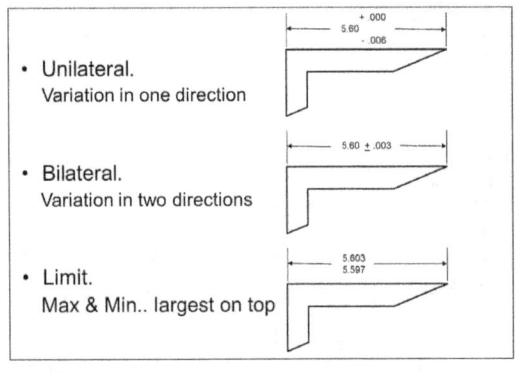

FIGURE B-92 Forms of Linear Tolerances

FIGURE B-93 Example General Tolerance Note on a Drawing

Table B-55 Typical Tolerances Possible for Various Manufacturing Processes	
Manufacturing Process	Typical Tolerance, ± inches
Machining	
—Turning	0.002
—Milling	0.003
—Drilling	0.002
Abrasive Processes	
—Grinding	0.0003
—Lapping	0.0002
Plastic Molding	0.010
Die Casting	0.005
Sand Casting	
—Aluminum	0.020
—Cast Iron	0.050
—Steel	0.060

B.4.22.2 Tolerances for 3-D Printing

In machining, the parts can be progressively processed to obtain smaller tolerances. 3-D printing, on the other hand, is a single automated production process. The tolerances of a 3-D printed part can't be refined beyond what the printer can produce without the use of subtractive methods. The designer needs to know what 3D printing capabilities are available and how printing costs vary as a function of the precision and accuracy needed. A good 3-D printer should be able to give a surface precision of ±0.005 inches.

B.4.22.3 Deciding What Tolerance to Use

The following three factors influence the decision on what tolerance to use:

- Functional requirements of mating parts
- Cost of production
- Available manufacturing processes

Table B-56 lists the two key rules for selecting a tolerance.

Table B-56 Two Key Rules for Selecting a Tolerance

- Choose as coarse a tolerance as possible without compromising functional requirements.
- Use proper balance between cost and quality of parts.

B.4.22.4 Fits of Mating Parts

Fit is the general term used to signify the range of tightness (or looseness) in the design of mating parts. As shown in Figure B-94, there are four types of fit that should be considered when working with tolerances: (1) **clearance fit**, (2) **interference fit**, (3) **transition fit**, and (4) **line fit**.

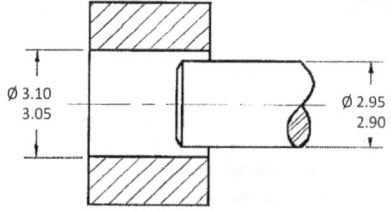

Clearance Fit- have limits of size so prescribed that a clearance always results when mating parts are assembled.

Max Clearance = 3.10-2.90= 0.20
Min Clearance = 3.05—2.95 = 0.10

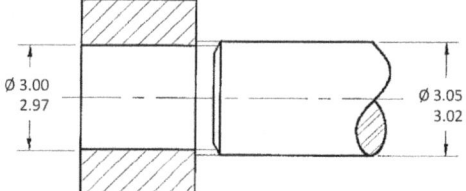

Interference Fit- have limits of size so prescribed that an interference always results when mating parts are assembled.

Max Interference= 3.05-2.97= 0.08
Min Interference = 3.02—3.00 = 0.02

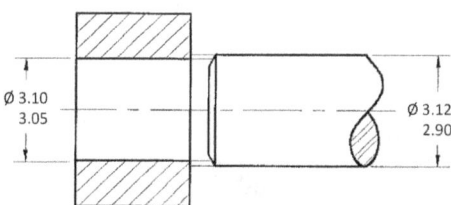

Transition Fit- have limits of size indicating that either a clearance or an interference may result when mating parts are assembled.

Max Clearance = 3.10-2.90= 0.20
Max Interference = 3.12-3.05 = 0.07

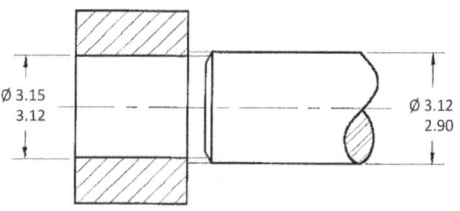

Line Fit: is the condition in which the limits of size are such that a clearance or surface contact may result between the parts.

Max Clearance = 3.15-2.90= 0.25
Surface Interference = 3.12-3.12 = 0.07

FIGURE B-94 Types of Fits

An interference fit produces a joint that is held together by friction after the mating parts are pushed together. An interference fit is also called a friction fit or press fit. Pressing together the mating parts is often accomplished with a hydraulic ram. Press fits have the potential for damaging one or both parts. If the parts must not sustain damage, then a shrink fit can be obtained by either heating or cooling one of the mating parts. When the treated part returns to normal temperature, the mating surfaces are held together by compressive forces.

B.4.22.5 Surface Finish

Surface finish or surface roughness is the deviation of the normal vector of a real surface from its ideal surface. If these deviations are large, the surface is rough; if they are small, the surface is smooth. Often surface roughness is measured by comparison to a sample of known surface roughness. A more precise method is to use either a contact or optical profilometer.

Figure B-95 shows the surface finish capabilities of various manufacturing operations. Figure B-96 shows the symbol used for various micro-inch finishes and the applications for them. The key design concept is to specify the roughest finish that will provide the needed

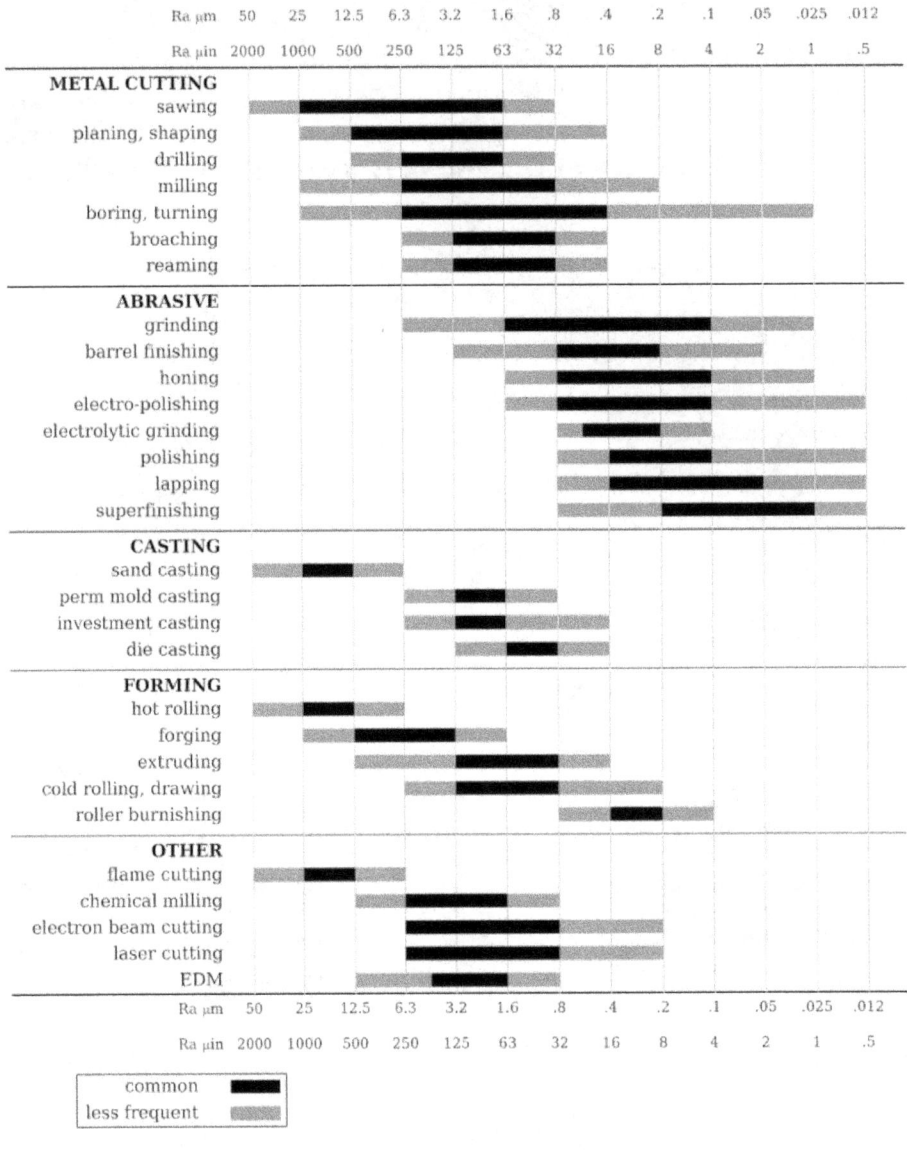

FIGURE B-95 Surface Finishes Possible for Various Manufacturing Processes

functionality because manufacturing costs increase greatly when smoother surface finishes are required.

Micro-inches Rating	Application
1000	Rarely used, suitable for unmachined areas on construction items.
500	Suitable for clearance features on machinery
250	Course production surface where tool marks are not objectionable.
125	Roughest surface for parts subject to vibration and high stresses. Lightly loaded and infrequently used bearing surfaces.
63	For close fits, stressed parts, bearing surfaces with light and infrequent loads.
32	For parts where stress concentration is present. For lightly loaded and non-continuous bearings.
16	Where smoothness is of primary importance such as extreme tension members, high speed or loading shaft bearings.
8	Surfaces for rings or packings to slide or withstand pressure, hydraulic cylinder surfaces, sensitive valve surfaces, bearings where lubrication not dependable.
4	Fine or sensitive instrument parts, certain gauge surfaces and low friction surfaces where lubrication is not dependable.
2, 1	Fine or sensitive instrument parts and certain gauge surfaces such as precision gauge blocks.

FIGURE B-96 Applications for Various Surface Finishes

B.4.22.6 Tolerance Stacking

An important task during detail design is to make sure that tolerance stack-ups allow proper clearances. Figure B-97 gives an example of tolerance stacking. Part A is a plug that is inserted first through the Part B spacer and then into a pocket in Part C. Parts A, B, and C are held together with two bolts. The tolerance concern is that the plug may be too long for the pocket or the gap between the end of the plug and the pocket end may exceed 0.100 inches. As shown in the referenced figure, the minimum gap between the end of the plug and the end of the pocket is 0.004 inches, and the maximum gap is 0.026 inches. This stackup meets the design requirements.

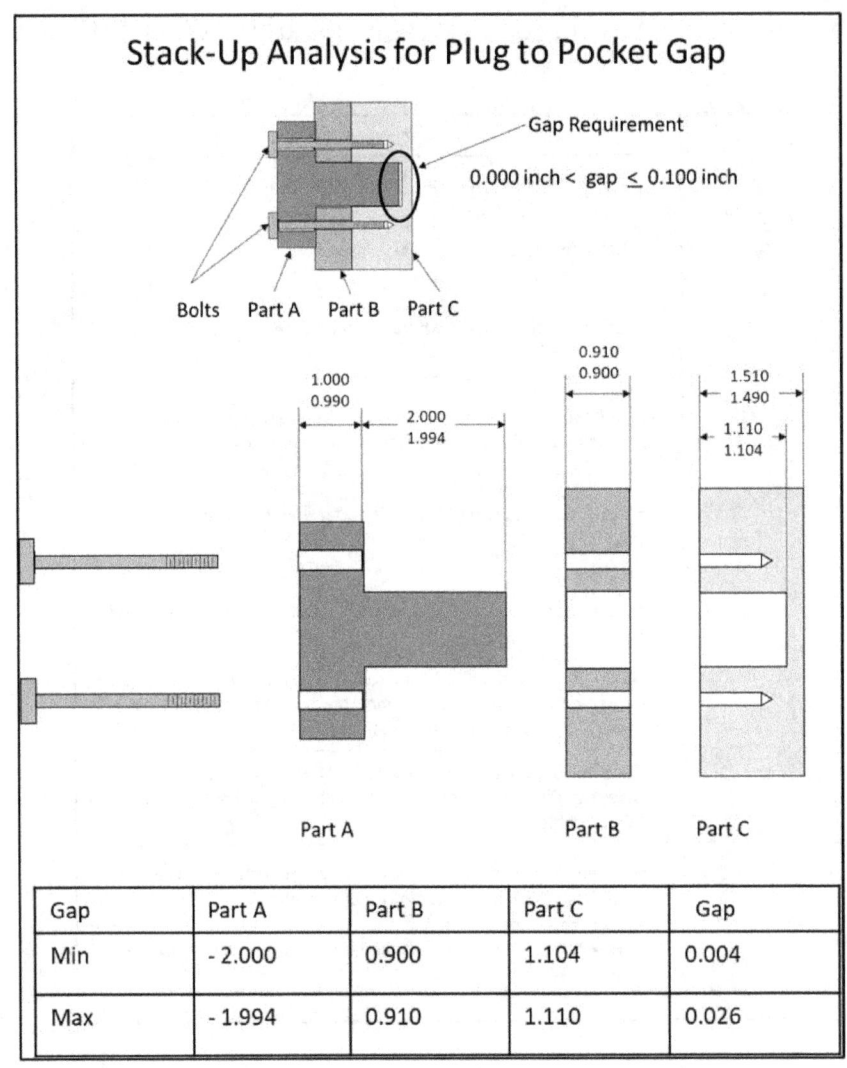

FIGURE B-97 Stack-Up Analysis for Plug to Pocket Gap

B.4.22.7 Questions to Ask When Considering Tolerances

The following are some key points that the team should address when deciding how to apply tolerances to a drawing:

- In general, the tighter the tolerances are, the more costly it is to fabricate the part.
- Use tight tolerances only when needed to ensure functionality.
- Be sure there is a general tolerancing note on the drawing for all nominal dimensions appearing on the drawing.
- Consult with the manufacturing experts to determine the costs associated with various levels of tolerance for key dimensions.
- Consider alternate design configurations that avoid the need for tight tolerances.
- Avoid tolerance stacking, if possible.

B.4.23 Manufacturing Plan

The team must have a plan for how they will manufacture the engineering prototype. Planning starts by defining the schedule, the available resources, and any special constraints. Resources include team labor, team manufacturing expertise, monetary budget, a place to assemble and inspect the engineering prototype, available suppliers, and their locations, etc. An example of a special constraint is a commercial-off-the-shelf (COTS) item that is available only from an overseas Chinese firm. The team then must arrive at a strategy that overcomes any special constraints and allows the manufacture of the engineering prototype to meet the schedule with the available resources.

Too often, teams forget to consider what might go wrong during the manufacturing process and how the team will resolve any problems and still stay on schedule and on budget. The first step in addressing this issue is to make sure that there is a team member assigned to each component. This liaison person keeps in contact with the supplier and helps resolve any issues that may arise. It is a good idea for the team to make a list of potential problems and discuss how the team would resolve each of them. An example issue is that a specialty supplier has fabricated a part that is out of print. The subject of resolving manufacturing issues will be discussed in more detail in further sections of this handbook.

Figure B-98 provides a flowchart for the manufacture of an engineering prototype. This chart is helpful in preparing the manufacturing plan. Table B-57 provides a suggested outline for the manufacturing plan.

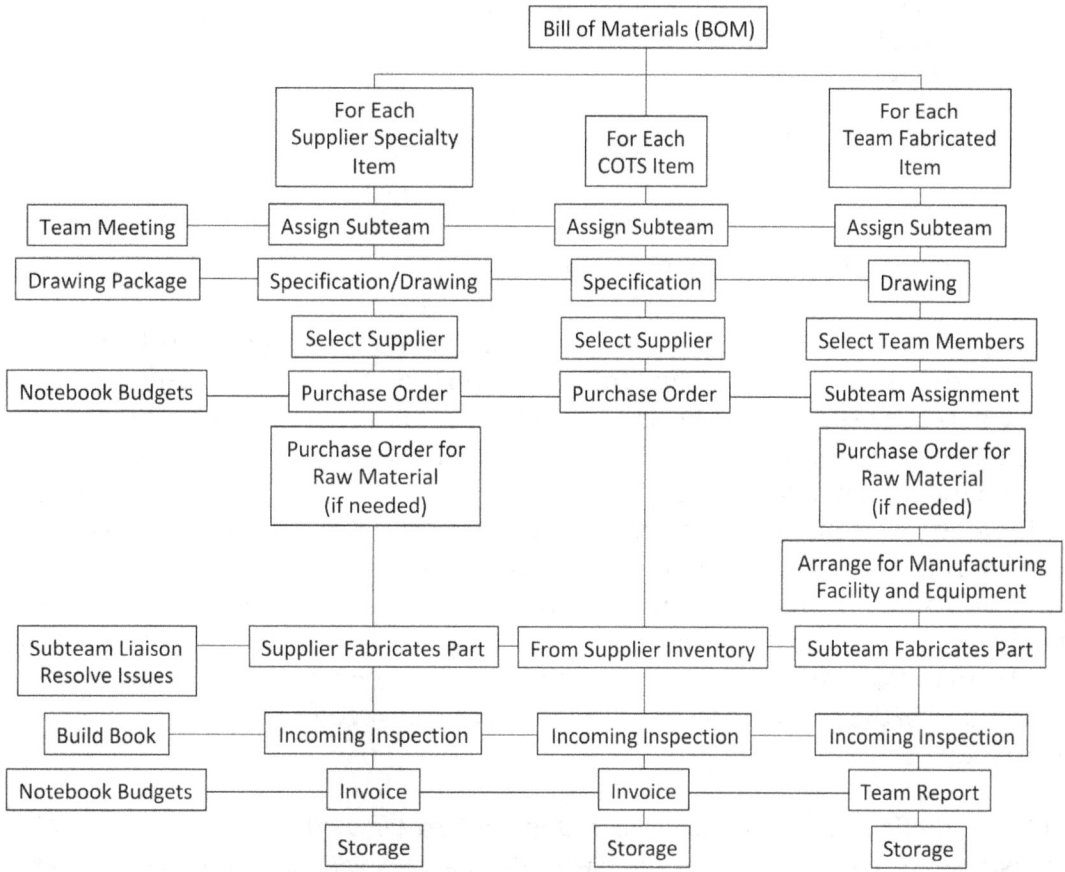

FIGURE B-98 Example Manufacturing Plan Flowchart

> **Table B-57 Suggested Outline for Manufacturing Plan**
>
> Manufacturing plan outline
> 1. Introduction
>   a. Purpose
>   b. Description of prototype and BOM
>   c. Resources
>   d. Schedule and constraints
>   e. Manufacturing strategy
> 2. Team processes
>   a. Selecting subteams for each component
>   b. Tracking changes to the drawing package during manufacturing
>   c. Tracking orders and expenses
>   d. Tracking changes with build book
>   e. Risk mitigation and resolving issues plan
>   f. Incoming inspection procedures, facilities, and equipment
>   g. Storage of completed components prior to assembly
> 3. Team-fabricated items
>   a. Facilities and equipment
>   b. Raw material procurement
>   c. Team member training
>   d. Part marking
> 4. Specific Plan for Each Component
> 5. Assembly
>   a. Facilities and equipment
>   b. Procedures including subassembly inspections
>   c. Assembly adjustments and calibrations
>   d. First article inspection
>   e. Final assembly marking
>   f. Final assembly storage

B.4.24 Manufacturing Process Planning

One of the goals of the capstone project is to give students the experience of making one or more of the project components. This is often done in the university's student machine shop. Before the part can be made, the process must be planned. This section covers this subject.

Most engineering products are made from individual parts that are joined in subassemblies and final assembly, and all these parts are manufactured individually using one or more of a wide variety of manufacturing processes. These processes differ with respect to their technical and economical capabilities. Manufacturing process planning covers the essential steps to manufacture an individual part or component in a job shop environment. Steps of manufacturing process planning include a) the analysis of the detailed part design drawing(s) first to identify information related to the processes, machine tools, tooling, and process parameters and b) the preparation of a routing sheet that specifies the operations sequence in the job shop.

B.4.24.1 Analysis of the Detailed Part Design Drawing(s)

In this task, the design team reads and carefully analyzes the part drawing to identify the following items:

- Material(s) from which the part is to be made, including its properties, geometry, and dimensions
- Number of pieces to be produced
- Dimensional and geometric tolerances specified for each part dimension and feature
- Surface finish specified for various part surfaces
- Manufacturing processes and operations needed to produce the part
- Machine tool(s) associated with each operation
- Tooling associated with each process
- Process parameters for each operation
- Sequence of operations through the job shop

B.4.24.2 Preparation of a Routing Sheet for the Part

The routing sheet documents the part's manufacturing process plan. It includes an operation list, machine tool list, and a tooling list. The routing sheet also specify the sequence of conducting the operations in the given job shop setup. The header of the routing sheet contains basic part information on the part name, part number, drawing number, quantity, material, and date. The routing sheet development is usually preceded by selecting the blank of the material from which the part is made, as detailed below.

Selection of the Blank Size of the Part

Selecting the right material, shape, and dimensions of the blank size of a part is a crucial task. The two main characteristics of a raw material from which a part is to be made are its geometry (shape and dimensions) and its properties (mechanical, physical, and chemical). The type of raw material used for the part is usually indicated in the drawing. The shape and dimensions of the raw material from which the part is to be made are determined from the part drawing. Engineering raw materials from which a part is made can be obtained in any of the following forms and geometries:

- Flat sheets
- Solid or hollow round bars or rods
- Flat bars
- Angle, channel, or box-shaped cross-sections.
- Preformed parts, such as castings, forgings, and stamping blanks

The design team must pay attention to the following points when selecting the shape and amount of the material required for the part:

- If the material is to be made by casting, calculate the weight of metal from the approximate volume of the mold cavities and add to it an allowance for the gating system, runners, risers, etc.
- If the part is to be made from a flat or bar (solid or hollow), cut the blank slightly larger than the finished dimensions of the part from the material that has the same type of cross-section (i.e., round bar for round part, square for a square, rectangular for rectangular).
- One may be forced to choose a round rod from which to cut a blank of rectangular or square cross-section with the intention of machining out the excess material. This happens when there is an excess stock of round bars and the proper rectangular bars are not available.

Example 1: Selecting a Blank and Preparing the Routing Sheet for a Stepped Shaft
Problem:
   a. Find the right blank size (i.e., diameter and length) from which the stepped AISI 1045 steel shaft part shown in Figure B-99 can be machined. Dimensions are in millimeters (mm).
   b. Prepare a routing sheet for the stepped shaft manufacturing process plan.
   c. Prepare an operation sheet for manufacturing the part.

Solution:
   a. The design team has indicated that they selected the AISI 1045 round steel bar because it meets the desired mechanical properties of tensile strength (83,000 pounds per square inch [psi]), yield strength (65,000 psi), and hardness (84 hardness Rockwell B). The team now has the task of selecting the length and diameter of the blank bar from which the part will be machined. Because machining is required, the blank AISI 1045 rod must have a diameter larger than 75 mm and a length longer than the 210 mm.

From MetricMetal.com,[1] the design team selected the 80-mm diameter bar, which is the next larger diameter from the finished 75 mm diameter of the stepped shaft. This provides 5mm machining allowance for the diameter. The length of the blank rod will be equal to the length of the finished stepped shaft (200 mm) plus the machining allowance for facing its ends and plus the length to mount it in a three-jaw chuck during turning on the lathe. Assume the latter allowance is 40 mm. Also, a 5mm allowance for facing the ends of the shaft at the end of the operations is provided. This makes the length of the blank (including the finished length of the stepped shaft of 200 mm) 250 mm.

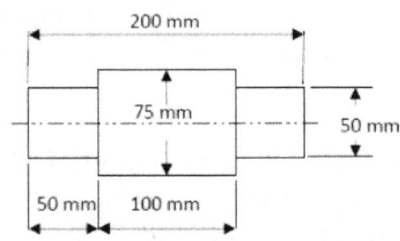

FIGURE B-99 Stepped AISI 1045 Round Steel Shaft Drawing

   a. After selecting and acquiring the blank of the stepped shaft, the design team has prepared the routing sheet in Figure B-100 for the manufacturing process plan.
   b. The routing sheet of a part is used for planning the operation, but the operation sheet documents the actual steps used in making the part. The part's operation sheet is prepared during part manufacturing by following the routing sheet operations sequence. It also includes the machine tool, cutting tool, process parameters (such as speed, feed, depth of cut) used during manufacturing, and a remarks column. The sheet is a record of all operations and has information that determines the machining time and number of tools used for the purpose of costing the part. The operation sheet elements vary according to the operation. An example of an operations sheet template is given in Table B-58.

---

1 MetricMetal.com. n.d. *Carbon Round Bar 1045 TGP*. https://www.metricmetal.com/product/carbon/round-bar-carbon/carbon-round-bar-1045-tgp/.

Routing Sheet for the Stepped Shaft				
Part Name:		Part No.		
Quantity:		Material: AISI 1045 steel bar		
Date:				
Operation No.	Operation		Equipment/ Machine	Tooling
10	Cut 250-mm long blank from the 80-mm dia. bar		Horizontal Band saw	Saw blade
20	Insert 40-mm length of one end of the bar in the 3-jaw chuck of the lathe		Engine lathe	
30	Face the free end of the bar		Engine lathe	Carbide insert tool
40	Turn the 80-mm diameter to 75-mm dia along the length of the shaft outside the chuck		Engine lathe	Carbide insert turning tool
50	Turn the free end of the 75 mm dia. shaft to 50 mm dia. for a length of 50 mm.		Engine lathe	Carbide insert turning tool
60	Remove the bar from the 3-jaw chuck		Manual	
70	Insert and secure the whole length of the 50 mm dia. machined end of the bar in the 3-jaw chuck		Engine lathe	
80	Center drill the unmachined end of the bar.		Engine lathe	Center drill
90	Advance the tailstock to the end of the bar to support it. Lock the tailstock.		Lathe	Tailstock
100	Part the shaft at a 152.5 mm length from the face of the chuck.		Engine lathe	Carbide insert
110	Face the end of the shaft by removing 2.5 mm length.		Engine lathe	Carbide insert facing tool
120	Turn the 75 mm dia. end of the bar to 50 mm dia. for 50 mm length.		Engine lathe	Carbide insert turning tool
130	Remove the machined stepped bar from the chuck		Manual	

FIGURE B-100 Routing Sheet for the Stepped Shaft

Example 2: Operation Sheet of the Stepped Shaft

A few sample elements of the operation sheet for the stepped shaft in Figure B-99 is shown in Table B-38, which provides a typical operation sheet template.

Table B-58 Operation Sheet Template								
Part No.:				Part No.:			
No	Operation	Machine tool	Cutting tool	Cutting speed	Feed rate	Depth of cut	Remarks
10	Cut bar	Hacksaw	Saw blade	-	-	-	-
20	Face end of bar	Engine lathe	RH facing tool	Selected from table	Select value	Select value	
..	..	..					

Example 3: Routing Sheet for the Galvanized Steel Bracket
The previous example of the routing sheet of for stepped shaft is for a machined part. This example is for a sheet metal formed part. The bracket shown in Figure B-101 has to be formed from a 3-mm thick galvanized steel sheet. The operations for making the part include blanking, drilling four holes, and bending. The routing sheet for the bracket is given in Figure B-102.

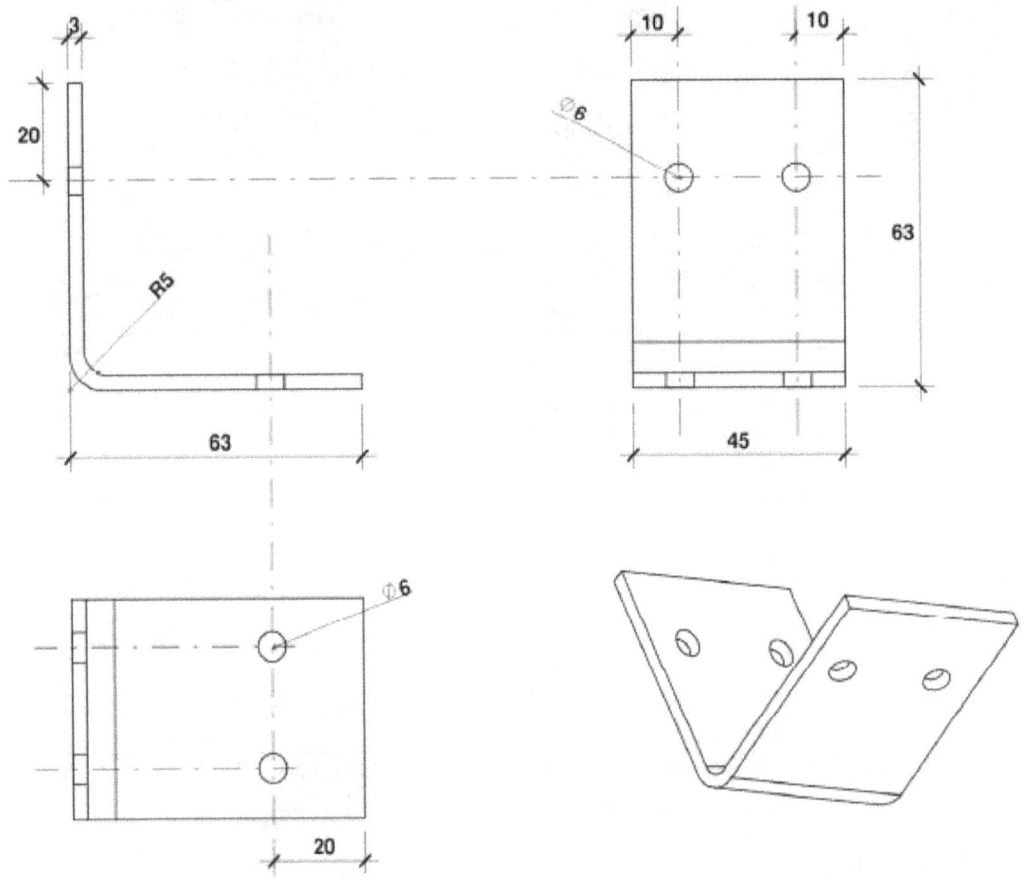

FIGURE B-101 Galvanized Steel Bracket[2]

Part No.:0001	Part Name: Bracket		Drawing No. 2xxx-03
Quantity:	Materia: 3-mm thick galvanized sheet		Update:
Operation No.	**Operation**	**Machine**	**Tooling**
100	Cut blank	Shearing (Guillotine)	Shear blade
200	Deburring	Manual	Vise and file
300	Drill 4 holes	Drill press	Twist drill (Φ6-mm)
400	Bending	Press break	V-die
500	Check dimensions	Measuring bench	Rule, set square

FIGURE B-102 Routing Sheet for Sheet Metal Bracket

---

2 All nominal dimensions are in millimeters.

B.4.25 Engineering Prototype Test Plan

In prior phases, the team prepared test plans for their proof-of-concept tests. In this phase, the team needs to prepare a more complete test plan that identifies each test to be performed during the development and validation of the engineering prototype. Plan preparation instructions are provided in Appendix BA4. The test procedures for each of these tests will be prepared as part of the next module.

B.4.26 Final Report Sections

During Phase 4P, the team is responsible for preparing the final report sections that cover this phase.

B.4.27 Preparation for Phase 3P and Phase 4P Design Review (DR4)

DR4 covers Phase 3P: Engineering Prototype Preliminary Design and Phase 4P: Engineering Prototype Detailed Design. The purpose of the design review is to make sure that the team is ready to enter Phase 5P: Engineering Prototype Development and Validation. DR4 occurs during Module 32, so the team needs to prepare for this during Module 31. The team prepared the Phase 3P portion of DR4 during Module 30. The team now includes the work done in Phase 4P as outlined previously in Table B-49.

B.4.28 Phase 4P Exit Criteria Checklist

Below is the phase-exit checklist for this phase. The checklist must be completed and approved by the sponsor before the team can move into Phase 5P: Engineering Prototype Development.

Phase 4P: Engineering Prototype Detailed Design Exit Criteria Checklist

Team Name: _____   Team Leader: _____
Members: _____   _____
Date Started: _____   Date Completed: _____

Done	EXIT CRITERIA	Comments
	1. Detailed analyses are complete.	
	2. Drawing package is complete.	
	3. Procurement, fabrication, and assembly instructions are complete.	
	4. Development and validation test plan and test procedures are complete.	
	5. All FMEA actions are complete.	
	6. Final cost estimates are complete.	
	7. The production and prototype designs meet their DTC targets.	
	8. First article inspection plan is complete.	
	9. Critical design review (CDR) is complete.	
	10. All corrective actions from CDR are complete.	
	11. Final report sections of the previous activities are complete and approved.	
	12. Notebook is checked and approved.	
Approved by	Name and Title	Date

B.4.29 Example Travel Iron Team Report for Week 18

B.4.29.1 Team Meetings
After the Module 31 lecture, the team met to review each team member's pre-lecture work. The team then created design teams to prepare the following: 1) prototype drawing package, 2) drawing package for any test rigs to be used during prototype development, 3) prototype manufacturing plan, and 4) prototype testing plan. Prior to the beginning of Module 32, the team conducted a second meeting to (1) review the drawings and manufacturing and testing plans and (2) prepare and practice the DR4 presentation to be given to the sponsor and others in Module 32. This design review covers Phase 3P: Engineering Prototype Preliminary Design and Phase 4P: Engineering Prototype Detailed Design.

B.4.29.2 Engineering Prototype Drawing Package
The team prepared a complete set of drawings.

B.4.29.3 Test Rig Drawing Package
A drawing package for the test rig was prepared. The test rig supports the iron while the sole plate temperatures are being determined.

B.4.29.4 Phase 5P Manufacturing Plan
Figure B-103 provides the manufacturing plan. A list of prototype components and their liaison team members is given in Table B-59.

B.4.29.5 Phase 5P Testing Plan
The team prepared a test plan following the instructions in Appendix BA4. The test matrix included in that plan is provided as Figure B-104. Several prototype requirements were deleted from the testing due to limitations in resources. The tests will be only conducted with 120 volts AC (VAC) power because 240 VAC power is not available in the garage where the tests will be conducted. The incline stability test was deleted because the shell was eliminated in the prototype design. The operating life test was reduced from 200 cycles to five cycles due to time limitations. These deleted requirements tests will be addressed in their entirety in the development of a second engineering prototype that is beyond the scope of this capstone course.

B.4.29.6 Preparations for DR4
The team prepared slides for DR4 following the outline provided previously in Table B-49. The team has only 10 minutes to present their work. A slide was prepared for each item listed in Table B-49. Care was taken to make sure each slide was readable and the presentation communicated the state of the prototype design. The team had to practice the presentation several times to ensure that they would not exceed the 10-minute time limit. To do this, some information had to be moved into backup slides that will probably not be shown during the slide presentation but will be included in the slide package provided to the sponsor.

B.4.29.7 Phase 4P Exit Criteria Checklist
The team completed the checklist, and it was approved by the sponsor.

Manufacturing Plan for the Travel Iron Engineering Prototype

1. **Introduction**
    a. **Purpose:** To plan how the prototype will be manufactured in a way to stay on schedule and budget.
    b. **Description of prototype and BOM:** See Team Project Notebook
    c. **Resources:** ASU Student Machine Shop, Team member A has a garage for assembly and testing.
    d. **Schedule and constraints:** Must be completed by the middle of Week 24. Budget is $500.
    e. **Manufacturing strategy:** Use ASU machine shop for sole plate and plastic cover, all other items COTS or team fabrication, assemble and inspect in Team member A's garage.
2. **Team Processes**
    a. **Selecting subteams for each component:** See table below
    b. **Tracking changes to the drawing package during manufacturing:** Any change to the BOM and/or drawings will be coordinated by Team member C. Team member B will update all drawings (including assembly and subassembly drawings) based on input from the liaison team member for each component.
    c. **Tracking orders and expenses:** The assigned liaison team member will communicate with the supplier to insure that the item will arrive on-time and be of high quality. The order, progress reports, cost estimates, invoices, etc. will be placed in the build book.
    d. **Tracking changes with build book:** Changes to any drawings will be recorded in the build book along with the necessary deviations and drawing changes.
    e. **Risk mitigation and resolving issues plan:**
      ---Student Machine Shop fabricated parts are the major risk. The liaison engineer will check daily to make sure that the parts are progressing as needed. The liaison engineer shall make any changes as necessary to keep the fabrication of the part on schedule and budget.
      ---If a COTS item incurs a schedule problem, the team will purchase an inexpensive iron and use it to provide the needed COTS item.
      ---The liaison engineer will make an effort to contact the team quickly to reach consensus on a manufacturing plan change; however, to stay on schedule, the liaison engineer is empowered to make the change if team members are unavailable.
    f. **Incoming inspection procedures, facilities and equipment:** Each component will undergo incoming inspection by the assigned liaison engineer. The inspection sheet will be placed in the build book. The procedures given in item e will be followed. The component will be stored in Team member A's garage. Storage, fabrication, assembly and first article inspection will be conducted in Team member A's garage which has all necessary tools. The electrical power of 120 VAC and 240 VAC will be available in Team member A's garage.
    g. **Storage of completed components prior to assembly:** See item f.
3. **Team Fabricated Items**
    a. **Facilities and equipment:** See item 2.f.
    b. **Raw material procurement:** The liaison team member for the fabricated part will also order the required raw material.
    c. **Team member training:** Team member A will provide any shop skills training.
    d. **Part marking:** At incoming inspection all components will be bagged or tagged with the part number.
4. **Specific Plan for Each Component:** See table below.
5. **Assembly**
    a. **Facilities and equipment:** See item 2.f.
    b. **Procedures including subassembly inspections:** References to assembly and inspection procedures will be noted on drawings. The procedures will be included in the build book.
    c. **Assembly adjustments and calibrations:** The adjustments and calibrations will be accomplished during the Phase 5P testing.
    d. **First article inspection:** See item 2.f.
    e. **Final assembly marking:** A part number tag will be affixed to the unit.
    f. **Final assembly storage:** See item 2.f.

FIGURE B-103 Manufacturing Plan for Travel Iron Engineering Prototype

Table B-59 List of Components and Assigned Liaison Team Members

No.	Part Number	Component	COTS	Fab	Liaison Member	Supplier and Comments
	100000X	Top Assembly Dwg			A	
1	100001X	Sole plate		X	A	Aluminum plate CNC machined in student shop
2	900001	Outer htr element	X		A	SWT Heater Supply
3	900002	Inner htr element	X		A	SWT Heater Supply
4	100002X	Well cover		X	B	Team-sheet metal fabrication
5	900003	Well filler	X		A	3M RTV
6	900004	Thermostat	X		A	SWT Heater Supply
7	900005	Thermostat screw	X		B	Ace Hardware
8	100003X	Black plastic cover		X	C	PEEK plate CNC machined in student shop
9	900016	LED	X		B	Fry's Electronics
10	900006	LED harness wahser	X		B	Ace Hardware
11	900007	LED harness screw	X		B	Ace Hardware
12	900008	Dual voltage switch	X		C	Fry's Electronics
13	900009	Power cord	X		C	Fry's Electronics
14	900010	Wire nut (2)	X		C	Ace Hardware
15	900011	Wiring	X		C	Ace Hardware
16	900011	Wiring	X		D	Ace Hardware
17	900011	Wiring	X		D	Ace Hardware
18	900011	Wiring	X		D	Ace Hardware
19	Deleted	Top shell		X	Deleted	
20	100005X	Temp/Off Knob		X	D	Team-plastic rod on lathe
21	Deleted	Covers (4)		X	Deleted	
22	Deleted	Screws (4)	X		Deleted	
23	Deleted	Clear plastic window		X	Deleted	
24	1000006X	Wood Handle		X	E	Team-Pine block cut on table saw
25	1000007X	Wood Transition		X	E	Team-Pine block cut on table saw
26	1000008X	Wood Plate		X	E	Team-Pine block cut on table saw
27	900012	Screws, Long (4)	X		D	Ace Hardware
28	900013	Screws	X		E	Ace Hardware
29	900014	Sleeve, aluminum	X		A	Ace Hardware
30	900015	Thick Copper Wire	X		B	Ace Hardware
	900016	Aluminum Plate	X		C	McMaster-Carr
	900017	PEEK Plastic Plate	X		D	McMaster-Carr
	900018	Thin Aluminum Sheet	X		E	McMaster-Carr
	900019	Plastic Rod, Large Dia	X		E	McMaster-Carr

Test No.	Engineering Requirement		Validation Method	Test Description	Time, Hr
1	Iron Operating Temp. Range	Three settings: 275 deg F, 375 deg F, 400 deg F	Test rig A with IR thermometer	With 110 VAC, Calibrate iron for the temperature settings by marking the thermostat knob alignment on the shell.	2
2	UL Reqt 1	automoatic temperature control--iron holds temperature in each of three settings ± 10 deg F	Test rig A and IR thermometer	With 110 VAC, Five trials for each setting.	2
3	UL Reqt 3	The sole plate on its highest setting cannot be higher than 662 deg F.	IR thermometer and test rig A where the sole plate is horizontal and supported by three metal pins.	Set at maximum power setting on thermostat	0.5
4	Press Time	< 2.0 minutes for pillow case	Test and measure time	With 110 VAC, standard size pillow case at 400 deg F. No wrinkles on either side.	0.5
5	UL Reqt 4	In the above heating test the following temperatures above ambient cannot be exceeded: a) insulation of supply cord and wiring is 60 deg K, b) non-use iron resting position surface is 65 deg K, c) non-metallic iron handle is 50 deg K, d) any non-metallic knob, grip, or other surface held for only a short period of time is 60 deg K including surfaces below handle.	See validation method above.	At the end of Test 4 take required temperature measurements. One reading per location.	1
6	Operating Life	> 200 cycles	Test by cycling test rig	For 110 VAC only. Not enough testing time to conduct complete test. Conduct 5 cycles of iron going from ambient temperature to setting S3 temperature.	2

FIGURE B-104 Travel Iron Prototype Test Matrix for Development and Validation Tests

Credits

Fig. B-95: Source: https://commons.wikimedia.org/wiki/File:Surface_Finish_Tolerances_In_Manfacturing.png.

Fig. B-96: "Applications for Various Surface Finishes," Adapted from www.cnccookbook.com/surface-finish-chart. Copyright © by CNC Cookbook, Inc.

MODULE 32

Starting Manufacturing, Prototype Testing Procedures, and Conducting DR4

OVERVIEW

Module 32 marks the beginning of Phase 5P: Prototype Development and Validation. This phase starts with the manufacturing of the prototype. In parallel with the manufacturing activities, the team prepares the prototype testing procedures and conducts the DR4 (this occurs in lieu of the lecture).

LEARNING OBJECTIVES

- Demonstrate the ability to start the manufacturing of the prototype by following the manufacturing plan previously prepared.
- Demonstrate the ability to prepare test procedures.
- Demonstrate the ability to conduct the Phase 3P and Phase 4P Design Review (DR4).

PRE-LECTURE ASSIGNMENT

- Read sections B.4.30–B.4.32.
- Each team member prepares test procedures for the first test in the test plan prepared in Module 31.
- Each team member prepares a paragraph describing their initial manufacturing tasks.
- Each team member practices their portion of DR4.

POST-LECTURE ASSIGNMENT

- Conduct a team meeting to complete the following tasks: 1) Review the feedback from DR4, 2) review the individual ideas on the test procedures for Test 1; 3) based on this review, the team decides on a format for the test procedures and assigns team members to prepare each procedure; and 4) each member reports on the status of their manufacturing plan assignments.
- Conduct additional meetings to finalize the test procedures and address any manufacturing issues identified by the liaison engineers.

TEAM DELIVERABLES

- Component orders
- DR4 presentation
- Finalized test procedures
- Minutes of team meetings
- Team activity as documented in the team notebook

B.4.30 Phase 5P: Engineering Prototype Development and Validation

EM-(f)

This module starts with the conduct of Phase 3P and Phase 4P Design Review (DR4). Following DR4 the team begins work on Phase 5P Engineering Prototype and Validation. Phase 5P has three sequential tasks: (1) Manufacture the engineering prototype and test rigs, (2) conduct development testing of the engineering prototype and rework the prototype as needed so that there is a high probability that the unit will pass all the validation tests, and (3) conduct prototype validation testing.

At the beginning of this phase, the team should review their plans for Phase 5P and make any modifications needed based on new information learned during Phase 4P. It is the responsibility of each team member to start their manufacturing tasks. To be successful, every team member needs to hold themselves accountable for completing their assigned work on time. If a team member encounters an issue, they must immediately bring this to the attention of the team so that the team can decide on a course of action to resolve the issue.

It is important to remember that the team is responsible for completing the manufacturing effort on time even if they encounter issues. Almost every project has manufacturing issues. Each team member must keep in contact with their suppliers to uncover any issues as soon as they occur and seek help from the rest of the team as needed to resolve these issues.

Maintaining the team's high performance is continued in this phase by conducting team 360-degree reviews, as was done in the prior phases.

B.4.31 Testing Procedures

In parallel with engineering prototype manufacturing, the development team needs to finalize test procedures for Phase 5P: Engineering Prototype Development. As a reminder, Appendix BA4 provides a test plan procedures template in Figure BA4-1. This template should be used to prepare the test procedures for Phase 5P. The test procedures are covered as Item 6 in this template.

B.4.32 Example Travel Iron Team Report for Week 19

The team started the week by conducting design review DR4. Feedback from the review were discussed during the first team meeting, and the team entered decisions into the team notebook. The team reviewed each team member's procedures for Test 1 and decided on a specific procedures format as shown in Figure B-105 for the travel iron development Test 1.

Each team member then reported on their progress in executing their part of the manufacturing plan. The liaison team member for the heating elements reported that a 48ohm inner heating element was not available for the prototype, although it could be provided by the manufacturer for the production unit. A 53.6ohm heating element was available, and the team decided to use this for the engineering prototype. The use of this new heating element results in a change in the expected performance for the engineering prototype, as shown in the analysis provided in Figure B-106. The maximum power for the engineering prototype is calculated to be 368 watts for a 120 VAC power input voltage.

The team held a second team meeting later in the week to review and consolidate the individual test procedures into a final set of test procedures.

Test Procedures for Test No. 1 Iron Operating Temperatures Calibration

1. Purpose: To mark the thermostat knob for the position that yields 51 = 275 deg F, 52 = 375 deg F and 53 = 400 deg F.
2. Approach: Use the test stand to support the sole plate of the unit. Find and mark the thermostat knob position for each of the temperature settings using the IR thermometer to measure the temperature of the bottom of the sole plate. 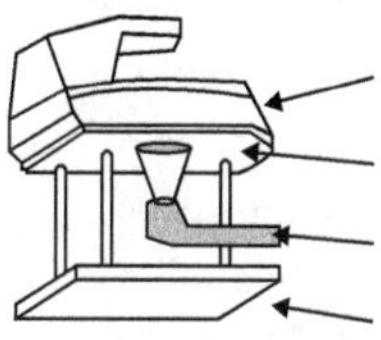
3. Test Article Description: The travel iron engineering prototype.
4. Description of the Test Setup: See figure at right.
5. Environment and Test Conditions: Room temperature in the garage of one of the team members.
6. **Safety Provisions:** Warning: Do not touch sole plate when hot.
   Warning: Do not look directly into laser on IR thermometer.
   Warning: Use caution when powering travel iron with electricity.
7. **Data Collection Sheets:** In this case, instead of a data sheet, take a picture of the thermostat knob with the three temperature positions marked.
8. **Step-by-Step Instructions:**
   a. Place prototype in test rig as shown in figure.
   b. Energize prototype with 110 VAC electricity.
   c. Adjust thermostat knob to achieve a sole plate temperature of 275 deg F using the IR thermometer. Measure sole plate at approximately the location in the figure. The temperature must be steady state for at least 1 minute.
   d. Mark body of prototype on its centerline next to knob and mark knob to line up with this mark and label knob with a 1 to represent setting 1. Use a felt pen to make these marks.
   e. Adjust thermostat knob to achieve a sole plate temperature of 375 deg F using the IR thermometer. Measure sole plate at approximately the location in the figure. The temperature must be steady state for at least 1 minute.
   f. Mark knob with a line to line up with the prototype body mark and label this knob mark with a 2. Use a felt pen to make these marks.
   g. Adjust thermostat knob to achieve a sole plate temperature of 400 deg F using the IR thermometer. Measure sole plate at approximately the location in the figure. The temperature must be steady state for at least 1 minute.
   h. Mark knob with a line to line up with the prototype body mark and label this knob mark with a 3. Use a felt pen to make these marks.
   i. De-energize prototype and let it cool to room temperature before removing it from the test rig.

FIGURE B-105 Procedures for Travel Iron Prototype Test 1

Travel Iron Calculations for Resistances of Heating Elements

Issue: The heating element resistances for travel iron design shown as Option B in Figure B-36 need to be selected.

Problem Statement: Select heating element resistances for travel iron such that one element operates at 240 VAC and that element in parallel with a second element of different resistance operates at 120 VAC. In both cases the iron produces 400 W of thermal energy.

Approach: Neglect the fact that 120 VAC and 240 VAC are only nominal voltages and there are acceptable tolerances allowed by the electricity provider. Assume the heating element can be procured for any resistance needed. Start by sizing the heating element for the 240 VAC case. Then, find the second heating element resistance.

Defining Equations:
AC Voltage = $V = V_{RMS}$ AC Current = $I = I_{RMS}$ Resistance = V/I
Electrical Power = VI = Thermal Power = $I^2 R = V^2/R$
For Travel Iron: V = 120 VAC, 240 VAC   P = 400 W   Inner, outer heating element resistances = R1, R2

Calculation Results:
1) For 240 VAC use R2:
    a) Find the current: I = P/V = 400 W/240 V = 1.67 A
    b) Find heating element resistance: R2 = P/I^2 = 400 W/ (1.67 A)^2 = $\boxed{144 \text{ Ohm} = R2}$
    c) Calculation Check: P = 400 W = (I)^2 * R2 = (1.67 A)^2 * 144 Ohm = 400 W <u>Check</u>
2) For 120 VAC use R2 and R1 in parallel:
    a) R2 has V = 120 VAC = I2 R2 → I2 = V/R2 = 120 V / 144 Ohm = 0.833 A
    b) P2 = V^2/R2 = (120 V)^2 / 144 A = 100 W
    c) Total current = I1 + I2 = I = P/V = 400 W/ 120 V = 3.333 A
    d) I1 = I − I2 = (3.333 − 0.833) A = 2.50 A
    e) R1 = V1/I1 = 120 V / 2.5 A = $\boxed{48 \text{ Ohm} = R1}$
    f) P1 = V^2/R1 = (120 V)^2 / 48 A = 300 W
    g) Calculation Check: P = 400 W = P1 + P2 = 300 W + 100 W = 400 W <u>Check</u>

Conclusions/Solution:
Travel Iron inner heating element resistance = 48 Ohm
Travel Iron outer heating element resistance = 144 Ohm

Recommendation:
Team should design the travel iron to use outer heating element when operating at 240 VAC. Team should design the travel iron to use the outer heating element in parallel with the inner heating element when the iron operates at 120 VAC.

FIGURE B-106   Engineering Prototype Calculations for Resistance of Heating Elements

MODULE 33

Engineering Prototype Build Book

OVERVIEW

In addition to other manufacturing tasks, the project team learns how to create an engineering prototype build book. This book documents all of the actions related to the prototype manufacturing. It is a structured document that presents the prototype assembly process procedures for use by those assembling the product and also contains all the information related to the engineering prototype. Its purpose is to archive all the information related to the prototype's manufacturing, assembly, and rework. The build book is a part of the project notebook.

LEARNING OBJECTIVES

- Gain knowledge and demonstrate the ability to complete all tasks necessary to prepare the build book documents.

PRE-LECTURE ASSIGNMENT

- Read sections B.4.33 and B.4.34.
- Each team member takes the build book outline given in Table B-60 and adds bullets to indicate what information they will be adding to the build book.

POST-LECTURE ASSIGNMENT

- Conduct a team meeting to 1) discuss the tasks of preparing the engineering prototype build book for the team's project, 2) agree on the tasks each team member will be responsible for in preparing the build book, and 3) modify the outline given in Table B-60 to document the decisions made by the team.

TEAM DELIVERABLES

- Build book outline for the project with assignments (see Table B-60)
- Updated project notebook table of contents showing where build book has been added
- Minutes of team meetings
- Team activity as documented in the team notebook

B.4.33 Engineering Prototype Build Book

The **engineering prototype build book** documents all the actions related to the manufacturing of the prototype, such as:

- COTS specifications, orders, invoices, and incoming inspection reports.
- Supplier special items, including specifications, orders, invoices, and incoming inspection reports.
- Raw material specifications, orders, invoices, and incoming inspection. Raw materials need to be tied to the parts that will be fabricated from them.
- First article inspection report.
- Deviations.
- Testing log.
- Failure analyses.
- Rework and repair documents.

The purpose of the engineering prototype build book is to archive all the information related to the prototype's manufacturing, assembly, and rework. This information is archived in a structured way that helps the team quickly access the information.

Table B-60 provides the build book structure. The build book can be either a hard-copy notebook or a computer file. However, if it is a computer file, then all hard-copy data, such as invoices and inspection sheets, must be scanned and placed in the computer file in a structured system of subfolders. The development and updating of the engineering prototype build book is the responsibility of the team member in charge of quality control. All team members must complete their assignments to support this effort.

B.4.34 Example Travel Iron Team Report for Week 20

Team members continued to work on their manufacturing tasks. During the team meeting, an outline for the prototype build book was created. The outline included all the tasks to be completed by each team member. The team decided to make the build book a computer document. Each team member agreed to scan and create a properly named file for each document they are responsible for in the build book.

Table B-60 Engineering Prototype Build Book Structure

Table of Contents:
Engineering prototype drawing package
For each COTS component:
-Order
-Supplier liaison documents
-Invoice
-Incoming inspection sheet
-Noncompliance decisions and waivers
-Inventory history
For each supplier-fabricated item:
-Order
-Serialization (if required)
-Supplier liaison documents
-Invoice
-Incoming inspection sheet
-Noncompliance decisions and waivers
-Inventory history
For each team-fabricated item:
-Fabrication description
-Serialization (if required)
-Incoming inspection
-Noncompliance decisions and waivers
-Inventory history
Assembly:
-Assembly process sheet
-First article inspection
-Noncompliance decisions and deviations
-Inventory history
Rework descriptions:
-Rework 1
-Rework 2
-Etc.

MODULE 34

Incoming Inspection and Manufacturing Troubleshooting

OVERVIEW

As components are received by the team, the team should inspect them to make sure that they meet the order specifications. During the incoming inspection, the team fills out an incoming inspection sheet and files it in the build book. Sometimes the design team receives components that, upon inspection, do not meet the specifications and/or drawings. As soon as a component discrepancy is identified, the team must decide on a course of action. This module addresses incoming inspection and how to resolve inspection issues.

LEARNING OBJECTIVES

- Gain knowledge of the incoming inspection process.
- Gain knowledge and demonstrate the ability to resolve incoming inspection issues, including manufacturing troubleshooting.

PRE-LECTURE ASSIGNMENT

- Read sections B.4.35–B.4.37.
- Each team member selects one COTS components and one fabricated component and prepares a paragraph for each on 1) how the component will undergo incoming inspection, 2) potential inspection issues, and 3) how these potential issues can be resolved.

POST-LECTURE ASSIGNMENT

- Conduct a team meeting to prepare an incoming inspection plan, including examples of how manufacturing issues resulting in failed incoming inspection will be resolved.
- Conduct a team meeting to 1) form three subteams and 2) assign the major parts of the engineering prototype to the subteams.
- Conduct additional meetings for the subteams to identify for the parts assigned to them: 1) issues that may make each part not meet the specifications and 2) inspections that must be performed on these parts.

TEAM DELIVERABLES

- List of what inspections are to be performed on the received parts
- Complete list of issues that may make each part not meet the specifications
- Minutes of team meetings
- Team activity as documented in the team notebook

B.4.35 Incoming Inspection

The purpose of incoming inspection is to determine how well the actual delivered component complies with the drawings and specifications for that item. For each item upon delivery, the liaison team member must perform an inspection of the item and determine if there are any noncompliance issues. Troubleshooting these issues is covered in the next section of the handbook.

To illustrate the incoming inspection procedure, the following example is given. A plate with a hole in it is received from the supplier. As shown in Figure B-107, the part drawing indicates that the plate must have a hole with a diameter of 0.500 ± .010 inches and the centerline of the hole must be located 2.000 ± 0.010 inches from the x-axis and 2.000 ± 0.010 inches from the y-axis.

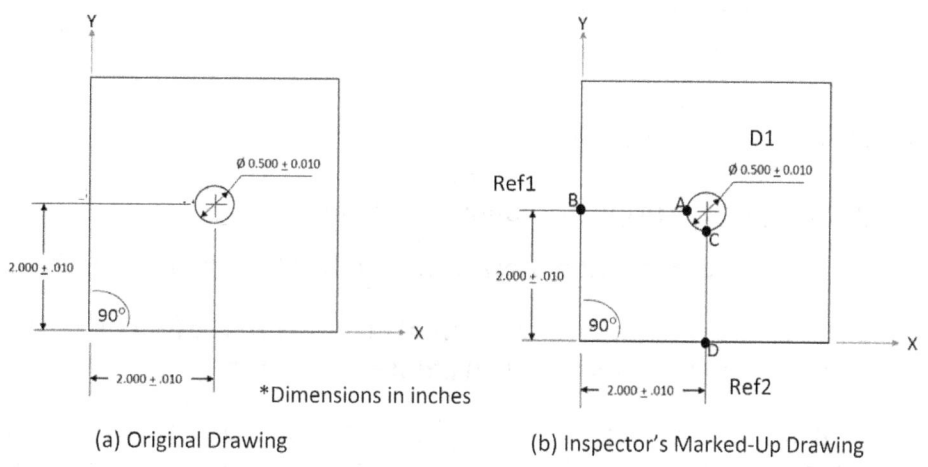

FIGURE B-107 Inspection Example

The first step in inspecting the part is to find the right measuring instruments or machines for the required measurements. The important part of selecting inspection equipment is to ensure that the instrument/gage capability meets the the rule of 10 to one. This means the measurement device, or working gage, should be 10 times more precise than the tolerance to be measured. The digital caliper shown in Figure B-108 was chosen because it has a scale that measures to 0.001 of an inch.

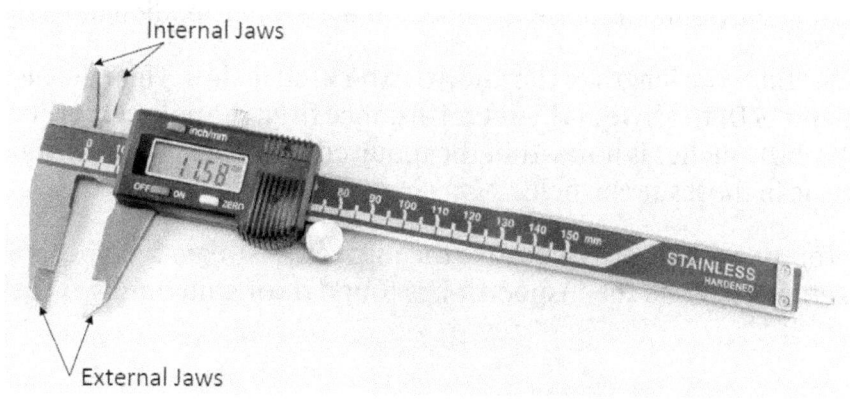

FIGURE B-108 Electronic Digital Caliper

Figure B-109 shows the inspection sheet for the hole-in-the-plate example. The steps followed by the inspector to fill out this inspection sheet are given below.

Inspection Report	Team No.: 45				Report No. IR-A-2		Date: 10-10-2020
Supplier: ACME Machining					Comments: Plate outer dimensions previously inspected.		
Part Name: Spacer							
Part No.: 100050							
Inspector: Mary Doe (Team member A)							
Inspection Type: Incoming							
Ref. No.	Zone	Qty	Feature	Tool	Requirement	Results	Issue No.	Calculations
D1	B-2	1	Hole	Caliper No. 1	0.500 ± 0.010 in	0.508 in	Pass	D1/2 = 0.254 in
Ref 1	B-2	1	Hole Cntrline	Caliper No. 1	2.000 ± 0.010 in	2.006 in	Pass	CD+ D1/2= 1.752 + 0.254 = 2.006 in
Ref 2	B-2	1	Hole Cntrline	Caliper No. 1	2.000 ± 0.010 in	2.014 in	A2	AB + D1/2 = 1.760 + 0.254 = 2.014 in

FIGURE B-109 Inspection Sheet for Hole-in-the-Plate Example

Using the digital caliper, the inspector follows these steps:

1. The inspector marks the inspection drawing by labeling the diameter of the hole as *D1*.

2. Using the caliper, the inspector finds the actual diameter of the hole to be 0.508. This is within the tolerance limits, so diameter D1 passes inspection.

3. The inspector marks the inspection drawing by (1) labeling the dimension of the hole centerline to the x-axis as *Ref1*, (2) labeling the edge of the hole nearest the x-axis as *C*, and (3) marking the point on the x-axis where line *CD* is perpendicular to the x-axis.

4. The inspector finds the length of the line *CD* to be 1.752 inches. This value is added to half of the diameter of D1. This results in the actual distance from the hole centerline to the x-axis. This value of 2.006 inches is within the required tolerance, so the inspector enters pass in the Issue No. field.

5. The inspector marks the inspection drawing by (1) labeling the dimension of the hole centerline to the y-axis as *Ref2*, (2) labeling the edge of the hole nearest the y-axis as *C*, and (3) marking the point on the y-axis where line *AB* is perpendicular to the y-axis.

6. The inspector finds the length of the line *AB* to be 1.760 inches. This value is added to half of the diameter of D1 to arrive at the actual distance from the hole centerline to the y-axis. This value of 2.014 inches is not within the required tolerance, so the inspector needs to label this issue in the Issue No. field.

7. The inspector uses the label *A2* because the inspector is Team Member A and this is the second discrepancy issue the inspector has found during incoming inspection on this project.

The general topic of inspection measurement is called *metrology*. This topic is covered in more detail in numerous manufacturing textbooks. Design teams should identify the types of inspection measurements required and acquire the competency to perform these necessary inspections.

B.4.36 Manufacturing Issues

In an ideal world, suppliers would always ship components that meet the specifications, and fabricated parts would always be made to print. Unfortunately, this is not the case. The team is often called upon to deal with issues such as Issue No. A2 discussed in the previous section. Fortunately, there are procedures for dealing with these issues. These procedures are part of a complete configuration control system.

Configuration control is an important function of configuration management as discussed in Section B.4.21. The primary objective of configuration control is to establish and maintain a systematic change management process that regulates life cycle costs. Configuration control tasks include initiating, preparing, analyzing, evaluating, and authorizing proposals for change to a system. The first step of configuration control is to identify and document the need for a change in a change request. This step is followed by analysis and evaluation of a change request and initiation of a change proposal. The change proposal is then approved or disapproved. The final step of configuration management is the verification and release of a change. Covering configuration control in any depth is beyond the scope of this handbook. For capstone students, configuration control is limited to preparing an incoming inspection plan for the engineering prototype that includes general guidelines for resolving manufacturing issues.

As purchased components are received by the team, they should be inspected to make sure they meet the order specifications. Some specification items may require testing that can be delayed until the overall prototype is tested; however, this introduces additional risk that there will be a hardware issue in the future. As soon as a purchased component discrepancy is identified, the team needs to decide whether the part can be used or must be returned to the supplier and another component purchased. Usually the short calendar time available for capstone prototype manufacturing results in the team accepting the out-of-specification part. It is important to document that this part deviates from its specification. This is done by using a deviation form that is logged into the prototype build book. Deviations will be addressed in the next module of this handbook.

A similar situation occurs with fabricated parts. Often the fabricator makes a mistake in some feature of the fabrication drawing. Then the team must decide how to proceed. Again, a tight schedule usually results in the team either accepting the part as is or reworking the part so that it can be used in the engineering prototype unit. These actions also need to be documented in a deviation form that is included in the build book. In this way, the team knows in what ways the as-built engineering prototype deviates from the manufacturing drawings.

Many potential deviations can be avoided by the team communicating often with their suppliers and fabricators. They can also be avoided by investigating the quality control characteristics of the supplier before the order is placed.

B.4.37 Example Travel Iron Team Report for Week 21

The team members continued to liaison with their assigned components during the manufacturing process. Incoming components were inspected, the project notebook was updated, and the team continued to work on the final report.

Credits

Fig. B-108: Copyright © by Jacek Halicki (CC BY-SA 4.0) at https://commons.wikimedia.org/wiki/File:2020_Suwmiarka_cyfrowa.jpg.

MODULE 35

Deviations and Mid-Course Adjustments

OVERVIEW

As discussed in the previous module, manufacturing errors that do not agree with the component drawing or specification are addressed by attaching a completed deviation form to the drawing. The procedure for filling out a deviation form is presented in this module. Often the prototype manufacturing process encounters issues that introduce the potential for not staying on schedule and/or exceeding the monetary and labor budgets. Mid-course adjustments in scope are often needed to keep the project on schedule and within budget. The process of making mid-course adjustments during the prototype development process is discussed in this module.

LEARNING OBJECTIVES

- Gain knowledge and demonstrate the ability to prepare drawing deviation forms.
- Gain knowledge and demonstrate the ability to make prototype development plan changes to stay on schedule and within budget.

PRE-LECTURE ASSIGNMENT

- Read sections B.4.38–B.4.40.
- Each team member makes a list of at least five potential issues for their team's prototype development that would result in not meeting schedule and/or not staying within the labor and/or monetary budgets.

POST-LECTURE ASSIGNMENT

- Conduct a team meeting to 1) review and update the process to resolve manufacturing issues developed in the previous module; 2) resolve any existing manufacturing issues; 3) prepare any deviation forms that are needed based on the decisions in Item 2; 4) review the current and expected progress on the prototype manufacturing effort and identify areas where mid-course adjustments are needed; and 5) prepare, document, and execute the needed mid-course adjustments to keep Phase 5P on schedule and within budgets.
- Continue the prototype manufacturing.

TEAM DELIVERABLES

- Deviation forms prepared by the team for resolving any existing manufacturing issues
- Mid-course adjustments, including current and planned progress to keep Phase 5P on schedule and within budget
- Minutes of team meetings
- Team activity as documented in the team notebook

B.4.38 Deviations

As shown in Figure B-110, each project component is inspected when received from the supplier or completed by the team. If the component fails the inspection, then there is an issue number for each failed inspection item. The team must then decide. For supplier-provided components, the component can be returned to the supplier, who can then send a new part. However, in many cases, this action would cause the project to slip schedule. In these cases, the issue then goes to a deviation decision, and the team's plan is captured on a deviation form.

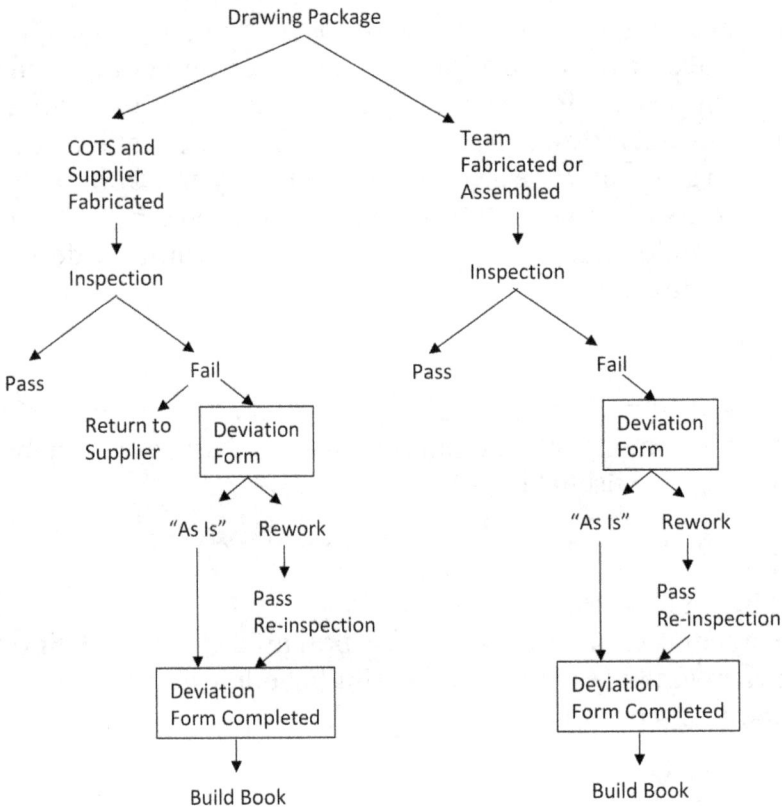

FIGURE B-110 Component Drawing/Specification Deviation Decision Flowchart

A deviation is a document that describes the noncompliance and rationale for accepting the item and using it even though it is not compliant at incoming inspection. It also describes what actions that should be taken for future manufacture of this item.

Granting a deviation during production is a serious issue, and most companies have a material review board to address these matters. For government programs, in general, deviations must comply with the requirements of MIL-HDBK-61A: Configuration Control.

For capstone projects, the form of deviation documentation is determined by the sponsor (instructor). Figure B-111 provides a recommended deviation form for use by capstone project teams.

Deviation Form				
Deviation No:		Supplier Name:		
Inspection Issue No.:		Purchase Order No.:		
Part Number:		Liaison Engineer		
Description:		Quantity:		
Revision:		Date:		

Current Requirement	
Requested Deviation	
Reason for Deviation	
Corrective Action	

Approval	Project Chief Engineer	Date	Comments
Corrective Action Completed	Item Liaison Engineer	Date	Comments

FIGURE B-111 Recommended Deviation Form Format

As shown in Figure B-110, the team has two options for deviating from the drawing and specification requirements. The first option is to accept the component as is. This option may keep the team on schedule, but it may limit the performance of the prototype. The second option is to rework the component, either by the supplier or the team. In most cases, the rework will still deviate from the drawing, but the component will now meet its functional requirements. An example would be a hole drilled in the wrong location. The rework could be as simple as adding another hole in the proper location. In other cases, the first hole would also have to be patched to achieve overall component functionality.

B.4.39 Mid-Course Adjustments

As described in Section B.4.36, there are often numerous manufacturing issues that pose either cost overruns or missed schedule milestones. The team needs to be flexible in dealing with these issues. Sometimes the scope of the engineering prototype design must be reduced to stay within schedule and/or budget. Changing the prototype hardware may also reduce the number of prototype tests that can be performed. Because the prototype must be assembled and ready for first article inspection by the middle of Module 37, it is important that the team review their progress in manufacturing the prototype now while there is time to make adjustments.

B.4.40 Example Travel Iron Team Report for Week 22

This week the team continued the prototype manufacturing. All of the COTS components have been acquired and passed incoming inspection. No deviations have occurred. The wooden simulated shell and handle are still being fabricated. Team Member E is leading the shell and handling manufacturing efforts, and all other team members are helping.

During the team meeting, the project schedule and budgets were updated and reviewed. The sole plate and plastic insulation plate will be completed in the student machine shop by the beginning of next week.

The team is planning to assemble the prototype and test rig near the end of Week 23. This gives a few days to rework any assembly issues and still meet the hardware fabrication cut-off date of the middle of Week 24.

MODULE 36

Assembly, First Article Inspection Form, and DR5 Preparations

OVERVIEW

The key concern for the team is to have an assembled and inspected engineering prototype by the middle of Week 24. This module covers those tasks. Guidance on how to prepare for the design review 5 (DR5) test readiness review scheduled for Week 24 is also covered in this module.

LEARNING OBJECTIVES

- Gain knowledge and demonstrate the ability to assemble and inspect the engineering prototype.
- Gain knowledge and demonstrate the ability to prepare for DR5.

PRE-LECTURE ASSIGNMENT

- Read sections B.4.41–.4.44.
- Each team member prepares their suggested first article inspection form for the engineering prototype.

POST-LECTURE ASSIGNMENT

- Conduct a team meeting to 1) review the manufacturing progress and resolve any schedule issues, 2) finalize plans for assembling and inspecting the prototype, 3) review the individual team first article inspection forms and finalize the forms, 4) prepare a plan for the DR5 presentation and make team assignments, and 5) agree on a time to review and practice the DR5 presentation before the start of Module 37.
- Continue the manufacturing of the prototype.

TEAM DELIVERABLES

- Team's plans for assembling and inspecting the prototype
- Completed first article inspection forms
- DR5 plan and presentation slides
- Minutes of team meetings
- Team activity as documented in the team notebook

B.4.41 Prototype Assembly

The plan for assembling the prototype is part of the manufacturing plan prepared in Module 31. The following assembly topics are covered in the plan:

- Facilities and equipment
- Procedures including subassembly inspections
- Assembly adjustments and calibrations
- First article inspection
- Final assembly marking
- Final assembly storage

The team should follow the subassembly and assembly procedures, take pictures of the various stages of assembly, and record all observations made during the assembly process. These photographs and written observations should be included in the prototype build book.

B.4.42 First Article Inspection

The term *first article inspection* is generally applied to the inspection of the first unit off of the production line. The term is applied to the prototype unit to emphasize the need to thoroughly document the as-is condition of the prototype before testing begins. This inspection should follow the general procedures described in Section B.4.35: Incoming Inspection, including archiving the inspection sheet in the prototype's build book.

B.4.43 DR5 Test Readiness Review Preparation

The purpose of DR5 is to verify that the team is ready to start development and validation testing of the prototype. The review should answer the following questions:

- Has the prototype (test article) been dimensionally inspected to verify that it conforms to the prototype drawing package, and have all inconsistencies been recorded on deviation forms that are archived in the build book?
- Is the testing plan doable with the resources available, including schedule and labor budget?
- Are there one or more tests to address each of the prototype design requirements?
- What steps have been taken to ensure that each test can be conducted in a safe manner?
- Is there enough time in the plan to allow each person involved in a test to be thoroughly clear on the procedures for that test?
- Do the procedures include how the test data are to be analyzed and documented?

Table B-61 provides an outline for DR5, assuming that there will be only 10 minutes for the team to present. The presentation slides must be clear, and the team must practice ensuring that they can stay within their 10 minutes while covering all of the pertinent test readiness information.

It is important for the team to remember that they must convince the sponsor to receive the resources to conduct the prototype testing. The sponsor must be convinced that there is a high probability that all the testing objectives will be met if the requested resources are provided.

Table B-61 DR5 Test Readiness Review List of Presentation Slides

1. **Introduction.** Include name of the presentation, team name, project name, and team member names. There should be a picture of the engineering prototype.
2. **Prototype description.** Describe what the prototype does. Include a simple graphic.
3. **Summary of first article inspection.** This must be a brief but effective slide.
4. **Description of test rigs.** Include a labeled graphic and brief description.
5. **Test plan summary.** Describe what is to be done during development testing and what will be done during validation testing. Describe the plans to accommodate test failures in both development and validation.
6. **Test matrix.** This table should include the following information for each prototype requirement: test name, facilities, testing time, number of persons required, equipment, very brief description of the test, and criteria for passing the test. Use more than one slide if necessary so that it can be easily read by the audience (no eye charts).
7. **Schedule.** The slide shows each test, including the tasks of preparation, testing, data analysis, and documentation.
8. **Labor budget.** Present a table listing each test and the labor required. Present the total hours required and the labor hours available in the labor budget.
9. **Safety provisions.** Describe how safety has been addressed for each test.
10. **Summary of a selected test.** The slides must cover the following:
    a. Purpose
    b. Prototype requirement
    c. Graphic of test setup
    d. Procedures (This chart should include the procedures to show the detailed used, but the audience must understand that they do not have to read the procedure because it is summarized on the slide.)
    e. Data sheet format
    f. Analysis and evaluation

B.4.44 Example Travel Iron Team Report for Week 23

B.4.44.1 Engineering Prototype Assembly
The travel iron team successfully assembled the prototype unit (see Figure B-112) and conducted first article inspection.

B.4.44.2 Test Rig Assembly
Figure B-113 shows the engineering prototype test rig. It was constructed from scrap wood, 3/32inch diameter Readibolt, nuts, and washers.

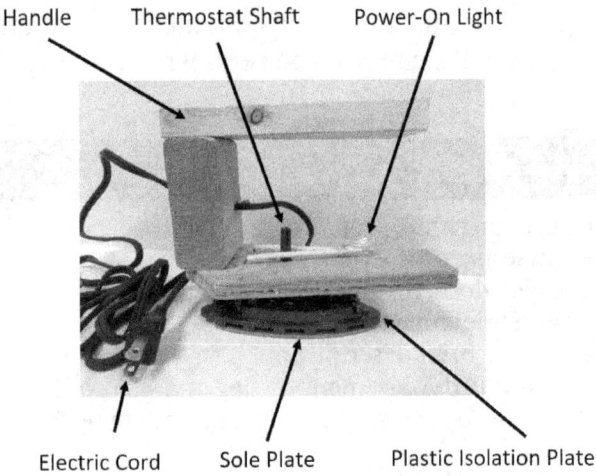

FIGURE B-112 Assembled Travel Iron Engineering Prototype

FIGURE B-113 Travel Iron Test Rig

B.4.44.3 DR5 Test Readiness Review Preparation

The team prepared the slides for the DR5 test readiness review as shown in Figure B-114.

B.4.44.4 Change in Temperature Measuring Device

The initial plan was to measure sole plate temperature with the LaserPro LP300 infrared thermometer. After the team received the device, they learned that its accuracy is dependent on using the correct emissivity for the surface being measured. The team was unsure about the emissivity of the sole plate. The device measures a spot larger than the laser pointer indicates. The size of this spot proved to be too large to distinguish between the sole plate temperature and the plastic isolation plate temperature.

Based on these findings, the team switched to the Minnesota Measurement Instruments LLC Model DM6801 K-Type thermometer. This device uses a surface measurement thermocouple wand that has a ¼-inch diameter measurement head. The device has two components: the digital thermometer and the surface probe. The accuracy of the digital thermometer is ±(0.3% + 2 °F). The accuracy of the surface probe is ±4.5 °F. The test procedures were updated to Rev 1 to incorporate the thermostat equipment change.

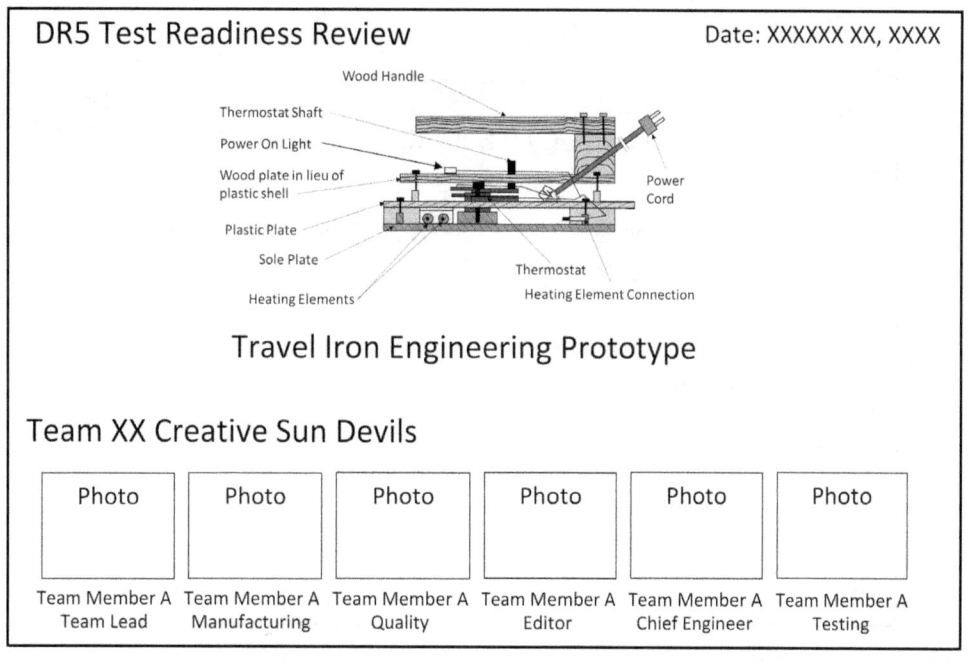

FIGURE B-114A Travel Iron DR5 Slides (1 of 10)

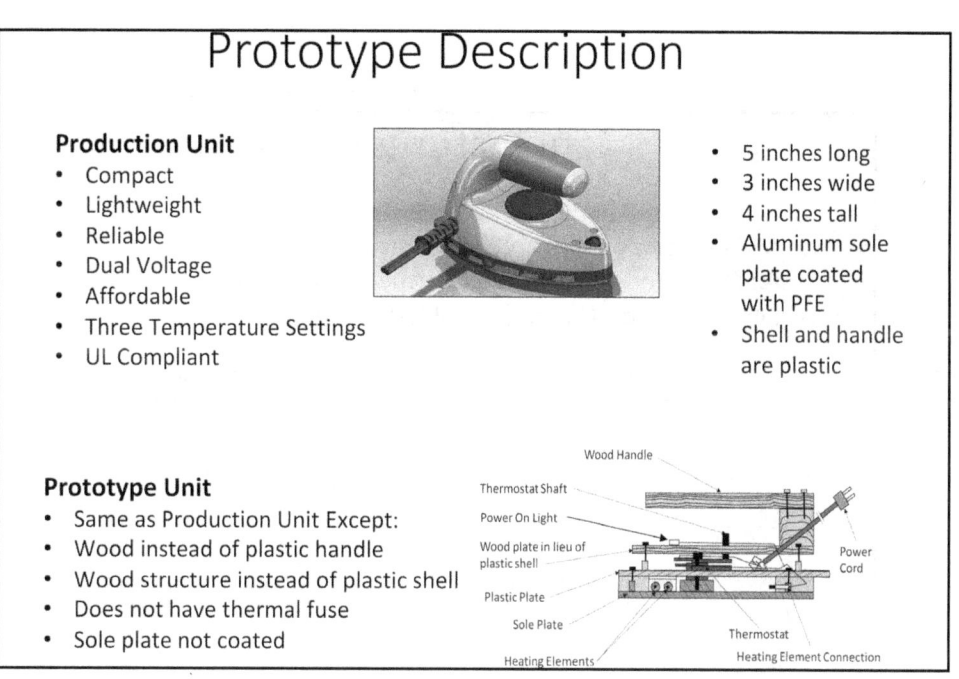

FIGURE B-114B Travel Iron DR5 Slides (2 of 10)

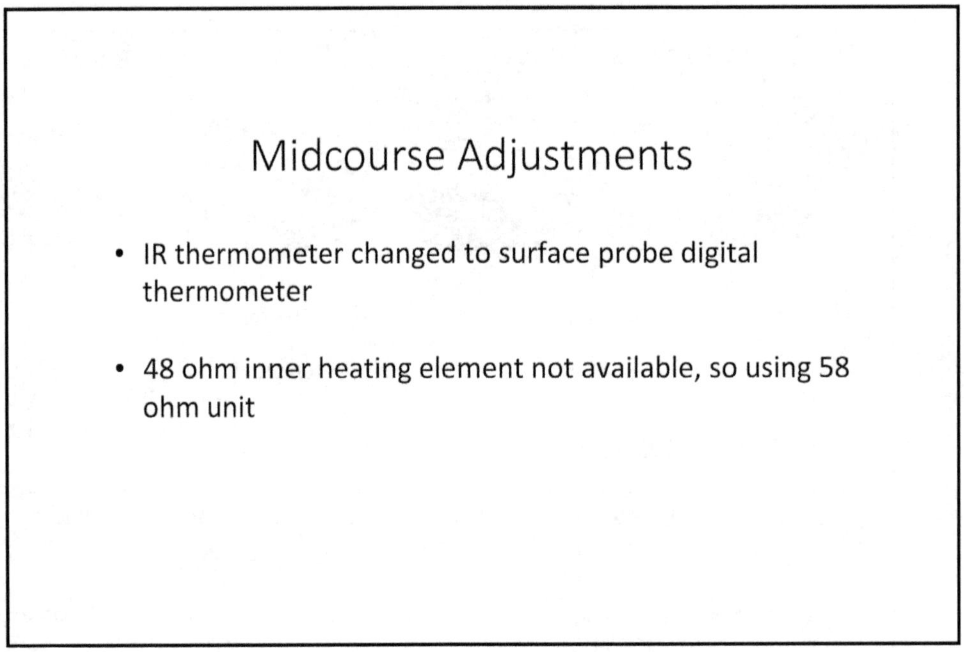

Midcourse Adjustments

- IR thermometer changed to surface probe digital thermometer

- 48 ohm inner heating element not available, so using 58 ohm unit

FIGURE B-114C Travel Iron DR5 Slides (3 of 10)

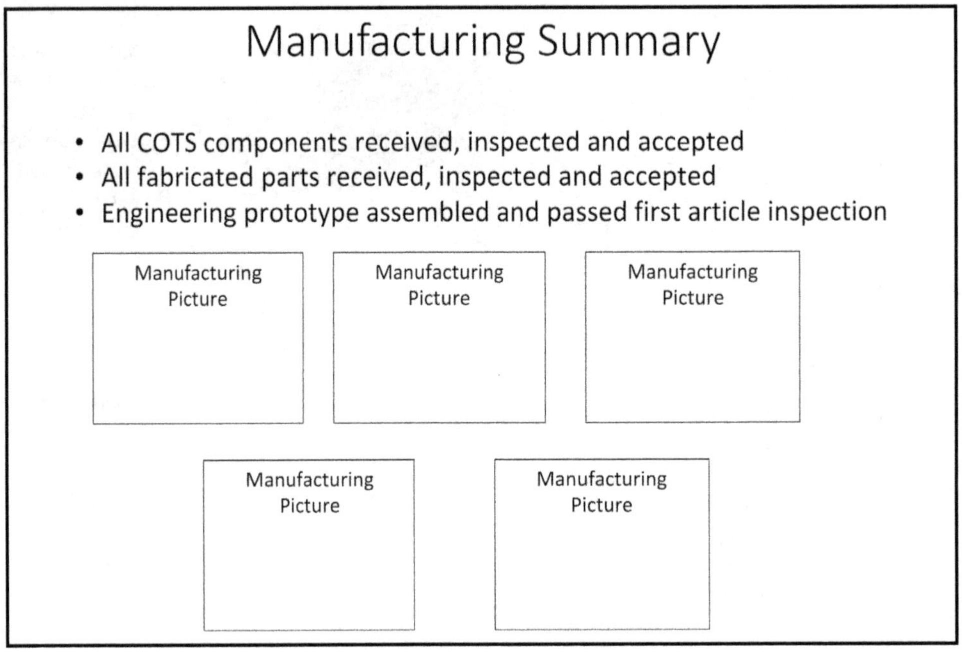

Manufacturing Summary

- All COTS components received, inspected and accepted
- All fabricated parts received, inspected and accepted
- Engineering prototype assembled and passed first article inspection

Manufacturing Picture	Manufacturing Picture	Manufacturing Picture
Manufacturing Picture	Manufacturing Picture	

FIGURE B-114D Travel Iron DR5 Slides (4 of 10)

Summary of First Article Inspection

Only two deviations

- E1 – Thermostat knob not on centerline
- E2 – Initially miss-located screw holes in wood structure plate. Extra holes still in plate.

Photo of Team Inspecting Unit

FIGURE B-114E Travel Iron DR5 Slides (5 of 10)

Test Rig, Facility and Equipment Ready

Testing at Team Member A's garage

- 120 VAC

Additional Equipment

- i-phone for timer and photos

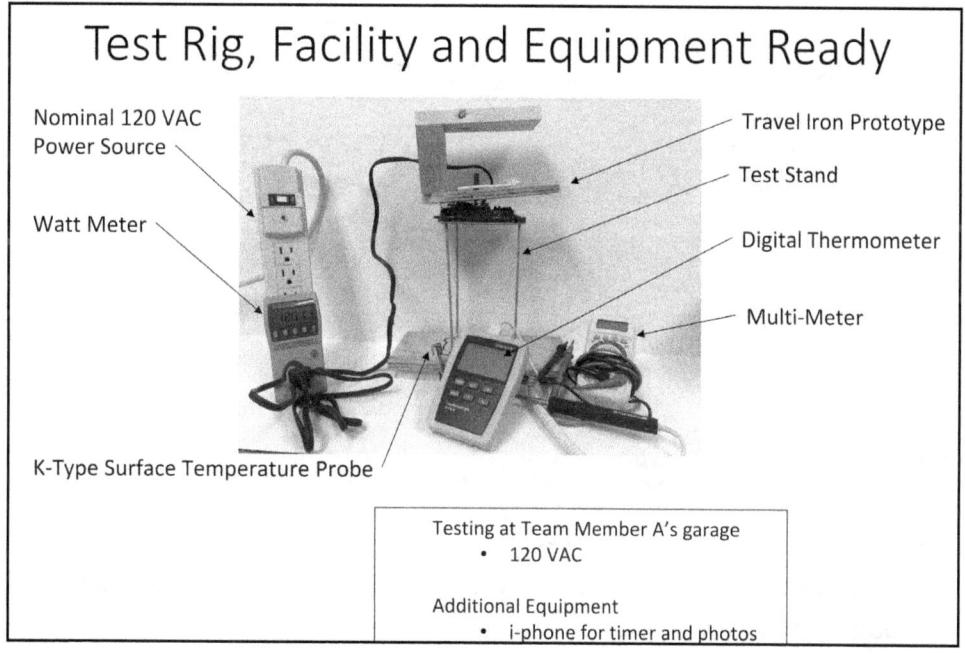

FIGURE B-114F Travel Iron DR5 Slides (6 of 10)

Test Plan Summary

- Development – test and rework until achieved if possible
- Validation – test once—if fail repair—complete other tests
- Risk Areas
  - Test 1 – change in inner heating element may preclude reaching 400 F
  - Test 2 – could be greater variation than ± 10 deg F

Test	Engineering Requirement		Validation Method	Power
1	Three Settings	275, 375, 400 deg F	Test Rig	With 110 VAC
2	UL Reqt 1	Hold settings ± 10 deg F	Test Rig	With 110 VAC
3	UL Reqt 3	< 662 deg F	Test Rig	With 110 VAC
4	Press Time	< 2.0 min. pillow case	Timer	With 110 VAC
5	UL Reqt 4	Temp - Ambient, deg K cord--60 resting surf--65 handle --50 touching surf--60	Test Rig	With 110 VAC
6	Operating Life	> 200 cycles Limited: 5 cycles	Test Rig, IR	With 110 VAC

FIGURE B-114G Travel Iron DR5 Slides (7 of 10)

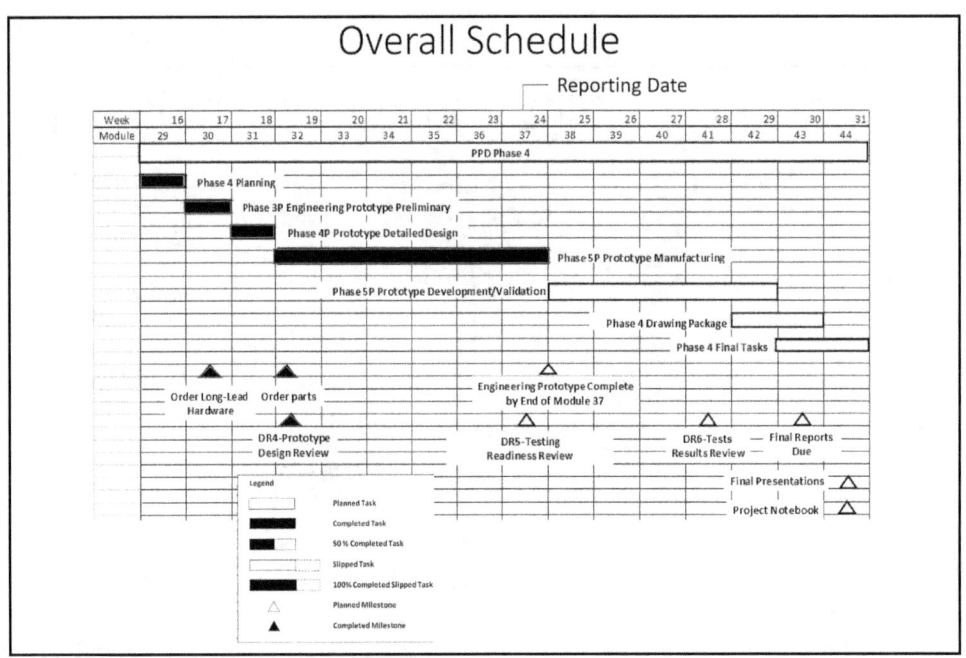

FIGURE B-114H Travel Iron DR5 Slides (8 of 10)

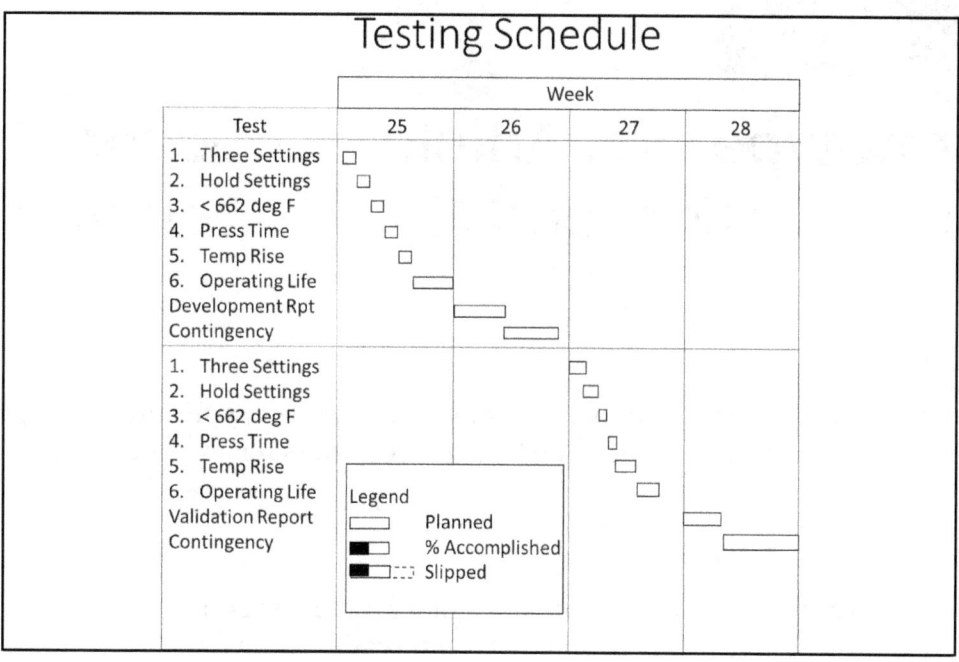

FIGURE B-114I Travel Iron DR5 Slides (9 of 10)

Safety Provisions

There are two major safety areas:

- Hot surfaces
  - The sole plate can reach a temperature of 410 deg F
  - A warning notice is included in the test procedures.

- Electrical shock
  - The unit operates on both 120 VAC electric power. that
  - A warning notice in assembly and test procedures to make sure the sole
    plate does not conduct electricity when unit is operating.
  - A warning notice regarding exposed electrical wires in prototype since shell is absent.

FIGURE B-114J Travel Iron DR5 Slides (10 of 10)

MODULE 37

Prototype First Article Inspection, DR5 Presentation, and Final Report Work

OVERVIEW

The team completes assembly of the engineering prototype and conducts its first article inspection. The team conducts the DR5 test readiness review and works on the final report this week.

LEARNING OBJECTIVES

- Demonstrate the ability to conduct a first article inspection.
- Demonstrate the ability to conduct a test readiness review.
- Demonstrate the ability to prepare portions of the project final report.

PRE-LECTURE ASSIGNMENT

- Read section B.4.45–B.4.47.
- Each team member prepares a progress report on their team-assigned final report writing.

POST-LECTURE ASSIGNMENT

- Conduct a team meeting to 1) review the feedback given during the DR5 test readiness review, (2) finalize preparations for the start of engineering prototype testing the following week, and 3) review the status and drafts of the final report sections already assigned.
- Each team member continues their assigned test preparation and final report writing and editing tasks.

TEAM DELIVERABLES

- Team's response to the feedback given during the DR5 test readiness review
- Summary of the preparations for the start of engineering prototype testing the following week
- Status and drafts of the final report sections already assigned
- Minutes of team meetings
- Team activity as documented in the team notebook

B.4.45 Completing Prototype Manufacturing

Prior to DR5, the team needs to complete the assembly of the engineering prototype and conduct its first article inspection. This inspection should follow the general procedures described in Section B.4.35: Incoming Inspection, including archiving the inspection sheet in the prototype's build book.

The team is responsible for reviewing and practicing the DR5 presentation prior to its formal presentation to the sponsor and other expert reviewers. The team must remember that they are communicating with reviewers who are dependent on the team for explaining the project. What is common knowledge among the team members will probably not be known by the reviewers unless the team presents that information. While giving the presentation, the team needs to watch the reviewers' body language to see if they are understanding what is being said. During the reviewers' feedback portion of the meeting, the team needs to be attentive, appreciative, and not argumentative. Feedback from the review must be recorded by the team. Following the presentation, the team needs to meet and review this feedback and decide what actions need to be taken. It is important to have customer feedback regarding the design and its expected performance. If possible, actual customers should attend. If this is not possible, then some of the experts should review the presentation from the customer's perspective.

EM-(f)

EM-(h)

EM-(m)

B.4.46 Working on Sections of the Project Final Report

In parallel with the prototype development testing effort, the team must focus on completing most of the sections of the final report. This includes assigning team members to the sections for both writing and editing the section drafts. A checklist for editing each section is given in Table B-62.

Table B-62 Checklist for Each Section of the Project Final Report

☐	1.	Does the section follow the outline in Section C of the handbook?
☐	2.	Is the section written for the reader (sponsor) and not the writer?
☐	3.	Are all facts, figures, etc. properly referenced?
☐	4.	Are all figures and tables properly labeled and introduced in the text before they appear?
☐	5.	Does the report section flow when read aloud?
☐	6.	Are there any spelling errors?
☐	7.	Are there any grammatical errors?
☐	8.	If there are checklists, is the backup for them easily found in the team notebook?
☐	9.	Have all missing items in this section been added?
☐	10.	Has this section been reviewed and approved by each team member?
☐	11.	Is there an email from each team member stating they approve of the section?

B.4.47 Example Travel Iron Team Report for Week 24

B.4.47.1 Engineering Prototype Manufacturing

Following final assembly of the engineering prototype, the unit underwent first article inspection. The unit passed all the inspection requirements and is ready to begin development testing.

B.4.47.2 DR5 Test Readiness Review

The team presented the slides shown previously in Figure B-114. The team was approved to begin development testing and acted upon review feedback. No additional tasks resulted from this feedback.

B.4.47.3 Progress on Final Report

To properly track the preparation of the final report, the team prepared the status table shown in Table B-63. The team is basically on schedule but must remain focused on finishing the writing and editing of this document.

Table B-63 Final Report Preparation Status as of End of Week 24

Section	Section Editor	Draft Due Date*	Final Due Date*	Date*	% Done	Status
Title Page	Member B	29	29	24	100	Done and approved
Team Member Page	Member B	29	29	24	90	Waiting for signatures of members
Executive Summary	Member B	29	29	24	50	Waiting for Validation results
Table of Contents	Member B	29	29	24	80	Need final page numbering
Table of Figures	Member B	29	29	24	80	Need final page numbering
Table of Tables	Member B	29	29	24	80	Need final page numbering
ABET Cross Reference Table	Member A	29	29	24	80	Waiting for Validation results
1. **Introduction**	Member B	5	29	24	100	Done and approved
2. **Final Design Description**	Member E	29	29	24	20	Waiting for Validation results
3. **Phase 1. Preconcept**	Member A	5	29	24	100	Done and approved
4. **Phase 2A. Requirements**	Member C	9	29	24	100	Done and approved
5. **Phase 2B. Conceptual Design**	Member C	9	29	24	100	Done and approved
6. **Phase 3. Preliminary Design**	Member C	15	29	24	100	Done and approved
7. **Phase 4. Detailed Design**	Member C	29	29	24	25	Waiting to complete product dwgs
8. **Phase 3P Engineering Prototype Preliminary Design**	Member D	17	29	24	100	Done and approved
9. **Phase 4P Engineering Prototype Detailed Design**	Member D	18	29	24	100	Done and approved
10. **Phase 5P Engineering Prototype Development/ Validation**				24		
10.1 Fabrication and Assembly	Member D	28	29	24	100	Done and approved
10.2 Development	Member D	28	29	24	50	Currently being finished
10.3 Validation	Member D	28	29	24	25	Waiting for Validation results
11. **Project Performance**	Member A	29	29	24	0	Behind schedule should be 40%
12. **Project Conclusions**	Member B	29	29	24	0	Behind schedule should be 40%
13. **Recommendations**	Member C	29	29	24	0	Behind schedule should be 40%
Appendices	All	29	29	24	0	Behind schedule should be 40%

End of Project Week Number

MODULE 38

Testing Issues and Repair/Rework

OVERVIEW

This module discusses how to address testing issues to obtain the needed design information while keeping the project on schedule and within budgets.

LEARNING OBJECTIVES

- Demonstrate the ability to address testing issues in order to maximize the information gained from the engineering prototype subproject while staying within the available budget and schedule resources.

PRE-LECTURE ASSIGNMENT

- Read sections B.4.48–B.4.49.
- Each team member prepares (1) a progress report on their testing work and (2) a hard copy of their final report write-ups and brings both items to the team meeting.

POST-LECTURE ASSIGNMENT

- Conduct a team meeting to 1) review the testing progress and resolve any issues and 2) review the final report write-ups and provide editing as needed.
- Each team member continues their assigned testing and final report writing and editing tasks.

TEAM DELIVERABLES

- Report on the testing the team conducted and the issues the team has resolved
- Compiled sections of the completed final report
- Minutes of team meetings
- Team activity as documented in the team notebook

B.4.48 Engineering Prototype Testing Purpose and Issues

The purpose of development testing is to uncover design issues that were, for a variety of reasons, inadequately addressed during design analysis. These testing issues and their resolution result in new knowledge about the project. The entrepreneurial engineer will look at these issues as opportunities to add more value to the customer and/or the producer.

EM-(a)

When the engineering prototype fails a test, the team uses their technical skills and knowledge to decide how to proceed. There are three options. In order of desirability, the options are as follows: (1) Redesign the prototype and rework the hardware so that it will pass the test, (2) accept the failed test and repair the prototype so that it can proceed with the other scheduled tests, or (3) accept the failed test and proceed directly to the next test without repairing the unit. The goal of development and validation testing is for the prototype to pass all the engineering requirements. However, the engineering prototype subproject still provides valuable results even if one or more tests fail. It just means that more design and testing work are needed before the program moves into building the production unit prototype. The choice of which option to follow is based on the available schedule and budget resources.

EM-(g)

Product development is risky. It is important for the team to appreciate the fact that their decisions on how to address testing issues can have both positive and negative outcomes. There are always risks associated with any action. The goal is not to avoid risk altogether, but rather to make decisions based on weighing the benefits versus the risks.

EM-(l)

B.4.49 Example Travel Iron Team Report for Week 25

B.4.49.1 Development Testing Setup

The test setup for travel iron prototype testing is shown in Figure B-115. A list of the equipment with their accuracies is given in Table B-64. It was noted that the power outlet voltage varied from 120 to 122 VAC.

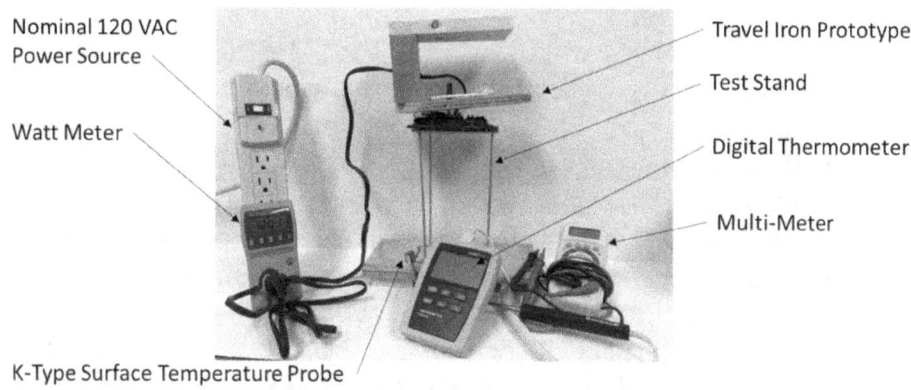

FIGURE B-115 Travel Iron Engineering Prototype Test Setup

B.4.49.2 Exploratory Testing

The team decided to perform some exploratory tests (tests 1A, 1B, 1C, and 1D) before starting the test plan development test matrix. As described below, the team learned that when the thermostat is at its maximum set point, the iron becomes energized and reaches a nominal 400 °F at the top location of the iron. When this maximum temperature is reached, the thermostat de-energizes the heating elements, and the top location temperature begins to decrease. When the temperature reaches a lower limit, the thermostat again energizes the iron until the top location temperature again reaches its maximum value. The team decided that the settings listed in Test 1 of the test plan (see Slide 7 of Figure B-114) will be for the maximum temperature at the top location for each of the three settings. It should be noted that the temperature at other locations on the sole plate will be lower than the temperature of the top location.

To summarize, at Setting 1, the maximum temperature of the top location should be nominally 275 °F. Likewise, for Setting 2, the temperature should be nominally 375 °F, and for Setting 3, the temperature should be nominally 400 °F.

B.4.49.3 Exploratory Test 1A: Maximum Sole Plate Temperature and Power Requirement

In preliminary design, the team researched at least 10 existing production irons that are capable of heating the sole plate to 400 °F. By taking into account iron sole plate area and rated power for each iron, the travel iron required power was estimated to be 400 watts (W) in order to provide a sole plate temperature of 400 °F. The purpose of Test 1A is to determine if the travel iron can actually achieve sole plate temperature of 400 degrees Fahrenheit. During preliminary design, it was determined that for a power source supplying 120 VAC, the outer heating element needed to be 144 ohms, and the inner heating element needed to be 48 ohms. For the engineering prototype, only a 58ohm inner heating element was available. The analysis presented previously in Figure B-106 showed that the use of this new heating element would result in the engineering prototype having a maximum power usage of only 348 W. The team was concerned that this lower power level would not yield a sole plate temperature as high as 400 °F. The objectives of this test are to (1) compare the actual power with the predicted 348 W of power and (2) compare the actual maximum sole plate temperature with the desired value of 400 °F.

Test 1A was done when the outlet voltage was 120 VAC. The thermostat was placed in the maximum set point position. This means turning the thermostat from the off position clockwise until it reaches the other stop. This is an arc of approximately 350 degrees. Five runs were conducted, with each consisting of turning the thermostat to the maximum position and waiting for the middle of the sole plate to reach its maximum temperature. The maximum temperature and input power were measured and recorded. The thermostat automatically turned off the electrical power, and the sole plate began to cool. At some temperature, the thermostat again allowed power to the heating elements. This low temperature was recorded, and the sole plate temperature began to increase again. The five cycles (runs) of the thermostat were measured and recorded as shown in Table B-64. It was noted that when the sole plate was at its maximum temperature and the thermostat turned off the power, it took approximately 20 seconds for the sole plate to cool to a temperature where the thermostat would turn power on again. When the power came on, it took only about two seconds for the sole plate to heat back up to its maximum temperature and the power to turn off.

Table B-64 Test 1A Data Sheet

	Sole Plate		Supply	
	Temperature		Power	Voltage
	degrees F		Watts	VAC
	Max	Min		
Run 1	362	343	345	120
Run 2	361	343	344	120
Run 3	360	342	344	120
Run 4	361	343	344	120
Run 5	362	342	345	120

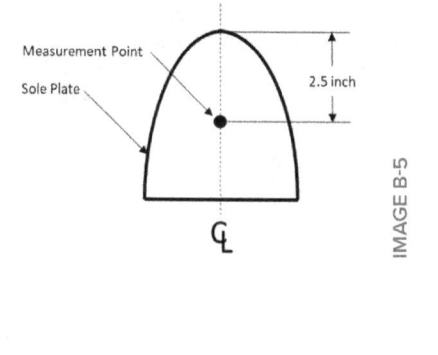

Figure B-116 documents the analyses conducted on the Test 1A data. The analysis process is based on the information provided in Appendix BA4. The conclusions of this test are that (1) the team was successful in predicting the engineering prototype's maximum electrical power and (2) the next engineering prototype for this project will need a lower resistance inner heating element to obtain a maximum sole plate temperature of 400 °F.

It should be noted that only one measurement location on the sole plate was used. The distribution of temperatures at other locations on the sole plate were not investigated in this test.

B.4.49.4 Exploratory Test 1B: Maximum Temperature at Other Locations on the Sole Plate Centerline

Based on the results of Test 1A, the team became curious regarding the distribution of temperatures on the sole plate. The team noted that the heating elements are U-shaped and cross the centerline nearer the top of the sole plate relative to the location investigated in Test 1A. Table B-65 shows the test data for three locations along the iron's centerline.

Problem Statement: (1) Compare the maximum travel iron prototype electrical power with the predicted value of 348 w, and (2) compare the measured maximum sole plate temperature witD the desired temperature of 400 degrees F.

Approach: Use the analysis procedures provided in Appendix B2.

Defining Equations:

AC Voltage = V = V_{RMS} AC Current = I = I_{RMS} I = V/R

Electrical Power = VI = Thermal Power = I^2R = V^2/R

Average X = $\Sigma X_i /n$ where n = number of measurements (runs)

Sample standard deviation $s = \sqrt{\dfrac{\sum_{i=1}^{n}(x_i - \bar{x})^2}{(n-1)}}$

Statistical Standard Uncertainty = u = s/sqrt(n)

Equipment Uncertainty for temperature probe = ui = ai / √3 where ai = manufacturer's accuracy data where u1 ~ temperature probe, u2 ~ temperature meter, u3 ~ power meter

Ut = total uncertainty = sqrt($u^2 + u1^2 + u2^2 + u3^2$)

To establish a confidence interval, U, at the 95% confidence level, the total standard uncertainty is then multiplied by a coverage factor of 2, U = 2Ut

Spreadsheet Analysis:

	Max Temperature, F		Min Temperature, F		Power, W	
	Xi	(Xi-X)^2	Xi	(Xi-X)^2	Xi	(Xi-X)^2
	362	0.64	343	0.16	345	0.36
	361	0.04	343	0.16	344	0.16
	360	1.44	342	0.36	344	0.16
	361	0.04	343	0.16	344	0.16
	362	0.64	342	0.36	345	0.36
Sum	1806	2.8	1713	1.2	1722	1.2
Average	361.2		342.6		344.4	
s		0.837		0.548		0.548
u		0.374		0.245		0.245
a1		4.5		4.5		0.0
u1		2.598		2.598		0.0
a2		3.1		3.1		0.0
u2		1.8		1.8		0.0
a3		0.0		0.0		6.9
u3		0.0		0.0		3.984
Ut		3.2		3.2		4.0
U		6.4		6.3		8.0
Value with tolerance		361.2 +/- 6.4		342.6 +/- 6.3		344.4 +/- 8.0
Confidence level		95%		95%		95%

Conclusions:

1) Power: 348 w predicted and 95% confident it is in the range 340.4 to 348.4 w. Prediction agrees with measurement.
2) Maximum Temperature: 95% confident max temperature is in the range 354.8 to 367.6 deg F. An inner heating element of lower resistance is needed to obtain 400 deg F. This should be addressed with the next engineering prototype.

FIGURE B-116 Data Analyses and Conclusions for Test 1A

Table B-65 Test 1B Data Sheet

Location	Maximum Temperature Degrees F		
	Top	Middle	Bottom
Run 1	380	362	338
Run 2	396	361	337
Run 3	400	360	337
Run 4	402	361	340
Run 5	402	362	339

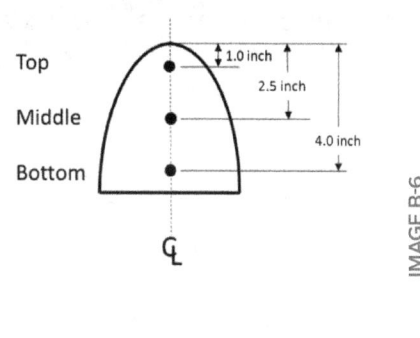

Figure B-117 displays the results of Test 1B with 95% confidence limits. The team is now 95% confident that the iron at maximum 120 VAC supply power has a maximum sole plate temperature at the top location somewhere in the range of 390.8 to 401.2 °F.

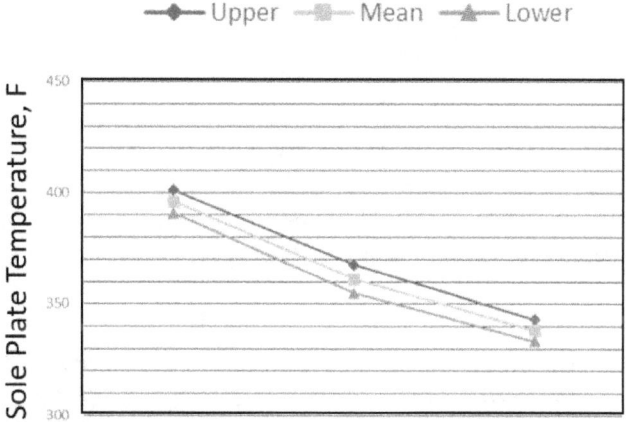

Location on Centerline	Top	Middle	Bottom
Value with tolerance	396.0 +/- 5.2	361.2 +/- 6.4	338.2 +/- 3.2
Confidence level	95%	95%	95%

FIGURE B-117 Test 1B Results: Temperatures Along Centerline with 95% Confidence Limits

The design intent is to iron cotton fabrics at the high temperature setting. These fabrics may scorch if they reach temperatures much above 400 °F. Because the team has high confidence that the current heating elements are able to achieve a sole plate temperature near 400 °F while not exceeding the 400-degree temperature by more than a few degrees Fahrenheit, the team has decided to change the production design to have an inner heating element of 56.3 ohms instead of 48 ohms.

The rated power of the production travel iron needs to be a value the team is confident that the iron will achieve. The actual power can be over but not below this value. Based on the 95% confidence limits for maximum power established in Test 1A, the team decided to rate the production iron, with the change in inner heating element, at 340 W instead of the current value of 400 W.

B.4.49.5 Exploratory Test 1C: Temperature Control Settings

In this test, the relationship between top location maximum temperature and thermostat setting was investigated. Based on these tests, marks for settings 1, 2, and 3 were made on the travel iron structure to align with the mark on the thermostat shaft.

B.4.49.6 Exploratory Test 1D: Plastic Insulation Plate Temperature

As shown in Figure B-118, with the iron at its maximum power setting, the temperature of the plastic insulation plate was only 250 °F. This is within the normal operating temperature of the plate's nylon 6 material.

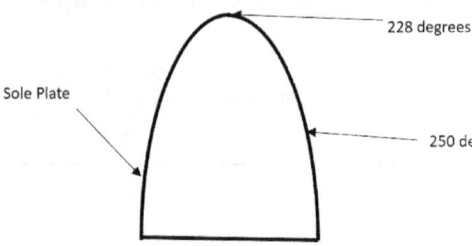

FIGURE B-118 Temperature of Plastic Insulation Plate at Maximum Power

B.4.49.7 Development Test 1: Three Temperature Settings

The purpose of this test is to demonstrate that when the iron is at one of its settings, it will have a maximum temperature indicated by this setting. Settings 1, 2, and 3 are to have maximum temperatures of 275, 375, and 400 °F, respectively. The data sheet for this test is provided in Table B-66.

	Table B-66 Development Test 1 Data Sheet		
	Setting 1	Setting 2	Setting 3
Run 1	270	370	380
Run 2	276	372	396
Run 3	274	375	400
Run 4	274	372	402
Run 5	273	371	402

The test data were analyzed following the same procedure as shown previously in Figure B-116. Table B-67 shows that Development Test 1 passed the test because the desired setting temperature falls within the tolerance bands of the measured data.

Table B-67 Development Test 1 Results			
Setting	1	2	3
Setting Temperature, F	275	375	400
Test Value with Tolerance, F	273.4 +/- 6.6	372.2 +/- 6.5	396.0 +/- 10.4
Confidence Level	95%	95%	95%
Test Pass or Fail	PASS	PASS	PASS

B.4.49.8 Development Test 2: Temperature Control Tolerances

This test addresses the UL requirement that the temperature control must hold the set point temperature ±18 °F (±10 °C). Table B-68 shows the data for Setting 3 (nominal 400 °F).

Table B-68 Development Test 2 Data Sheet					
Run	1	2	3	4	5
Max Temp, deg F	380	396	400	402	402
Min Temp, deg F	351	341	355	354	358

The team pondered how to interpret these test results. Table B-69 shows the difference between the measured maximum temperature and the measured minimum temperature for each of the five runs. The difference in all but Run 1 are greater than the allowed 36 °F spread. If the uncertainty of the measurements are taken into account, these differences are even greater. Based on this analysis, the team decided all runs except Run 1 failed. The team was uncertain how to score Run 1. Because at least one of the runs failed, the team considers Test 2 to be failed.

Table B-69 Development Test 2 Results					
Run	1	2	3	4	5
Max Temp, deg F	380	396	400	402	402
Min Temp, deg F	351	341	355	354	358
Difference	29	55	45	48	44
Pass or Fail	?	FAIL	FAIL	FAIL	FAIL

B.4.49.9 Summary of Engineering Prototype Development Tests

Table B-70 summarizes the results of the engineering prototype development testing. All six tests passed except Test 2. The major conclusions from these development tests are as follows:

- The current heating elements could heat the sole plate to a maximum of 400 °F. Therefore, the production unit should have an inner heating element changed from 48 ohms to 56.3 ohms, and the iron should be rated at 340 W instead of 400 W.
- The current thermostat is not capable of meeting UL Requirement 1 and needs to be changed to one capable of a tighter tolerance around the set point. The team decided that this change should be made in a second engineering prototype and the current development testing should continue with the existing thermostat.
- UL Requirement 3 was passed because the maximum temperature of the iron was only 400 °F at its maximum setting.
- The iron is capable of pressing the pillowcase in less than two minutes.
- UL Requirement 4 was not passed in its entirety because the production unit shell was not simulated.
- The operating life requirement of 200 cycles was not accomplished due to resource limitations; however, the iron did operate properly for a five-cycle life test.
- The current engineering prototype is ready for validation testing.
- A second engineering prototype that has an improved thermostat and a shell similar to the production unit should be built and tested prior to moving to Phase 5.

Table B-70 Summary of Engineering Prototype Development Testing Results

Test	Engineering Requirement		Test Method	Power	Status
1	Three Settings	275, 375, 400 deg F	Test Rig, Surface Probe Thermometer	120 VAC	PASS
2	UL Requirement 1	Hold settings ±18F	Test Rig, Surface Probe Thermometer	120 VAC	FAIL
3	UL Requirement 3	< 662 deg F	Test Rig, Surface Probe Thermometer	120 VAC	PASS
4	Press Time	Pillow Case < 2 min	Timer and Pillow Case	120 VAC	PASS
5	UL Requirement 4	Temperature Rise above Ambient: cord < 108 F	Test Rig, Surface Probe Thermometer	120 VAC	PASS
6	Operating Life	> 200 cycles— Limited to 5 cycles	Test Rig, Surface Probe Thermometer	120 VAC	PASS

B.4.49.10 Progress on Final Report

The team continued to work on the final report following the schedule previously given in Figure B-74.

MODULE 39

Test Analyses/Reporting and Updating Engineering Prototype Drawings

OVERVIEW

The ability to analyze test data is one of the Accreditation Board for Engineering and Technology (ABET) outcome requirements. This module discusses this issue along with procedures for test reporting. The task of updating the engineering prototype drawings is also discussed.

LEARNING OBJECTIVES

- Demonstrate the ability to properly analyze test data.
- Demonstrate the ability to prepare test reports.
- Demonstrate the ability to update engineering drawings based on test results.

PRE-LECTURE ASSIGNMENT

- Read sections B.4.50–B.4.53.
- Each team member prepares and brings to the team meeting (1) a progress report on their testing work and (2) their final report write-ups.

POST-LECTURE ASSIGNMENT

- Conduct a team meeting to 1) review the testing progress and resolve any issues and 2) review the final report write-ups and provide editing as needed.
- Each team member continues their assigned testing and final report writing and editing tasks.

TEAM DELIVERABLES

- Report on testing progress and any related issues that were resolved
- Edited final report write-ups
- List of the testing and final report writing and editing tasks that were assigned to team members
- Minutes of team meetings
- Team activity as documented in the team notebook

B.4.50 Test Analyses

An important part of test planning is deciding how the data will be analyzed to determine whether the unit being tested has met the design requirement. In general, the confidence that the test data represent the actual physical characteristic of the unit under test increases as the number of test data points increases. However, the absence of time, labor, and or money may limit the amount of data obtained. The team must be proficient in statistics to determine the confidence level in the conclusions drawn from test data. Appendix B2 provides some statistical approaches for development and validation testing.

A key success factor for product development is to budget and schedule adequate resources for the analysis of test data. This is a challenge in a fast-paced, two-semester capstone project course. If the team has followed the planning processes presented so far in this handbook, then they should be able to analyze the development and validation test data in a timely manner.

B.4.51 Test Reporting

EM-(c)

Initially, each product test is documented in the project notebook. This information is then formalized in the project final report. The format for test reporting is discussed in Appendix BA4: Product Development Testing and Section C: Final Report Outline. The key outcomes of the engineering prototype subproject are recommendations in the test reports for retaining or changing design features of the production unit.

B.4.52 Updating Engineering Prototype Drawings

The engineering prototype drawings must be updated with any changes that were made to the engineering prototype during development testing repair or rework.

B.4.53 Example Travel Iron Team Report for Week 26

B.4.53.1 Development Testing
As shown in Figure B-119, the team completed development testing this week.

B.4.53.2 Final Report Writing
During the team meeting, each member reported on the progress of their final report assignments. In addition, they presented their new sections to the team for final editing and approval. The team reviewed the final report schedule and concluded that the team was on schedule to complete the final report during Week 29.

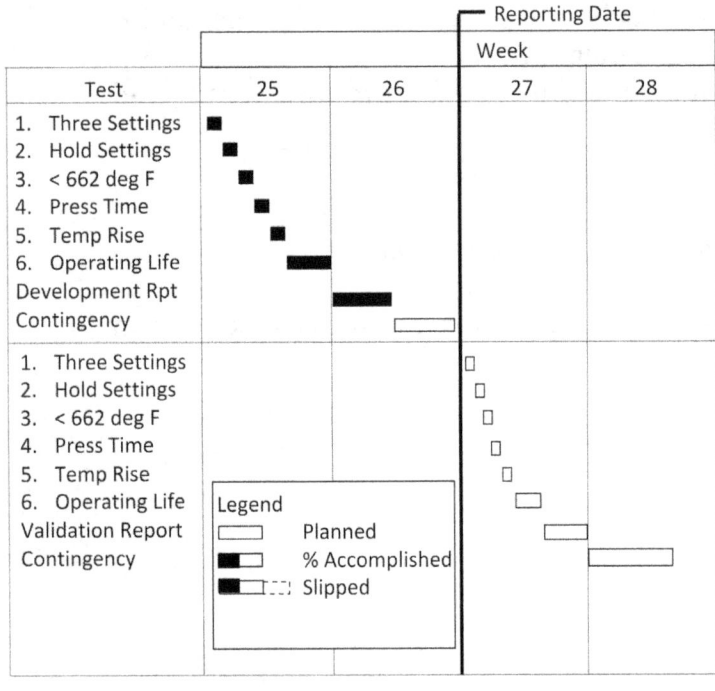

FIGURE B-119 Engineering Prototype Testing Schedule for End of Week 26

MODULE 40

Validation Testing and Test Results Design Review 6 (DR6) Preparations

OVERVIEW

Many of the engineering prototype requirements are validated through testing. This module covers this subject and the preparations for design review 6 (DR6; the test results design review).

LEARNING OBJECTIVES

- Demonstrate the ability to conduct validation testing.
- Demonstrate the ability to prepare a test results review.

PRE-LECTURE ASSIGNMENT

- Read sections B.4.54–B.4.56.
- Each team member prepares and brings to the team meeting a progress report on their testing work.

POST-LECTURE ASSIGNMENT

- Conduct a team meeting to 1) review the testing progress and resolve any issues, 2) prepare the slides for DR6 and decide who will present each slide, and 3) schedule a team meeting to practice the DR6 presentation prior to the beginning of Module 41.
- Each team member continues their assigned testing and practices their part of the DR6 presentation.
- The team conducts a second meeting to practice for DR6.

TEAM DELIVERABLES

- Report on the testing progress and the issues the team has faced and how they are resolved
- Slides for DR6 and who will present each slide
- Minutes of team meetings
- Team activity as documented in the team notebook

B.4.54 Validation Testing

The purpose of development testing is to ensure that the test article will pass all the validation tests. The validation tests provide a formal means of validating that the prototype will meet the design requirements.

Because the capstone project is resource- and schedule-constrained, the team will sometimes find that they are not able to rework the engineering prototype design so that it will pass all the scheduled validation tests. It is important to understand that if this is the case, the engineering prototype development and validation subproject is still a success because it has increased knowledge regarding the production design.

B.4.55 Preparing for Test Results Design Review (DR6)

EM-(i) In most cases, the process of testing the engineering prototype will reveal new information about the design. Prior to DR6, the entrepreneurial team will study these new discoveries to find additional value that can be integrated into the production unit design.

The purpose of DR6 is to review the engineering prototype test program results and to summarize what changes need to be made in the production unit design. An outline for the DR6 presentation is provided in Table B-71. As with all team presentations, the team should work together to outline the presentation, make assignments for the preparation and presentation of each slide, and practice the presentation as a team.

Table B-71 Outline for DR6 Review of Engineering Prototype Test Results

1. **Title page**—Call it "DR6: Test Results Review" and include date, name of project, picture of project, team picture, team names, and sponsor name.

2. **Issue and brief problem statement**—Provide a brief description of the issue or societal need the design team is addressing; succinct and brief, solution-neutral problem statement

3. **Final prototype solid model**—Include a solid model for the prototype assembly; use sectioned assembly drawing to show internal details and parts, exploded view of labeled CAD model and BOM. Show all prototype components.

4. **Team's response to DR5 feedback**—Provide feedback given during the DR5 test readiness review and the team's response to the feedback

5. **Completed prototype assembly**—Include clear photos of the completed prototype and its parts taken from different angles; videos may be included to demonstrate how the prototype works.

6. **Development testing**—For development tests, provide the test procedure, equipment, and instruments used; data collected; analysis procedure; graphical plots and estimated error bars; and discussion of results and conclusions. For failed development tests, describe how failures were resolved; more than one slide can be used.

7. **Validation testing**—Include the engineering prototype requirements validation matrix for which validation testing was conducted; note that every validation test should have unofficially been conducted during development and passed; summarize the key validation test procedures.

8. **Failed validation tests**—List and describe the validation tests that failed and the design issues that led to the failure; how the team resolved the issues that led to the failed validation test; the team's plan to go forward with the prototype; if the prototype fails a validation test, whether it will still be used for further validation.

9. **Example of testing with data analysis**—Include detailed description of one of the most important tests; show test procedure, data collected, and analysis.

10. **Problem-solving example**—Show an example of a problem that was faced by the team during development and how it was solved; the problem must be a significant technical issue (not, for example, replacing a hardware with another one or a type of material with another).

11. **Identified changes in production design**—Describe the suggested production unit design changes based on the testing and validation results.

12. **Project schedule status**—Use the format given in the handbook; schedule must be readable; include one-sentence summary.

13. **Labor loading budget status chart**—Include reporting date, budget line, actual line, ETC line and one sentence summary.

14. **Financial budget status**—Show the team financial budget utilization.

15. **Final report writing status**—Show the status of the final report sections writing.

16. **Lessons learned**—Include bulleted list of top five things the team has learned so far about the phased product development process as applied to their project.

17. **Go-forward plan**—Include remaining tasks, schedule of expected completion date.

B.4.56 Example Travel Iron Team Report for Week 27

B.4.56.1 Validation Testing

Table B-72 summarizes the engineering prototype validation testing. The results of these tests validated the results and conclusions reported in the section on development testing. The validation test report was also prepared and archived in the project notebook.

Table B-72 Summary of Engineering Prototype Validation Testing Results

Test	Engineering Requirement		Test Method	Power	Status
1	Three Settings	275, 375, 400 deg F	Test Rig, Surface Probe Thermometer	120 VAC	PASS
2	UL Requirement 1	Hold settings ±18F	Test Rig, Surface Probe Thermometer	120 VAC	FAIL
3	UL Requirement 3	< 662 deg F	Test Rig, Surface Probe Thermometer	120 VAC	PASS
4	Press Time	Pillow Case < 2 min	Timer and Pillow Case	120 VAC	PASS
5	UL Requirement 4	Temperature Rise above Ambient: cord < 108 F	Test Rig, Surface Probe Thermometer	120 VAC	PASS
6	Operating Life	> 200 cycles– Limited to 5 cycles	Test Rig, Surface Probe Thermometer	120 VAC	PASS

B.4.56.2 Preparations for DR6

The team prepared the slides and practiced their presentation several times to make sure that they stayed within the time limit imposed by the instructor.

B.4.56.3 Production Unit Detailed Design Drawing Package

The first version of the production unit detailed design drawing package was created. The team did not have enough time to select a new thermostat for the production design. The BOM, assembly drawing, and thermostat control drawing were changed to show the thermostat to be a to-be-determined (TBD) item.

B.4.56.4 Preparation of the Final Report

The team made a special effort to complete at least a draft of all the final report sections. The team was successful in this effort, and a complete draft of the final report was sent to each team member for review. During this process, the team met to collectively work on Section 2: Final Design Description and the Executive Summary.

MODULE 41

Conducting DR6, Updating the Commercialization Plan, and Completing the Final Report

OVERVIEW

There are several project deliverables during the last three weeks of the capstone project. The team needs to be especially organized to manage all of these activities. Each team member must hold themselves and the other team members accountable for completing tasks on time.

LEARNING OBJECTIVES

- Demonstrate the ability to conduct a test results review.
- Demonstrate the ability to update the commercialization plan.
- Demonstrate the ability to summarize the final results of the capstone project as Section 2 of the project final report.
- Demonstrate the ability to complete the final report editing as a team.

PRE-LECTURE ASSIGNMENT

- Read sections B.4.57– B.4.62.
- Each team member prepares their assigned portions of the final report and brings them to the team meeting.
- Each team member prepares and brings to the team meeting a progress report on their final edits for the final report.

POST-LECTURE ASSIGNMENT

- Conduct a team meeting to 1) review the feedback from DR6 and integrate it into the final report, (2) update the production CAD model with the results of the Phase 5P development and validation tasks, (3) review and update the commercialization plan, (4) review/discuss team members' suggestions for Section 2 of the final report and prepare the team's final version of this section, (5) review and discuss the progress on completing the project final report and hold team members accountable for finishing their assignments, and (6) review the checklist for Phase 5P and assign any outstanding tasks that need to be completed before Module 42, when it is due to the sponsor.

TEAM DELIVERABLES

- Updated production CAD model based on the results of the Phase 5P development and validation tasks
- Updated commercialization plan
- Team's final version of final report Section 2
- Statement of progress on completing the project final report
- Status of the items in the checklist for Phase 5P
- Minutes of team meetings
- Team activity as documented in the team notebook

B.4.57 Conducting Test Results Design Review (DR6)

The purpose of DR6 is to present the test results to the sponsor (i.e., capstone instructor) and obtain approval to make the recommended changes to the Phase 4 production drawing package. The team should note all feedback received during this review.

B.4.58 Updating the Commercialization Plan

This updated commercialization plan will be used by the team that picks up the project and takes it the rest of the way into commercialization. The plan should follow the Goldsmith Commercialization Model and restate the value propositions for both customer and producer. Value streams from multiple perspectives should be included.

The production and use of this product will have an effect on the ecosystem. Describe how the product is being designed to be environmentally friendly.

EM-(e) During engineering prototype subproject, the team learns a great deal about what is required to complete the product development process. Sometimes this subproject identifies additional tasks that must be done before the project can move into Phase 5: Production Unit Prototype Development. For example, a second engineering prototype may be required to address all engineering require-

EM-(j) ments. Changes such as this will alter the overall commercialization timeline.

In parallel with the engineering prototype testing, the team has the oppor-

EM-(k) tunity to monitor the overall sociopolitical environment and identify any changing customer and market trends. An example is the fast-paced solar energy market. These changes also need to be addressed in the updated commercialization plan.

B.4.59 Preparing Section 2 of the Project Final Report

Section 2 of the final report is the key section to describe the (1) team's final production unit design and (2) results of engineering prototype testing. Section C outlines what should be included in this section of the final report.

B.4.60 Final Editing of the Final Report

Before the final report is submitted, each team member needs to make a final review of the report and list any needed changes. These changes must gain team approval before they are incorporated into the final report. The previously presented Table B-62 provides a checklist for this final review of the report.

EM-(p) The team needs to spend time reflecting on their performance. The final report should include this reflection. It should also include the results of the team's 360-degree reviews and how they were used to move the team into a high-performance level.

EM-(q) The team had to improve their marketing and business acumen skills to integrate all the pertinent customer and producer value streams into the product. The team needs to document this improvement process in both the project notebook and final report.

B.4.61 Phase 5P Exit Criteria Checklist

Below is the Phase 5P phase-exit checklist. The team must complete this list and submit it to the sponsor by the end of Week 29.

Phase Exit Criteria Checklist: Phase 5P: Development and Validation Testing

Team Name: _____	Team Leader: _____
Members: _____	_____
_____	_____
_____	_____
Date Started: _____	Date Completed: _____

Done	EXIT CRITERIA		Comments
	1.	Development testing has been completed, and the engineering prototype drawings have been updated to reflect any rework that occurred.	
	2.	The hardware has been prepared for validation testing.	
	3.	The unit has completed validation testing with any deviations from the requirements noted.	
	4.	DR6 review of engineering prototype test results has been presented.	
	5.	Feedback from DR6 has been reviewed, and appropriate actions have been taken.	
	6.	Applicable sections of the final report have been written, edited, and approved by the team.	
	7.	Notebook is checked and approved.	
Approved by	Name and Title		Completion Date

B.4.62 Example Travel Iron Team Report for Week 28

B.4.62.1 DR6 Review of Engineering Prototype Test Results

The team successfully presented their testing results during DR6 review of engineering prototype test results. The sponsor agreed that (1) the thermostat should be changed to one that is more precise, (2) a second engineering prototype should be accomplished before beginning Phase 5, and (3) the initial Phase 4 drawing package should be prepared with the specific thermostat being a TBD item. The Phase 4 production unit drawing package will again be updated after the second engineering subproject, which is beyond the scope of the capstone course.

B.4.62.2 Final Report

The team updated the production unit CAD model with the results of the Phase 5P testing. The team then used the outline in Section C to prepare a draft of Section 2 of the final report, which summarizes the final production unit design and engineering prototype.

The team reviewed drafts of the other final report sections by using the checklist presented in Table B-73. Then, the team prepared a statement of progress for completing the report and submitted it to the sponsor.

		Table B-73 Final Report Checklist
☑	1.	Does the section follow the outline in Section C of the handbook?
☑	2.	Is the section written for the reader (sponsor) and not the writer?
☑	3.	Are all facts, figures, etc. properly referenced?
☑	4.	Are all figures and tables properly labeled and introduced in the text before they appear?
☑	5.	Does the report section flow when read aloud?
☑	6.	Are there any spelling errors?
☑	7.	Are there any grammatical errors?
☑	8.	If there are checklists, is the backup for them easily found in the team notebook?
☑	9.	Have all missing items in this section been added?
☑	10.	Has this section been reviewed and approved by each team member?
☑	11.	Is there an email from each team member stating they approve of the section?

B.4.62.3 Phase 5P Exit Criteria Checklist

The team completed the Phase 5P checklist and had it approved by the sponsor.

B.4.62.4 Project Notebook

The team reviewed the status of the project notebook and updated it as needed.

B.4.62.5 Updating the Commercialization Plan

The team began work on updating the commercialization plan for the final report. A second engineering prototype will be added to the plan.

MODULE 42

Preparation of Phase 4 Drawing Package and Project Final Presentation

OVERVIEW
The Phase 4 drawing package is due during Week 30. The team starts this week to finalize this deliverable and prepares an outline for the final report presentation, which is also due this week. The slides used by the example travel iron project team based on this outline are provided.

LEARNING OBJECTIVES
- Demonstrate the ability to prepare a detailed design drawing package.
- Demonstrate the ability to prepare the final report presentation.

PRE-LECTURE ASSIGNMENT
- Read sections B.4.63–B.4.65.
- Each team member selects at least two of the slides described in Table B-72 and prepares rough drafts of these slides to present at the team meeting.

POST-LECTURE ASSIGNMENT
- Conduct a team meeting to (1) prepare rough drafts of the final presentation slides and assign a team member to each slide for its final preparation and presentation and (2) prepare an outline of items to be included in the Project 4 detailed drawing package and assign these items to team members.
- Team members work on the final presentation and Phase 4 drawing package.

TEAM DELIVERABLES
- Rough drafts of the final presentation slides
- Team member assignments to prepare and present each slide
- Outline of items to be included in the Phase 4 detailed design drawing package
- Team member assignments to prepare the drawing package
- Minutes of team meetings
- Team activity as documented in the team notebook

B.4.63 Final Report Presentation Outline and Rubric

Table B-74 provides a rubric for the final capstone project presentation. The rubric lists each slide that should be included in the final presentation and provides a scale of compliance for determining how well each slide describes the required content.

Table B-74 Final Capstone Presentation Slides and Rubrics

Slide	Item	Compliance (1–5, with 5 best)
1	**Introduction slide:** name of the presentation, team name, project name, team photo and member names, and picture of engineering prototype	
2	**Societal issue and problem statement:** societal issue to be solved; solution-neutral problem statement supported by appropriate graphic images as needed	
3	**Basic physics involved:** display basic physics involved in design and development; use graphics and key equations and concepts	
4	**Need with value statements:** value for the user; value for the sponsor	
5	**Production unit requirements validation matrix from the voice of the customer (VOC):** how quality function deployment is used to identify engineering requirements from the VOC; list the requirements identified from the customer need; target values, method of validation, and status of validation	
6	**Phase 2 final conceptual design:** display the final product conceptual design selection using morphological chart; use labeled sketches to show all parts	
7	**Phase 3 production unit CAD model Rev 3:** the updated CAD model Rev 3 from the preliminary design analyses results (i.e., RMS, FMEA, DMFA, DTC analyses); label all parts; use sectioned assembly (or exploded) drawing to show internal details	
8	**Engineering prototype scope and requirements:** engineering prototype scope and prototype requirements selected from the production unit requirements	
9	**Phase 3P engineering prototype CAD model:** how the engineering prototype differs from that of the production unit; design of any test rigs needed during prototype development	
10	**Phase 4P engineering prototype detailed design drawings:** show examples of fabricated part drawings with tolerance and surface requirements	
11	**FMEA analysis:** Top five failure modes and how the risk priority numbers were improved	

(continued)

Slide	Item	Compliance (1–5, with 5 best)
12	**Problem-solving example:** technical problem; impact on the design; approach for solving the problem; impact of the selected solution	
13	**Fabrication and assembly:** photos with captions showing fabrication processes for selected parts	
14	**Completed prototype assembly:** clear photos of the completed prototype and its parts taken from different angles; if possible, videos to demonstrate how the prototype works and meets the customer requirements	
15	**Development:** development tests you did prior to validation; use pictures; describe rework	
16	**Validation:** final requirements validation matrix; methods and results of validation tests	
17	**Example testing with analysis:** a sample of prototype testing with a lot of analysis and planning	
18	**Phase 4: Changes to production CAD model and drawing tree:** list changes in the production unit based on the prototype results; production package drawing tree	
19	**Commercialization slides:** value proposition for customer and producer; how the Goldsmith commercialization model was used to prepare the plan; the key remaining production issues that need to be addressed in the go-forward efforts; a brief business plan summary	
20	**Project schedule with key milestones:** use Gantt chart format given in the handbook	
21	**Labor chart budget line:** display chart and rationale with budget and actual lines	
22	**Team performance:** how well the team met the schedule, budget, and worked as a team	
23	**Accreditation Board for Engineering and Technology (ABET) outcomes:** show in a table evidence of how the project design team achieved each of the ABET 1–7 outcomes criteria	
24	**Entrepreneurial mindset (EM):** indicate the degree at which your project team met the 17 Arizona State University Fulton Engineering EM@FSE 2.0 indicators	
25	**Top lessons learned:** three thoughtful and hard-hitting items for the Capstone project experience	

(continued)

Slide	Item	Compliance (1–5, with 5 best)
26	**Summary:** prototype picture and five-bullet summary	
27	**Acknowledgements:** acknowledge those who provided support to the team during the project execution	

*Compliance score:

1. Poor: None of the expectations are met
2. Fair: Some of the expectations are met
3. Good: Meets most of the expectations
4. Very good: Meets all expectations
5. Excellent: Exceeds expectations

B.4.64 Phase 4 Production Unit Detailed Design Drawing Package

If the team decides that the engineering prototype subproject has provided all of the information needed to complete the production unit detailed design, then the team is ready to complete Phase 4 by creating a production unit detailed drawing package. In many cases, the team may decide on a project to conduct more engineering prototype design, development, and validation subprojects before the production unit detailed design is finalized.

Most capstone projects find that additional engineering prototypes are needed before the team can commit to a final production detailed design. However, a goal of the capstone course is to have teams prepare a Phase 4 drawing package. Therefore, even if more engineering prototypes are planned in the future, the capstone team will integrate the results from their engineering prototype subproject into their production CAD model for Phase 4 and then create a detailed design drawing package. A list of this drawing package is provided in Table B-75.

Table B-75 Phase 4 Production Unit Detailed Drawing Package Outline

- **Outline drawing** that provides the overall shape and dimensions of the product, identifies interfaces, and lists the product's weight. (Here the product is the engineering prototype.)
- **Assembly drawing** that labels each component of the product and includes notes on how these components are to be joined to make the final assembly. For relatively simple products, the assembly drawing can use cross-sections to show all the parts. For complex products, an exploded-parts drawing is generally used.
- **Subassembly drawings.**
- **Detail drawings**, which are provided for all parts that are fabricated either in house or by a supplier.
- **Control drawings or supplier specification sheets**, which are provided for all standard components.
- **Bill of Materials (BOM)**, which is a centralized source of information used to manufacture a product. Items included in a BOM are the part number, part name, quantity, unit of measurement, assembly references, method of parts construction, and additional notes.
- **Process sheets**, which are also included if there are any specialized processes required to fabricate or assemble the product.

B.4.65 Example Travel Iron Team Report for Week 29

B.4.65.1 Final Presentation
The team continued to practice the final presentation.

B.4.65.2 Phase 4 Detailed Design Drawing Package
During the first team meeting of the week, the team prepared the drawing package outline and made team assignments. A major effort for the team this week was completing the drawing package.

B.4.65.3 Project Notebook
The team updated the project notebook to include the results of the week's activities.

MODULE 43

End-of-Project Deliverables

OVERVIEW

This module covers the final work to complete the capstone product design and development course. The final report and end-of-project deliverables are due. The team meets after the class lecture to practice their project final presentation.

LEARNING OBJECTIVES

- Demonstrate the ability to deliver project deliverables on schedule.
- Demonstrate the ability to prepare a Phase 4 production unit detailed drawing package.
- Demonstrate the ability to practice a team project final presentation.

PRE-LECTURE ASSIGNMENT

- Read sections B.4.66–B.4.68.
- Each team member practices their portion of the project final project presentation.

POST-LECTURE ASSIGNMENT

- Conduct a team meeting to (1) agree on the tasks needed to complete the Phase 4 drawing package and assign tasks to the team members and (2) practice the team's project final presentation to be given in the next and final module.
- Conduct a second team meeting for the team to review and edit the final version of the Phase 4 production unit detailed drawing package.

TEAM DELIVERABLES

- List of tasks that are needed to complete the Phase 4 drawing package and the assigned tasks to the team members
- Proof that team practiced final presentation
- Final version of the Phase 4 production unit detailed drawing package
- Minutes of team meetings
- Team activity as documented in the team notebook

B.4.66 End-of-Capstone-Project Activities

The capstone project ends with the archival and submittal of the deliverables listed in Table B-76.

Table B-76 Capstone End-of-Course Deliverables

- Project final report (submitted the prior week)
- Phase 4 exit criteria checklist
- Team project notebook
- Engineering prototype build book
- Engineering prototype hardware
- Slides from project final presentation

B.4.67 Phase 4 Exit Criteria Checklist

Below is the Phase 4 exit criteria checklist.

B.4.68 Example Travel Iron Team Report for Week 30

The team continued to practice their project final presentation. Initially, the presentation exceeded the time limit, but some slides were modified to just summarize the subject, and the details were put on a separate backup slides.

The team filled out the Phase 4 exit criteria checklist and obtained the sponsor's approval. The team completed a final review and edit of the team project notebook, which includes the engineering prototype build book. The team then submitted this document to the sponsor.

The team reviewed the Phase 4 detailed design drawing package. A few drawing issues were identified, and team members were assigned to resolve these issues and have the complete drawing package ready for submittal in Week 31.

The team submitted the engineering prototype unit and test rig(s) to the sponsor.

Exit Criteria Checklist: Phase 4: Detailed Design of Production Unit

Team Name: _____	Team Leader: _____
Members: _____	_____
_____	_____
_____	_____
Date Started: _____	Date Completed: _____

Done	EXIT CRITERIA	Comments
	1. Engineering prototype phases 3P, 4P, and 5P checklists are complete and approved.	
	2. Production unit CAD model has been updated with changes resulting from engineering prototype subproject.	
	3. Detailed analyses are complete for the production unit.	
	4. Drawing package is complete for the production unit.	
	5. Final report Section 2 is complete.	
	6. All sections of final report are complete and submitted.	
	7. Slides for presentation of final report results are complete.	
	8. Final report presentation is completed.	
	9. Team project notebook is complete and submitted.	
Approved by	Name and Title	Date

Yes or No	Entrepreneurial Mindset Indicators Exit Criteria	Questions Related to Criteria
	a. Critically observes surroundings to recognize opportunity	The testing of an engineering prototype results in new knowledge. Did the team identify new opportunities based on this new learning, and are these opportunities discussed in the testing section of the project notebook?
	b. Explores multiple solution paths	During the Phase 4 planning activities, did the team make a list of unexpected events that could impact their plan, did they then identify a number of potential solutions, and were these risks and solutions documented in the team notebook?
	c. Gathers data to support and refute ideas	Were production unit design ideas supported or refuted by the engineering prototype testing data, and were these activities documented in the team notebook?
	d. Suspends initial judgement on new ideas	To manage unplanned events, the team must consider new ideas. Did the team (1) consistently monitor themselves in team meetings to suspend judgement of new ideas until they are explored, and (2) was this documented in the team minutes? Provide an example.
	e. Observes trends about the changing world with a future-focused orientation/perspective	Does the commercialization plan presented in Section 2 of the final report include the integration of changing trends?
	f. Collects feedback and data from many customers and customer segments	Did the team arrange for customers to attend the design reviews? If not, were other experts tasked with the role of reviewing the presentations from the customer's perspective? Did the team take actions to respond to the customer's feedback and record those actions in the project notebook?
	g. Applies technical skills/knowledge to the development of a technology/product	Did the team document in the project notebook how they applied their technical skills/knowledge to resolve test issues?
	h. Modifies an idea/product based on feedback	Did the team review the DR5 feedback, make the necessary changes, and document this work in the project notebook?
	i. Focuses on understanding the value proposition of a discovery	Prior to DR6, did the team increase the value of the production design by integrating new test discoveries into the CAD model for the production unit?

(continued)

Yes or No	Entrepreneurial Mindset Indicators Exit Criteria	Questions Related to Criteria
	j. Describes how a discovery could be scaled and/or sustained, using elements such as revenue streams, key partners, costs, and key resources	Did the team integrate the Phase 5P testing discoveries into the updated commercialization plan?
	k. Defines a market and market opportunities	Did the team create an updated commercialization plan during Phase 4 that included updated market information and opportunities?
	l. Engages in actions with the understanding that they have the potential to lead to both gains and losses	When deciding on how to address issues that arose during testing, did the team consider the potential benefits versus risks, and were these considerations recorded in the team minutes?
	m. Articulates the idea to diverse audiences	The design reviewers had various backgrounds and knowledge of the project as compared to the team members. Did the team list in the team minutes the specific actions taken to ensure that the design reviews were effective to these diverse audiences?
	n. Persuades why a discovery adds value from multiple perspectives (technological, societal, financial, environmental, etc.)	Did the team complete a comprehensive update of the commercialization plan that included additional multiple value streams identified during Phase 4?
	o. Understands how elements of an ecosystem are connected	Did the team make sure that the updated commercialization plan describes how the product's design promotes sustainability of the ecosystem?
	p. Identifies and works with individuals with complementary skill sets, expertise, etc.	Did the team demonstrate that they were a high-performing team? Did they conduct and document team member assessments?
	q. Integrates/synthesizes various kinds of knowledge	Did the team document in the project notebook how they improved their marketing and business acumen to properly integrate these considerations into the product's technical design?
Approved by	Name and Title	Completion Date

MODULE 44

End-of-Project Deliverables Due

OVERVIEW

The capstone project concludes with this module. The Phase 4 drawing package is submitted to the sponsor, and the project final presentation is given to the sponsor and their review experts.

LEARNING OBJECTIVES

- Demonstrate the ability to prepare and submit a Phase 4 production unit detailed design drawing package.
- Demonstrate the ability to give a team project final presentation.

PRE-LECTURE ASSIGNMENT

- Read sections B.4.69–B.4.71.
- Each team member practices their portion of the project final project presentation.

POST-LECTURE ASSIGNMENT

- Conduct a team meeting to (1) agree on the tasks need to complete the Phase 4 drawing package and assignee tasks to the team members and (2) practice the team's project final presentation to be given in the next and final module.
- Conduct a second team meeting for the team to review and edit the final version of the Phase 4 production unit detailed drawing package.

TEAM DELIVERABLES

- Phase 4 production unit detailed drawing package
- Final report
- Team project notebook (includes build book)
- Engineering prototype hardware
- Slides from project final presentation
- Minutes of team meetings
- Team activity as documented in the team notebook

B.4.69 Final Report Presentation

The team makes the final project presentation to the sponsor and subject experts. The slides are then archived and submitted to the sponsor.

B.4.70 Phase 4 Detailed Design Drawing Package

The team archives the Phase 4 detailed design drawing package and submits it to the sponsor.

B.4.71 Project Closure

After ensuring all of the deliverables have been submitted and approved by the sponsor, the team can consider this project completed.

It should be noted that the capstone course ends in Phase 4 product detailed design. For industry projects, the team will continue to complete any additional engineering prototype subprojects and the corresponding updates to the Phase 4 drawing package. The team will then move into Phase 5, which is the production unit prototype's manufacture, development, and validation. At the successful completion of Phase 5, the team will take the production unit into Phase 6: Production.

Section B

Phase 5: Production Prototype Development

B.5 Phase 5: Production Prototype Development

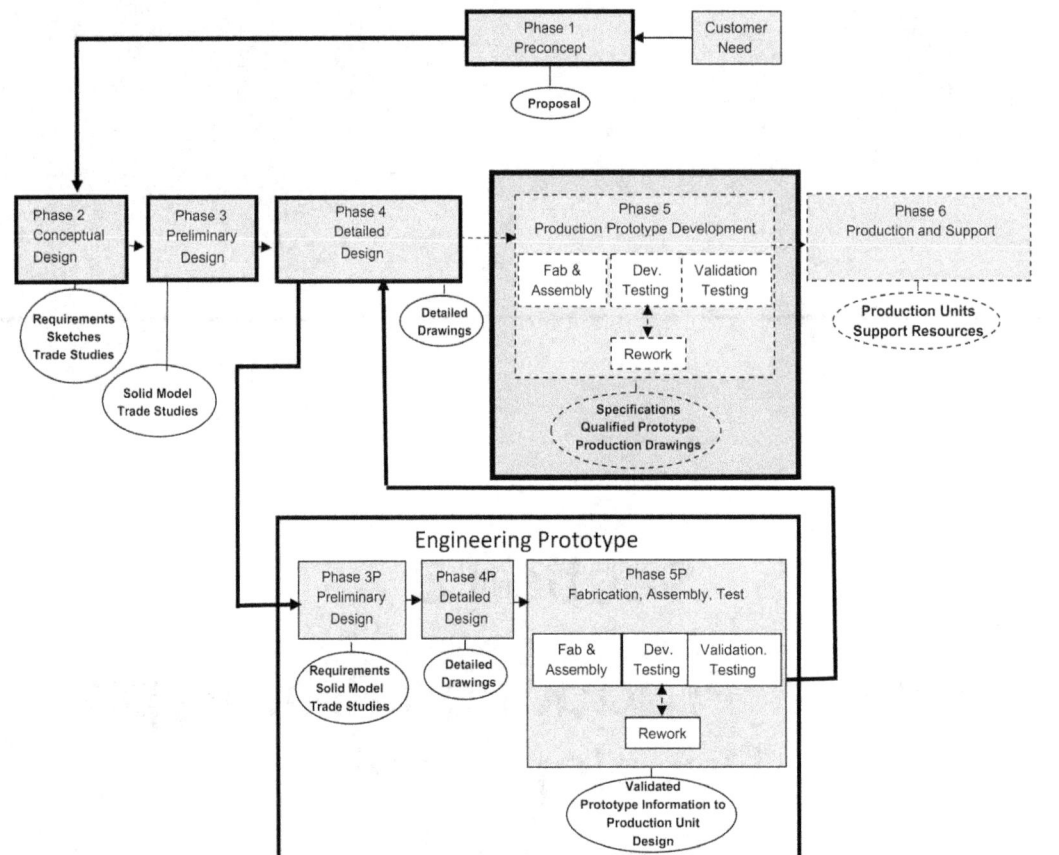

Phase 5 is for the development of the production unit prototype by the design team in industry. It was mentioned in the previous phases that due to financial and labor budget constraints, capstone projects stop at Phase 4 of the phased product development (PPD) process. It is similar to Phase 5P, during which the engineering prototype was developed (see sections B.4.30 through B.4.62). At the conclusion of Phase 5, the production unit drawing package is updated with any changes prior to release for production in Phase 6.

One of the goals of the PPD process is to minimize the time and resources needed to conduct Phase 5. This is accomplished by using analyses and testing in prior phases to minimize Phase 5 rework.

The completion of Phase 5 is a key contractual milestone between the sponsor and team. The sponsor wants to be sure that the design is ready for production. The team wants the sponsor to be satisfied so that the team can be rewarded for their efforts. Many upsets occur between the team and sponsor when the criteria for declaring Phase 5 complete are not clear. The production unit engineering requirements must be measurable, and the specific process for validating each of these requirements must be documented and approved by the sponsor prior to starting the work in Phase 5.

Section B
Phase 6: Production and Support

B.6 Phase 6: Production

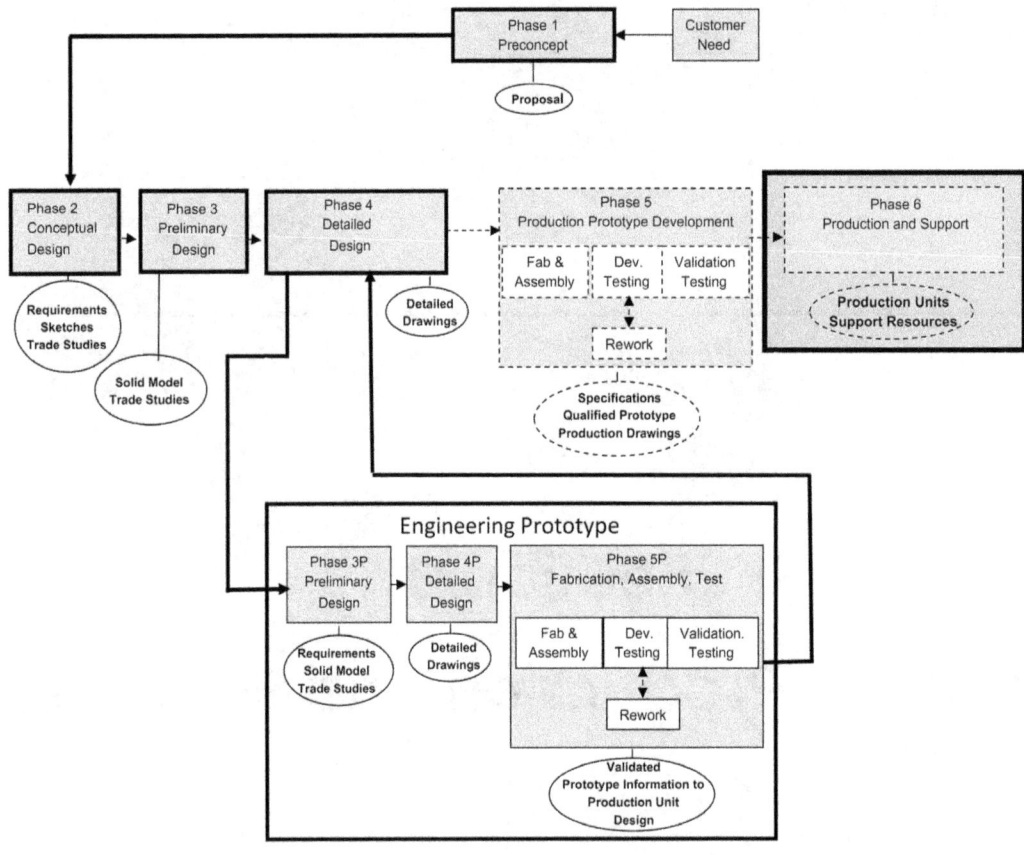

1. Acquire, install and commission the equipment and facilities, and acquire and train the labor force necessary to produce the product.
2. Conduct a low-rate initial production run to verify that the production process is functioning.
3. Produce a high-quality product that meets the drawing package and production cost goals.
4. Market and distribute the product in a way that meets sales objectives.
5. Maintain a service after the sale program that provides quality customer and warranty support.
6. Provide on-going product improvement and product growth programs.
7. Maintain an updated market exit strategy that addresses customer and environmental needs.

FIGURE B-120 Phase 6 Key Objectives

Market conditions are not stagnant. Customer needs, regulations, production costs, and supplier shortages are some of the many issues that must be addressed during the product's production life. Most of these issues require engineering input. Engineering is an important service during the entire production life of the product. The business plan needs to provide adequate funding for these engineering services.

Often the product development team is replaced by a different team during production. Having a well-written product development final report and team project notebook is vital to the production team as they tackle the issues listed above.

Section B

Product Development Final Thoughts

B.7 Product Development Final Thoughts

This innovative, entrepreneurial, and structured product design and development handbook, based on industry practice, is useful to both engineering students and practicing engineers.

In order to provide structure, specific formats have been provided for deliverables such as design reviews and reports. These formats may vary from organization to organization, but the content required is the same.

One of the innovative features of the handbook is the use of learning modules to facilitate both classroom and independent learning. Although this handbook evolved from materials developed for a mechanical engineering capstone program over the past eight years, it is readily applicable to other engineering disciplines involved in product design and development.

Much of this material was used for three years in a transdisciplinary product development capstone program that had team members majoring in business, industrial design, and sustainability. The engineering students came from the electrical, civil, computer sciences, chemical, and mechanical engineering fields. These materials have also been used in a solar energy graduate program where multidisciplinary teams designed and built projects for industry customers.

The handbook enables capstone design teams to meet Accreditation Board for Engineering and Technology (ABET) outcome criteria at the stated levels of mastery. Likewise, the entrepreneurial mindset (EM) framework as developed by Arizona State University and the Kern Entrepreneurial Engineering Network is an integral part of the handbook. The phase exit criteria include ABET criteria and EM indicators.

The handbook has also emphasized the use of the commercialization process, through the Goldsmith Commercialization Model, which combines technical product development with the requirements of marketing and business activities.

The authors through Trimble Consulting Inc. have tailored these materials to train and lead teams in the aerospace, utility, energy, automotive, industrial, and consumer products industries.

As stated above, the specifics of this handbook can be tailored to integrate into an organization's existing product development process. However, this integration needs to include the actions listed in Table B-77.

Table B-77 Key Product Design and Development Actions

1. Use a product development process with phase-exit criteria.
2. Create a high-performing team through commitment, accountability, and diversity.
3. Develop an entrepreneurial mindset (EM) within the team.
4. Focus on the voice of the customer (VOC).
5. Understand the basic physics of the issue.
6. Translate the VOC into measurable engineering requirements and maintain an engineering requirements matrix.
7. Explore the entire design space and identify multiple options.
8. Establish and update as needed the value propositions for both the customer and the producer.
9. Partner with the sponsor and understand and address their needs.
10. Plan the work, monitor the work, and be ready to make mid-course adjustments.
11. Manage the project scope to be always on schedule and within budget.
12. Use analyses and testing early in the project to minimize rework during prototype development.
13. Maintain a project notebook and document all activity.
14. Remember that reports and design review presentations are for the audience and not the writer or speaker.
15. Act with professionalism.

The authors are committed to continuous improvement. The authors invite both students and practicing engineers who use this handbook to contact them with case histories of its use and ideas on how to improve it.

Engineering is a noble profession. It is based on service to society. Engineers act with integrity to not only meet the needs of their customers and employers but also protect the welfare of society. The authors encourage both graduating engineers and practicing engineers to join the Order of the Engineer organization and take the obligation given in Figure B-121.

> "I am an Engineer. In my profession I take deep pride. To it I owe solemn obligations. As an Engineer, I pledge to practice integrity and fair dealing, tolerance and respect; and to uphold devotion to the standards and the dignity of my profession, conscious always that my skill carries with it the obligation to serve humanity by making the best use of the Earth's precious wealth. As an Engineer, I shall participate in none but honest enterprises. When needed, my skill and knowledge shall be given without reservation for the public good. In the performance of duty and in fidelity to my profession, I shall give my utmost."
>
> (Source: https://order-of-the-engineer.org/about-the-order/obligation/)

FIGURE B-121 Obligation of an Engineer

Credit

Fig. B-121: "Obligation of an Engineer," https://order-of-the-engineer.org/about-the-order/obligation. Copyright © by Order of the Engineer.

Section B

Appendices

APPENDIX BA1

360-Degree Teammate Review Form

Each team member periodically completes the 360-degree teammate review form to provide feedback on the performance of other team members, as well as providing a self-evaluation. Please rate the performance of your team members and yourself by providing a score for each of the listed performance items.

360-Degree Review Form					
Reviewer:					
Member 1:		Overall Score:			
	Excellent	Very Good	Good	Fair	Poor
Meeting attendance	5	4	3	2	1
Completed work on time	5	4	3	2	1
Practiced presentations	5	4	3	2	1
Reviewed others' work	5	4	3	2	1
Other teamwork elements (attention during meetings, general attitude, willingness to listen, etc.)	5	4	3	2	1
Member 2:		Overall Score:			
	Excellent	Very Good	Good	Fair	Poor
Meeting attendance	5	4	3	2	1
Completed work on time	5	4	3	2	1
Practiced presentations	5	4	3	2	1
Reviewed others' work	5	4	3	2	1
Other teamwork elements (attention during meetings, general attitude, willingness to listen, etc.)	5	4	3	2	1

(continued)

Member 3:		Overall Score:			
	Excellent	Very Good	Good	Fair	Poor
Meeting attendance	5	4	3	2	1
Completed work on time	5	4	3	2	1
Practiced presentations	5	4	3	2	1
Reviewed others' work	5	4	3	2	1
Other teamwork elements (attention during meetings, general attitude, willingness to listen, etc.)	5	4	3	2	1

Member 4:		Overall Score:			
	Excellent	Very Good	Good	Fair	Poor
Meeting attendance	5	4	3	2	1
Completed work on time	5	4	3	2	1
Practiced presentations	5	4	3	2	1
Reviewed others' work	5	4	3	2	1
Other teamwork elements (attention during meetings, general attitude, willingness to listen, etc.)	5	4	3	2	1

Member 5:		Overall Score:			
	Excellent	Very Good	Good	Fair	Poor
Meeting attendance	5	4	3	2	1
Completed work on time	5	4	3	2	1
Practiced presentations	5	4	3	2	1
Reviewed others' work	5	4	3	2	1
Other teamwork elements (attention during meetings, general attitude, willingness to listen, etc.)	5	4	3	2	1

(continued)

Self-Assessment:		Overall Score:		
	Excellent	Very Good	Good	Fair	Poor
Meeting attendance	5	4	3	2	1
Completed work on time	5	4	3	2	1
Practiced presentations	5	4	3	2	1
Reviewed others' work	5	4	3	2	1
Other teamwork elements (attention during meetings, general attitude, willingness to listen, etc.)	5	4	3	2	1

APPENDIX BA2

ASU Design Project Checklist

Team Information

Project Name:

Primary Contact:	Team Member:	Team Member:	Team Member:
Email:	Email:	Email:	Email:
Date:		Instructor:	

Instructions

Projects must meet certain minimum requirements to be considered acceptable capstone design projects. Please complete the checklist and project description to ensure compliance with the minimum standards. The project must be reviewed and approved by your instructor.

Project Checklist

☐ 1. Your project must fall within the generally accepted areas of mechanical engineering, such as thermal systems or mechanical systems (please specify).

☐ 2. Your project must utilize concepts from several (at least four) technical disciplines within engineering, such as automatic control, rigid-body dynamics, radiation heat transfer, aerodynamics, signal processing, and material fracture. Individual team members must take responsibility for and develop or hone expertise in one or more of these areas. Please specify the technical disciplines.

☐ 3. Your project must require knowledge and skills learned in previous required courses (please specify courses and the particular knowledge or skill learned in that course).

☐ 4. Your project must require a sufficient amount and level of technical analysis. Please specify analysis required, including specific equations and/or numerical analysis tools.

☐ 5. Your project must have a clearly defined goal function, preferably couched in mathematical terms (maximize efficiency where η = power out/power in, minimize cost where $ = development cost + manufacturing cost + operating cost, etc.). Please specify your project's goal function.

☐ 6. Your solution must be subject to multiple realistic constraints, such as cost, size, weight, required power, manufacture, environment, and government regulation. Please state at least four realistic constraints on your design.

☐ 7. In your design analysis, you must consider applicable engineering standards, such as SAE International, American Society for Engineering Education, or military standards. (Your project must be one for which engineering standards apply). Please list the engineering standards that apply. design.

☐ 8. The capstone courses (MEE 488 and MEE 489) require that you build and test your designed product, component, or system. Please describe anticipated components (at least two custom components) that can be manufactured using the School for Engineering of Matter, Transport & Energy machine shop.

Project Description

Please provide a 150-to-200-word description of your project. Please include 1) a discussion of how the final product will meet a recognized and realistic societal need with a definition of the product's customer and 2) an explanation of why you think the scope of the project is realistic for your team to complete in two semesters (approximately 160 hours per team member per semester).

I certify that the information provided is complete and correct. I acknowledge the time commitment to this course as outlined in the project description.

X_____
Team Member Name and Signature

X_____
Team Member Name and Signature

X_____
Team Member Name and Signature

X_____
Team Member Name and Signature

X_____
Team Member Name and Signature

X_____
Team Member Name and Signature

X_____
Approved by Instructor

APPENDIX BA3

Engineering Analysis for Design

1. Introduction

Engineering design requires both creativity and the application of scientific principles. This discussion considers the latter requirement. Most engineering courses are focused on the application of scientific principles to engineering problems in fields such as statics, dynamics, thermodynamics, etc. As one progresses through the curriculum, these basic principles are combined to solve more complex problems such as those encountered in machine design. Unfortunately, most textbook problems do not fully prepare the student for design analysis for the reasons given in Table BA3-1. This appendix addresses these aspects of analysis as it pertains to the design process.

Table BA3-1 Reasons Design Analysis Differs from Textbook Problems

- The designer often does not need an exact answer to proceed with a design decision, and the project manager does not want to spend time and money on a detailed analysis when a more top-level one is sufficient.
- Most design problems are ill-defined. A major part of the analysis is understanding the problem and making assumptions to arrive at a simplified problem that is amenable to analysis.
- The designer is usually not interested in just one answer (i.e., a point estimate). Instead, the designer wants to know how the answer is affected by changes in the input variables (i.e., the designer wants sensitivity studies).
- The design sponsor wants the analysis to be well-documented so that it can be evaluated and, if needed, replicated by others.

2. Analysis Planning: Selecting the Right Issues to Analyze at the Right Level

Several years ago, some capstone project assessment reviewers complained that the capstone projects reviewed lacked engineering analyses. The response the following year was a large amount of analyses, but most of it was not needed either at all or not at the detailed level, such as a finite element analysis (FEA), when a simple "P over A" analysis would suffice. Analysis planning is an important part of the design process. Analysis requires precious project resources. The benefits must justify the expense.

Most design analyses are used to determine whether the design will meet the design requirements. Once a failure mode and effects analysis (FMEA) is performed, additional analyses are often identified as ways to reduce the probability of a failure mode occurrence. At the beginning of each design phase, the team should create an analysis plan that identifies what analyses will be done during that design phase and at what level each analysis will be completed.

Once the analysis issues are identified, then the team needs to decide what level of analysis is needed for each issue. An analogy to a baseball park can be used to illustrate the level of analysis required. Some problems need a very exact answer—that is, they need to be at home plate. Other problems need to be tightly bounded—that is, they need to be in the in-field. Other issues just need to be in the ballpark.

For the latter issues, a back-of-the-envelope approximation is often sufficient. Detailed hand calculations are used to obtain a more precise answer. For difficult issues involving complex geometries, multiple loadings, or many variables, the use of modern engineering tools such as finite elements analysis (FEA) or computational fluid dynamics (CFD) may be required. Obviously, the resources needed from the project increase as the level of analysis increases.

When preparing the analysis plan, the skills and abilities of the team must also be considered. Accomplishing an FEA may take a large amount of calendar time and project labor if no one on the team is familiar with using FEA to solve the problem being considered. On the other hand, the same analysis may be a relatively easy task for a team member who has experience with these types of problems and ready access to the required FEA tools.

3. The Importance of Starting with a Graphic

A fundamental approach to engineering analysis is to present the issue in the form of a graphic. For a mechanics problem, a free-body diagram is usually the first step. For a thermodynamics problem, it is usually a system block diagram with temperatures, pressures, flows, and energy transfers indicated. Other types of engineering analyses call for similar graphics. These graphics help both the analyst and the reader understand the problem.

4. Documenting Analyses

Most engineering students—and practicing engineers—don't like to document analyses. However, it is a necessary task. If one has ever had to try and duplicate another engineer's poorly documented analyses, the importance of clear and detailed documentation is apparent. Without proper documentation, the reader is unable to understand the basis for the answers provided. Without understanding, the reader can put little faith in the results and therefore the design quality.

Most engineers don't solve a problem by starting with a clean sheet of paper and carefully writing out the solution to a given format. Instead, most engineers start by making a quick graphic and writing down some equations. They look at the problem from many perspectives and search for a good approach. They try several ideas until a proper approach becomes clear. Along the way, they must make assumptions and simplifications. Once the approach is clear, the right assumptions are known, and the results have been found, it is the proper time for the engineer to prepare the final documentation. Now is the time to bring out a clean sheet of paper and organize the work that has already been done. The format for this documentation is given in Table BA3-2.

Many students initially reject this two-step process, feeling that it is redundant. They see it as having to do the work twice. However, as the student gains experience in solving complex engineering analysis problems, it becomes clear that this two-step process is really the only effective way to arrive at a well-thought-out and documented analysis.

Table BA3-2 Analysis Documentation Format

- **Issue:** The issue is described, with a graphic included.
- **Problem statement:** Proper problem-solving techniques are used to arrive at a well-articulated problem statement that effectively addresses the issue. A graphic should be included.
- **Approach:** The approach to solving the problem in terms of simplifying assumptions and equations to be used should be covered. The approach also describes what kind of answer is needed (i.e., a ballpark solution, a very precise and accurate home plate answer, or something in between). The method of analysis, such as hand calculations or computer-aided analysis, should be included. Often a flowchart is helpful to show the analysis steps.
- **Defining equations:** The defining equations are listed, and each variable is defined.
- **Assumptions:** The assumptions are listed, and the rationale for each assumption must also be included.
- **Calculations:** The calculations must be documented. If there are several calculations that use the same equations, the top-level documentation should include a set of sample calculations. The remaining calculations should be archived in an appendix or the project notebook.
- **Results:** The results must be displayed in one or more tables and/or graphs. These graphics must be clearly and completely explained.
- **Conclusions:** The conclusions drawn from the results are listed and explained.
- **Solution:** The solution to the problem statement is based on the conclusions.
- **Recommendations:** What should the team do as a result of the analysis? Should the solution be implemented? If so, how and when?

5.   Modeling

A model is a schematic description of a system, theory, or phenomenon that accounts for its known or inferred properties and may be used for further study of its characteristics. Based on this definition, a model could be a mathematical equation, a finite element mesh, a Solid-Works three-dimensional (3-D) drawing, or any number of similar descriptions.

Both mathematical and physical models can be used to design a prototype. Math models are used to perform analysis, while physical models are used for testing. Both approaches yield information the designer needs to arrive at a prototype that optimally meets the design requirements.

6.   Optimization

Optimization is the process of finding not only a solution to a problem, but also the best solution. When optimizing a system, it is important to remember that component design choices must be made not to necessarily optimize the component, but to optimize the system as a whole. Too often, the designer treats the system as a grouping of independent components and attempts to optimize the system by optimizing the design of each component. This often results in a suboptimum system design. Some simple optimization examples follow.

Example 1: Optimizing Volume from a Metal Sheet

A manufacturer needs to make a large, open-topped box for an electronics enclosure. The largest sheet of metal available is 24 inches by 24 inches. The manufacturer can use square cut-outs at the corners of the sheet to form the bottom and sides of the box. The sides can then be folded to form the box. The manufacturer needs to know how large the side of the cut-outs should be to yield as large a volume as possible.

Two methods can be used to solve this problem. The exact solution is found by using calculus, as shown in Figure BA3-1. The numerical solution that estimates the answer to within an acceptable error is given in Figure BA3-2.

Issue: A manufacturer needs to make a large, open-topped box for an electronic enclosure. The largest sheet of metal is 24 in × 24 in. The manufacturer can use square notches (cut-outs) at the corners to form the bottom and sides of the box. The manufacturer needs to know how large the sides of the cut-outs should be to yield as large a volume of the box as possible.

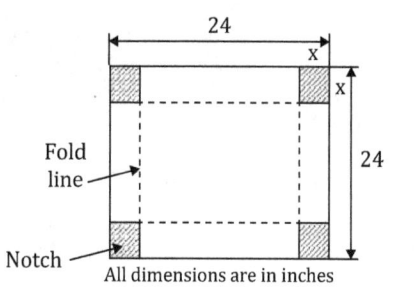

All dimensions are in inches

Problem statement: Find the value of x to obtain a maximum volume of the box.

Constraints: This is an unconstrained optimization problem because no limit was set neither on the values of x or the value of the volume.

The approach: Write equations for the volume of the box and take derivative $dV/dx = 0$ for maximum volume.

Defining equations:

Volume, V = area of the bottom × notch height = $(24-2x)^2 \times x$
$V = 4x^3 - 96x^2 + 24^2 x$
$dV/dx = 12x^2 - 192x + 24^2 = 0$
The eqaution has two roots: $x = 4$ or 12
The solution is $x = 4$, because $x = 12$ will result in Volume = 0

Conclusions: $x = 4$ gives the maximum volume of the box, but there is no sensitivity.

Solution: $x = 4$
Recommendations: Make each notch (cut-out) 4 in × 4 in

FIGURE BA3-1 Calculus Solution to Electronics Box Volume Optimization Problem

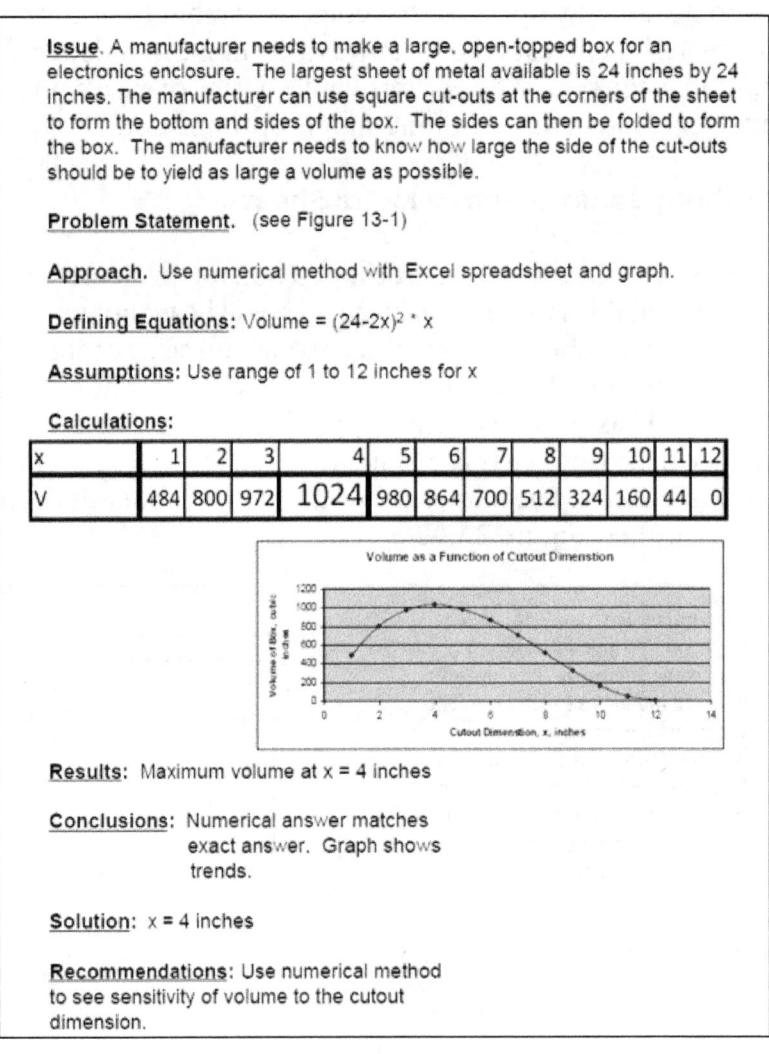

FIGURE BA3-2 Numerical Solution to Electronics Box Volume Optimization Problem

Example 2: Optimizing a Function of Two Variables

In this example, there is a dependent variable, z, that is a function of two independent variables, x and y. The equation for z is given below.

$$z = -2x^2 + 12x + 2xy - y^2 + 6y + 4$$

The constraint is that each independent variable has a range between 1 and 15.

Figure BA3-3 provides an exact solution using calculus. Figure BA3-4 provides a numerical solution using a Microsoft Excel spreadsheet. The answer is shown to be the same, although it is difficult to find the maximum value on the 3-D graph. However, the graph does show the sensitivity of the maximum to changes in both the x and y variables. This subject of sensitivity studies is presented in the next section.

> **Issue:** Show how to find an optimum for a function of two variables using an exact solution.
> **Problem statement:** Use the following function and find its maximum over the range 1 to 15 for both variables.
> $$Z = -2x^2 + 12x + 2xy - y^2 + 6y + 4$$
>
> **Approach:** Take the partial derivatives and set them = 0.
>
> **Defing equations:** Maximum: $\partial z/\partial x = 0$, and $\partial z/\partial y = 0$
> **Assumption:** There is a maximum in the allowed ranges.
> **Calculations:** Take the partial derivatives and set them = 0.
> Take the first derivative of z with respect to x and y:
> $$\partial z/\partial x = -4x + 12 + 2y = 0$$
> $$\partial z/\partial y = 2x - 2y + 6 = 0$$
> **Results:** On solving the two simultaneous eqautions, we obtain
> $$x = 9 \text{ and } y = 12.$$
> **Conclusions:** The method works; the answer is within the allowed range.
> **Solution:** Optimum is at $x = 9$ and $y = 12$.
> **Recommendations:** Use this answer and compare to numerical solution.

FIGURE BA3-3 Exact Solution to the Optimization of the Example Function of Two Independent Variables

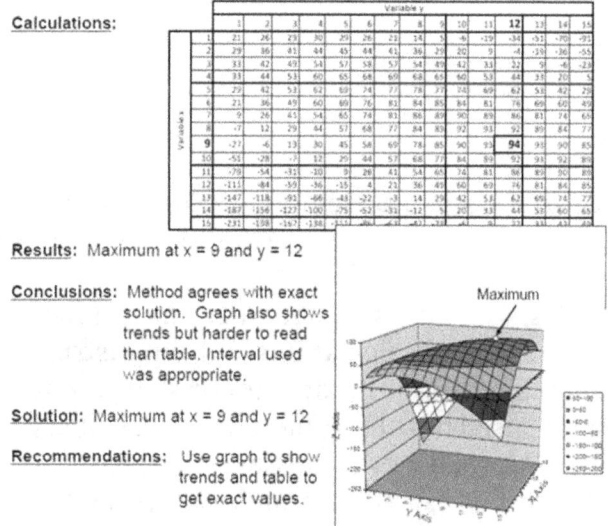

FIGURE BA3-4 Numerical Solution to the Optimization of the Example Function of Two Independent Variables

Appendix BA3 Engineering Analysis for Design | **369**

7. Single-Value Parametric Studies

In general, the designer wants not only a point estimate output, but also a feeling for how sensitive that answer is to its input variables. For the two example problems given in the previous section, the numerical solution graphs provide the designer with this sensitivity information.

Another type of sensitivity study is the single-value parametric study. In this method, there is a dependent variable, U, that has a number of independent variables, $x_1, x_2, x_3 \ldots x_n$. The analyst starts by determining the dependent variable answer when point values are provided for each of the independent variables. Then the analyst wants to know how sensitive the dependent variable is to changes in each of the independent variables when all other independent variables are kept at their point values. This is called a *single-value parametric analysis*.

A good way to conduct this type of analysis is to vary each variable over a percentage change range. Then the value of the dependent variable can be plotted as a function of the percentage change in each of the independent variables. The general format for this is shown in Figure BA3-5.

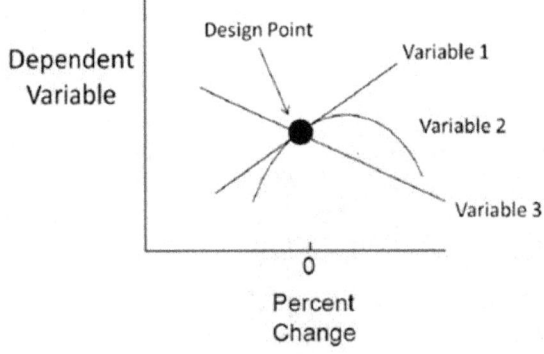

FIGURE BA3-5 Graph Format for a Single-Value Parametric Study

8. The Importance of Readable Hand Calculations

The myth among engineering students is that engineering is done solely by computer and there is no need for well-written hand calculations in the industrial setting. The truth is the opposite. Most engineering calculations for decision-making are still done with paper and pencil. These calculations are used by others; therefore, they must be readable. If a student currently has poor handwriting, then now is the time to improve. Most poor handwriting is the result of not taking the time to carefully form the letters and numbers. Team members must hold their teammates accountable for providing handwritten and readable analysis reports to the team project notebook.

9. Use of Modern Engineering Computer Tools

When the analysis issue involves complex geometries, multivariable optimization, etc., the use of modern, computer-based tools are often needed. Care should be taken by the team in deciding to use these tools. Modern engineering computer tools are valuable in many circumstances; however, they require considerable effort to use them properly. Before using one of these tools, the team must clearly understand its assumptions and limitations. Hand calculations to verify its results are mandatory. Analyses accomplished with these tools must be clearly documented in the project notebook to a degree that another engineer not on the project could read the report and then replicate the results. A diagram of a part's colorful stress field is meaningless without the proper documentation to explain the interpretation of the graphical result.

One of the Accreditation Board for Engineering and Technology (ABET) criteria is demonstrating the proper use of modern engineering tools. Moreover, many potential employers are expecting recent engineering graduates to have as a minimum the capabilities to prepare 3-D computer drawings, analyze stress and vibration problems with an FEA package and conduct analyses in MATLAB or similar computing software.

The capstone project report is an excellent document for showing one's computer capabilities to a potential employer. However, their use must be justified. Teams should try to define a problem statement that requires the use of these tools. The team must then allocate adequate resources to ensure that the analyses are conducted and documented in the proper manner.

A common practice by student teams is to delegate the use of the computer tools to one person in the team who is already skilled in their use. Unfortunately, this means that the rest of the team does not have an opportunity to develop and use their capabilities in this area. Teams are encouraged to find an efficient way to assign the work so that all team members have an opportunity to use one or more of these tools.

10. Documenting Created or Modified Analysis Software

During the process of analyzing the design, the team will utilize software tools such as MATLAB, Excel, Ansys, or Fluent. When tools such as these are used, they must be documented such that another team could use this information to replicate the results.

APPENDIX BA4

Product Development Testing

1. Testing as an Integral Part of the Design Process

Because so much of engineering tool development happens in computer simulation, many students are under the impression that testing now plays a small role in product design. In general, this is not the case. Testing should be prominent during all phases of design. In the early stages of design, there are often times when the designer wants to integrate a new concept into the design, but its viability is in question. Rather than just performing analyses and waiting until system development testing, the designer should conduct a simple proof-of-concept (POC) test without the entire system in existence. These POC tests are an excellent method of retiring development schedule and budget risk.

Once the prototype is built, it must be tested to see if it can meet the design requirements. These tests should be planned to best use the schedule, labor, and monetary resources available. It is almost a certainty that the prototype will not meet all the requirements the first time. When this happens, the design and its associated hardware need to be reworked before development testing can be completed. Once all of the rework has been completed and the final prototype hardware can meet all the design requirements, then validation testing can begin. Validation testing does not allow any further product development. This testing is done to verify that the current hardware meets all the design requirements.

The development and validation testing should be conducted in accordance with the quality assurance provisions that accompany the requirements list.

2. Types of Testing

Testing is done for two reasons: (1) to determine if a requirement has been met and (2) to determine the effect of certain factors on a particular outcome. This second form of testing is called an *experiment*. Both types of testing are needed during the design and development process. Proper documentation is needed for both types of testing.

When testing, the engineer wants to be sure that the test results are valid—that is, the test conclusions reflect the reality of the situation. In the first type of testing, the test procedures are developed with the customer to ensure that the test creates the conditions needed to fulfill the requirement. In the second type of testing, the engineer must design the test to explore the problem being investigated.

3. Basic Steps in Conducting an Experiment

Table BA4-1 lists the steps in conducting an experiment.

Table BA4-1 Steps in Conducting an Experiment
1. Define the problem by asking a question that is testable with the resources available.
2. Form a hypothesis about what will happen to the dependent variable when the selected independent variable is changed.
3. Design the experiment in terms of the apparatus and procedures such that only the independent variable of choice is changed while all other independent variables are held constant.
4. For the independent variable to be changed, determine the number of values to be investigated and the number of repeated trials to be performed at each value. (More repeated trials will reduce the probability of random errors.)
5. Use a data sheet to record both qualitative and quantitative data as the experiment is conducted.
6. Analyze the data to determine whether or not the hypothesis is supported.
7. Review the test procedures and data and determine what sources of error may be present and quantify these if possible.
8. Based on the work done, draw conclusions about the experiment.
9. Based on the conclusions, make recommendations on what should be done next to further the design effort.
10. Document the experiment and file it in the project notebook.

4. Areas Where Students Have Trouble with Testing

An important ABET outcome is the ability to design and conduct experiments as well as analyze and interpret the data. Industry reviewers often complain that capstone students show a low level of mastery in this area.

A major reason for this criticism by the reviewers is that there is generally little testing information covered in the capstone project final report or final presentation. Part of this may be due to the lack of reporting testing that was completed. Another contributor may be the fact that many teams take most of the second semester to build their project, and there is little time for testing. A third reason may be that teams perform only development tests and do not complete the design process with validation testing. A fourth reason is that uncertainty of the measurements is not considered when planning the test nor when the resulting test data are analyzed and reported. Based on these reasons, Table BA4-2 provides a list of actions teams should follow to demonstrate their mastery of testing.

Table BA4-2 Steps to Ensure Proper Testing on the Project

1. When preparing the project plan, include adequate calendar time and resources to conduct proof of concept (POC), development, and validation testing.
2. Conduct simple POC tests during the project's proposal, conceptual design, preliminary design, and detailed design phases.
3. Prepare a detailed testing plan during detailed design and present it at the detailed design review.
4. Prepare test procedures for all tests. These documents should be concise but complete. They should provide all the information necessary for another engineer or technician unfamiliar with the program to replicate the test.
5. Display the test results in properly created graphs with adequate labeling.
6. Create error bars for each data point graphed to indicate the range of error in the measuring of the data. The method used to establish this range must be included in the test report.
7. Prepare test reports for all tests that include the test procedures, data analysis, conclusions, and recommendations.
8. File all test reports in the project notebook.
9. Include the test plan and information on key tests in the final report.

5. Test Plan Template for Phased Product Development Phase 5

A test plan template for Phase 5 testing is provided in Figure BA4-1. This plan should be presented at the Phase 4 detailed design review.

Test plan for _____ Project

1. Purpose: This document describes the tests to be conducted on the _____ test article during both the development and validation phases of the program.

2. Approach: During the development phase, the test article will undergo testing to identify any rework that is necessary prior to final validation testing. Validation testing will not be initiated until all necessary rework has been performed and all development testing has been completed. If the test article fails a validation test, then validation testing shall be suspended and the test article shall be reworked as necessary to pass the test that was failed. Before validation testing is resumed, the team and the instructor shall meet and determine if any prior validation testing must be repeated due to the changes in the test article. The minutes of these meetings shall be documented and filed in the Project Notebook.

3. Requirements and tests:
Table 1 lists the test article requirements and the tests planned for each requirement. Table 2 provides more information about each test including the following: 1) Test number, 2) Features to be tested, 3) Acceptance criteria, 4) Expected results, 5) Test conditions, 6) Test setup and test rigs, 7) Summary of test procedures. Table 3 provides a list of any test article features that will be excluded from testing. It should be noted that some of the tests will be accomplished at a subsystem or component level during development.
Add Tables 1, 2 and 3 here.

4. Environmental and test rig needs: This section provides a description of the test environment and test rig needs for each test. It also lists all tools and instrumentation needed.

5. Schedule and personnel assignments: An estimated schedule for performing the tests for both the development and validation testing phases of the project are given in Figure 1. The team members needed for each test are also indicated in the referenced figure. (Add Figure 1 here.)

6. Test procedures:
6.1 Each test procedure will be organized as follows:

1. Title of the test	4. Description of test article	7. Safety provisions
2. Purpose	5. Description of the test setup	8. Data collection sheets
3. Approach	6. Environment and test conditions	9. Step-by-step instructions

6.2 Test Reports: Each test report will be organized as follows:

1. Title of the test	4. Test procedures	7. Results
2. Purpose	5. Data collection sheets	8. Conclusions
3. Approach	6. Data analysis	9. Recommendations

FIGURE BA4-1 Test Plan Template

6. Determining Testing Uncertainty

Importance of Determining Uncertainty

For every measurement, there is a region of doubt as to its exact value. An important part of the test planning process is to estimate this region of doubt. Generally, when the data are shown on a graph, the average of the data taken are shown along with uncertainty bars to indicate the region of uncertainty. For tests to determine requirements compliance, not only the averaged result but also the uncertainty bars must be within the compliance window. For investigative experiments, the uncertainty bars must be small enough to ensure that the trends found are actual trends and not just the result of testing uncertainties. Uncertainty bars are a function of the ability to observe and measure the data. It is also a function of the number of trials that are repeated. All test data used in the capstone project should have uncertainty bars. The example given below shows the procedure for determining uncertainty.

Procedure Example

An experiment results in the information given in Figure BA4-2, where the effect of Variable B on Variable A is measured. The accuracy of the devices to measure variables A and B are also given. The result of these measurements is to be expressed as an average value with a statement of the uncertainty of the measurement.

		\multicolumn{5}{c}{Value of Variable B}
		2	4	6	8	10
Value of Variable A	Run 1	9.8	19.6	29.4	39.2	49
	Run 2	9.85	19.7	29.55	39.4	49.25
	Run 3	9.9	19.8	29.7	39.6	49.5
	Run 4	9.95	19.9	29.85	39.8	49.75
	Run 5	10	20	30	40	50
	Run 6	10.05	20.1	30.15	40.2	50.25

Measurement Device for Variable B: Range 0 to 20 with full scale + 0.1 and -0.1
Measurement Device for Variable A: Range 0 to 100 with full scale + 0.2 and -0.2

FIGURE BA4-2 Experiment Results

To establish the uncertainty of the measurements, a spreadsheet is used. The procedure for building the spreadsheet is given below.

1. Assume variation in sample values are Type A uncertainties where measurement errors are distributed normally.

2. For each set of B variable trials, list the values of Variable B and calculate the sum and average of each column. Once the average of Xi is known, then the columns $(X-Xi)^2$ can be filled in. The results are shown in Figure BA4-3.

	B = 2		B = 4		B = 6		B = 8		B = 6	
	Xi	(Xi-X)^2	Xi	(Xi-X)^2	Xi	(Xi-X)^2	Xi	(Xi-X)^2	Xi	(Xi-X)^2
	9.8	0.015625	19.6	0.0625	29.4	0.140625	39.2	0.25	49	0.390625
	9.85	0.005625	19.7	0.0225	29.55	0.050625	39.4	0.09	49.25	0.140625
	9.9	0.000625	19.8	0.0025	29.7	0.005625	39.6	0.01	49.5	0.015625
	9.95	0.000625	19.9	0.0025	29.85	0.005625	39.8	0.01	49.75	0.015625
	10	0.005625	20	0.0225	30	0.050625	40	0.09	50	0.140625
	10.05	0.015625	20.1	0.0625	30.15	0.140625	40.2	0.25	50.25	0.390625
Sum	59.55	0.04375	119.1	0.175	178.65	0.39375	238.2	0.7	297.75	1.09375
Avg, X	9.925		19.85		29.775		39.7		49.625	

FIGURE BA4-3 Calculation of Average X and $(X_i - X)^2$

3. As shown in Figure BA4-4, for each set of trials, estimate the standard deviation, s, of the sample using the equation below, where n is the number of trials and x bar is the average of the sample measurements.

$$s = \sqrt{\frac{\sum_{i=1}^{n}(x_1 - \bar{x})^2}{(n-1)}}$$

	B = 2		B = 4		B = 6		B = 8		B = 6	
	Xi	(Xi-X)^2	Xi	(Xi-X)^2	Xi	(Xi-X)^2	Xi	(Xi-X)^2	Xi	(Xi-X)^2
	9.8	0.015625	19.6	0.0625	29.4	0.140625	39.2	0.25	49	0.390625
	9.85	0.005625	19.7	0.0225	29.55	0.050625	39.4	0.09	49.25	0.140625
	9.9	0.000625	19.8	0.0025	29.7	0.005625	39.6	0.01	49.5	0.015625
	9.95	0.000625	19.9	0.0025	29.85	0.005625	39.8	0.01	49.75	0.015625
	10	0.005625	20	0.0225	30	0.050625	40	0.09	50	0.140625
	10.05	0.015625	20.1	0.0625	30.15	0.140625	40.2	0.25	50.25	0.390625
Sum	59.55	0.04375	119.1	0.175	178.65	0.39375	238.2	0.7	297.75	1.09375
Avg, X	9.925		19.85		29.775		39.7		49.625	
s		0.093541		0.187083		0.280624		0.374166		0.467707

FIGURE BA4-4 Calculation of the Standard Deviation

The following final calculations are made and shown in Figure BA4-5:

4. For each set of trials, determine the standard uncertainty, $u1$, using the following equation:

$$u1 = s/\sqrt{n}$$

5. For equipment uncertainty, assume a rectangular distribution and that the semirange, a, is 0.2. Therefore, $u2$ can be determined as follows:

$$u2 = a/\sqrt{3}$$

6. The total standard uncertainty, ut, is then the root sum of the squares of $u1$ and $u2$:

$$ut = \sqrt{(u1^2 + u2^2)}$$

7. To establish a confidence interval, U, at the 95% confidence level, the total standard uncertainty is then multiplied by a coverage factor of 2:

$$U = 2ut$$

8. The measurement is then reported as ±U for 95% coverage.

Problem: Determine the average value of Variable A with uncertainty bars for each value of Variable B given the Hypothetical Measurement Data.

		Value of Variable B				
		2	4	6	8	10
Value of Variable A	Run 1	9.8	19.6	29.4	39.2	49
	Run 2	9.85	19.7	29.55	39.4	49.25
	Run 3	9.9	19.8	29.7	39.6	49.5
	Run 4	9.95	19.9	29.85	39.8	49.75
	Run 5	10	20	30	40	50
	Run 6	10.05	20.1	30.15	40.2	50.25

Measurement Device for Variable B: Range 0 to 20 with full scale + 0.1 and -0.1
Measurement Device for Variable A: Range 0 to 100 with full scale + 0.2 and -0.2

	B=2		B=4		B=6		B=8		B=10	
	Xi	(Xi-X)^2	Xi	(Xi-X)^2	Xi	(Xi-X)^2	Xi	(Xi-X)^2	Xi	(Xi-X)^2
	9.8	0.015625	19.6	0.0625	29.4	0.140625	39.2	0.25	49	0.390625
	9.85	0.005625	19.7	0.0225	29.55	0.050625	39.4	0.09	49.25	0.140625
	9.9	0.000625	19.8	0.0025	29.7	0.005625	39.6	0.01	49.5	0.015625
	9.95	0.000625	19.9	0.0025	29.85	0.005625	39.8	0.01	49.75	0.015625
	10	0.005625	20	0.0225	30	0.050625	40	0.09	50	0.140625
	10.05	0.015625	20.1	0.0625	30.15	0.140625	40.2	0.25	50.25	0.390625
Sum	59.55	0.04375	119.1	0.175	178.65	0.39375	238.2	0.7	297.75	1.09375
Avg X	9.925		19.85		29.775		39.7		49.625	
s		0.093541		0.187083		0.280624		0.374166		0.467707
u1		0.038188		0.076376		0.114564		0.152753		0.190941
a		0.2		0.2		0.2		0.2		0.2
u2		0.11547		0.11547		0.11547		0.11547		0.11547
ut		0.121621		0.138444		0.16266		0.191485		0.22314
U, k=2		0.243242		0.276887		0.32532		0.382971		0.446281
Result	9.925 +/- 0.243		19.850 +/- 0.277		29.775 +/- 0.325		39.700 +/- 0.383		49.625 +/- 0.446	

Error bars indicate that there is a 95% confidence level that the mean of the population is within the +/- range.

FIGURE BA4-5 Example Uncertainty Problem and Solution Spreadsheet

APPENDIX BA5

FMEA Methodology

1. Introduction

The methodology presented is based on information from the following sources. This information has been tailored to meet the needs of the capstone design course:

- *Failure Mode, Effects, and Criticality.* Analysis. *MIL-STD-1629A.*
- *Society of Automotive Engineers (SAE) Standard SAE J1739.*
- Reliasoft.com.
- Weibull.com.

2. Definition, Purpose, and Practices

The FMEA is a systemized group of activities designed to:

- Recognize and evaluate the potential failures of a product and its effects
- Identify actions that could eliminate or reduce the chance of potential failure occurring
- Document the process

The FMEA should be initiated early in the design process, but no later than preliminary design. It must be accomplished by a team that possesses the background experience and necessary skills to successfully manage the process.

The FMEA process is summarized as follows:

1. Assemble a team with the necessary skills and product information.
2. Define the system through schematics, drawings, and specifications.
3. Assemble additional information such as customer input and performance of similar systems.
4. List the functions the system must provide.
5. Utilize a worksheet to document the analysis
6. For each component, identify the ways it could fail (i.e., failure modes).
7. For each failure mode, list the effects on the component and on the system.
8. Establish the criticality of each failure mode by determining its risk priority number (RPN)—that is, the product of the severity (S), probability of occurrence (O), and ability to detect (D).

9. For each failure mode, list the action items that must be accomplished to reduce the RPN.

10. Assign a person to be responsible for each action item.

11. Establish a date for the action item to be completed.

12. Once the worksheets have been completed, list the failure modes by descending RPN.

13. Identify the critical failure modes by selecting those with a high RPN.

14. Designate a team leader to be responsible for ensuring that all actions recommended have been implemented or adequately addressed before the production readiness review milestone.

15. Make the FMEA a living document that always reflects the latest design level, as well as the latest relevant actions, including those occurring after the start of production operations.

3. Failure Modes

A potential failure mode can be described as the manner in which a component, system, or subsystem could fail to meet the design intent. These failure modes are described in "physical" or technical terms. Examples of failure modes are given in Table BA5-1.

Table BA5-1 Examples of Failure Modes

Bending	Deflection	Loosened	Short Circuit
Binding	Deformation	Misalignment	Spalling
Breakage	Fracture	Open Circuit	Sticking
Cracking	Leaking	Oxidized	Wear

Failure modes are defined as all the ways in which a product can fail to perform its intended function. More than one failure mode can exist for a given component of the product. Some failures may be gradual and/or partial, whereas others may occur immediately and completely.

4. RPN

The RPN is the product of the severity (S), occurrence (O), and detection (D) ranking:

$$RPN = (S) \times (O) \times (D)$$

The RPN is a measure of <u>design risk</u>. The values of S, O, and D can be defined in various ways. For the process reported here, each of these factors can take on a value from 1-to-10.

Therefore, the RPN in this method will be between 1 and 1,000. Failure modes with high RPNs should be addressed first by the team.

5. Severity (S)

S is an assessment of the seriousness of the effect of the potential failure mode. Severity applies to the effect only. A reduction in severity ranking index can be effected only through a design change. Severity is estimated on a 1-to-10 scale. Table BA5-2 provides the severity evaluation criteria.

Table BA5-2 S Evaluation Criteria

Effect	Criteria: Severity of Effect	Ranking
Hazardous—without warning	Very high severity ranking when a potential failure mode affects safe operation and/or involves noncompliance with regulations without warning.	10
Hazardous—with warning	Very high severity ranking when a potential failure mode affects safe operation and/or involves noncompliance with regulations with warning.	9
Very high	Product/item inoperable, with loss of primary function.	8
High	Product/item operable, but at reduced level of performance. Customer dissatisfied.	7
Moderate	Product/item operable but may cause rework/repair and/or damage to equipment.	6
Low	Product/item operable but may cause slight inconvenience to related operations.	5
Very low	Product/item operable but possesses some defects (aesthetic and otherwise) noticeable to most customers.	4
Minor	Product/item operable but may possess some defects noticeable by discriminating customers.	3
Very minor	Product/item operable but is in noncompliance with company policy.	2
None	No effect.	1

Note: Team members must agree on the criteria for ranking S and apply them consistently.

6. Occurrence (O)

O is the likelihood that a specific cause/mechanism will occur. Occurrence is estimated on a 1-to-10 scale. Table BA5-3 provides the O evaluation criteria. A method for estimating the failure rate is to use generic rates.

Table BA5-3 O Evaluation Criteria

Probability of Failure	Possible Failure Rates	Ranking
Very high: Failure is almost inevitable	1 in 2	10
	1 in 3	9
High: Repeated failures	1 in 8	8
	1 in 20	7
Moderate: Occasional failures	1 in 80	6
	1 in 400	5
	1 in 2,000	4
Low: Relatively few failures	1 in 15,000	3
	1 in 150,000	2
Remote: Failure is unlikely	1 in 1,500,000	1

Note: Team members must agree on the criteria for ranking O and apply them consistently.

7. Detection (D)

D is an assessment of the ability of the system to detect a failure mode before it occurs and prevent the failure from occurring. The detection could occur during a factory inspection for modes that occur during production or in the field for modes that occur as a result of operation. As shown in Table BA5-4, detection is estimated on a 1-to-10 scale.

Table BA5-4 D Evaluation Criteria

Detection	Criteria: Likelihood of D by Design Control	Ranking
Absolute uncertainty	Design control will not and/or cannot detect a potential cause/mechanism and subsequent failure mode; or there is no design control.	10
Very remote	Very remote chance the cesign control will detect a potential cause/mechanism and subsequent failure mode.	9
Remote	Remote chance the design control will detect a potential cause/mechanism and subsequent failure mode.	8
Very low	Very low chance the design control will detect a potential cause/mechanism and subsequent failure mode.	7
Low	Low chance the design control will detect a potential cause/mechanism and subsequent failure mode.	6
Moderate	Moderate chance the design control will detect a potential cause/mechanism and subsequent failure mode.	5
Moderately high	Moderately high chance the design control will detect a potential cause/mechanism and subsequent failure mode.	4
High	High chance the design control will detect a potential cause/mechanism and subsequent failure mode.	3
Very high	Very high chance the design control will detect a potential cause/mechanism and subsequent failure mode.	2
Almost certain	Design control will almost certainly detect a potential cause/mechanism and subsequent failure mode.	1

Note: Team members must agree on the criteria for ranking D and apply them consistently.

8. Sample Worksheet

IMAGE BA5-1

Failure Modes and Effects Analysis Worksheet

Part Name/No.		Customer Application	
Design Responsibility		Key Date	
Prepared by		FMEA Date (Orig) (Rev)	
FMEA Number		Core Team	

Find No.	Part Name	Function	Potential Failure Mode	Potential Failure Effect	SEV	Potential Causes Mechanisms of Failure	OCC	Current Design Controls	DET	RPN	Recommended Actions Responsibility/ Target Completion Date	Actions Taken	pS	pO	pD	pRPN

Appendix BA5 FMEA Methodology | **383**

9. Worksheet Category Definitions

The following is a description of the categories noted on the FMEA form:

1. Part name/no.
   Enter the name and number of the system, subsystem, or component being analyzed.

2. Design responsibility
   This is the design team.

3. Prepared by
   Enter the name and email of the team member responsible for the FMEA.

4. Customer/Application
   Enter the name of the external user and the application.

5. Key Date
   Enter the initial FMEA due date, which should not exceed the scheduled preliminary design due date.

6. FMEA Date (Orig.)   (Rev.)
   Enter the date the original FMEA was compiled and the latest revision date, when applicable.

7. Core Team
   List the names of the responsible individuals who have the authority to identify and/or perform tasks. (It is recommended that all team members' names and email addresses be included on a distribution list.)

8. FMEA Number
   Enter the FMEA document number, which may be used for tracking.

9. Item/Number
   Enter the name and number of the item being analyzed. Use the nomenclature and show the design level as indicated on the engineering drawing.

10. Potential Failure Mode
    Potential failure mode is defined as the manner in which a component, subsystem, or system could potentially fail to meet the design intent. The potential failure mode may also be the cause of a potential failure mode in a higher-level subsystem, or system, or be the effect of one in a lower-level component. Each potential failure mode for the particular item and item function should be listed. The assumption is made that the failure could but may not necessarily occur (e.g., corroded battery case).

11. Potential effect(s) of failure
    Potential effect(s) of failure are defined as the effects of the failure mode on functions, as perceived by the customer. Describe the effects of the failure in terms of what the customer might notice or experience, remembering that the customer may be an internal customer

as well as the ultimate end user. State clearly if the function could impact safety or noncompliance to regulations (e.g., deteriorated life of and impaired operation of the battery).

12. Severity (S)

    Severity is an assessment of the seriousness of the effect of the potential failure mode to the next component, subsystem, system, or customer if it occurs. S applies to the effect only and should be estimated on a "1-to-10 scale.

13. Potential cause(s)/Mechanism(s) of failure

    Potential cause(s)/mechanism(s) of failure is defined as an indication of a design weakness, the consequence of which is the failure mode. List, to the extent possible, every conceivable failure cause and/or failure mechanism for each failure mode. Typical failure causes may include but are not limited to incorrect material specified, inadequate design life assumption, and inadequate maintenance instructions. Typical failure mechanisms may include but are not limited to yield, fatigue, material instability, wear, and corrosion.

14. Occurrence (O)

    Occurrence is the likelihood that a specific cause/mechanism will occur. The likelihood of O ranking number has a meaning rather than a value. Removing or controlling one or more of the causes/mechanisms of the failure mode through a design change is the only way a reduction in the O ranking can be effected. The likelihood of occurrence of potential failure cause/mechanism is estimated on a 1-to-10 scale.

15. Current design controls

    List the prevention, design validation/verification (DV), or other activities which will ensure the design adequacy for the failure mode and/or cause/mechanism under consideration. Current controls, such as mathematical studies, laboratory testing, feasibility reviews, and prototype tests, are those that have been or are being used with the same or similar designs. There are three types of design controls/features to consider, those that: (1) prevent the cause/mechanism or failure mode/effect from occurring or reduce their rate of occurrence, (2) detect the cause/mechanism and lead to corrective actions, and (3) detect the failure mode.

16. Detection (D)

    Detection is an assessment of the ability of the proposed type (2) current design controls to detect a potential cause/mechanism (design weakness) or the ability of the proposed type (3) current design controls to detect the subsequent failure mode before the component, subsystem, or system is released for production. Ranking is on a scale of 1 to 10, dependent on the detection level.

17. Risk Priority Number (RPN)

    The *RPN* is the product of the S, O, and D ranking. The RPN is a measure of design risk and will compute between 1 and 1,000.

18. a) Recommended Action(s)

    When the failure modes have been rank-ordered by RPN, corrective action should be first directed at the highest-ranked concerns and most critical items. The intent of any recommended action is to reduce any one or all of the O, S, and/or D rankings.

18. b) Responsibility & Target Completion Date
    Enter the organization and individual responsible for the recommended action and the target completion date.

19. Actions Taken
    After an action has been implemented, enter a brief description of the actual action and effective date.

20. Potential Severity (pS)
    If the actions are completed, this is the potential S number.

21. Potential Occurence (pO)
    If the actions D completed, this is the potential O number.

22. Potential Detection (pD)
    If the actions are completed, this is the potential D number.

23. Potential RPN (pRPN)
    This is the product of pS, pO, and pD. The pRPN is the value of the RPN predicted once the actions are complete.

SECTION C

FINAL REPORT OUTLINE

The following is an outline for writing the project final report.

TITLE PAGE

- Include the following: project name, picture of team with names, computer-aided design (CAD) picture of production unit, sponsor's name, and date.

TEAM MEMBER PAGE

- A statement that each team member has contributed to and reviewed the final report and agrees with its contents.
- List of team members' names followed by dated signatures

EXECUTIVE SUMMARY (SEPARATE PAGE)

This is the most important part of the project final report. In one page, the entire report is summarized. This is a standalone item. It should be written after the entire report is written. Adequate time must be allotted to this vital task. It must cover customer need, requirements, technical approach, final design, and te key items relating to the development of this product, including the engineering prototype. It should finish with conclusions and recommendations for what the sponsor should do next (e.g., funding the production prototype). The lower one-third of the page should be a labeled graphic of the final production design.

TABLE OF CONTENTS

TABLE OF FIGURES

TABLE OF TABLES

ABET CRITERIA/FINAL REPORT CROSS-REFERENCE TABLE

- Use three columns. The first column has Accreditation Board for Engineering and Technology (ABET) criteria. The second column has target level of mastery. The third column has the best example of these criteria in the project final report along with the page numbers where they are located in the final report.

1. Introduction (start at the top of a new right-hand page)

The first paragraph should identify the following: purpose of the report, sponsor's need, customer need, product solution, project time frame, deliverables, and team member names. The second paragraph should tell the reader what will be covered in the remainder of this section.

1.1. Design Need

- Describe the customer need. The use of graphics is encouraged.
- Quantify the need (i.e., the solution needs to be a product with an estimated production rate of X units per year for X years at a sell price of $X per unit).

1.2. Problem Statement

- Present the problem statement and box it. Include the deliverables in the problem statement.
- Describe how the team went from the customer need to a specific project problem statement. Use a graphic to make this section more effective.

1.3. Physics Involved

- It is imperative that the team demonstrate that they understand the basic science and math involved in designing a product that meets the problem statement. To do this, the basic physics involved must be explained. Use graphics such as free-body diagrams, cycle block diagrams with state points, control volumes, mass volumes, and lumped-mass transient analysis. Be sure to label these graphics.
- Include the key defining equations (with variables defined) and other scientific factors that will affect the design.

1.4. Design Approach

- Phased product development (PPD) process divided into six phases.
- Report covers first four phases and ends with detailed design. Results from an engineering prototype were used as part of the detailed design process.
- The commercialization plan includes Phase 5: Production Unit Prototype Development and Phase 6: Production.
- The KEEN entrepreneurial mindset tools were used to help in demonstrating all ABET criteria.

1.5. Project Scope and Limitations

- It is important to define what is within the scope of this project. Do this in a clear and concise paragraph.
- It is equally important to define what is *not* within the scope of this project. Do this by listing what is not within the scope in a clear list with associated explanatory text.

1.6. Societal Impact

- The role of the engineer is to create products that benefit society. Describe how this device will benefit the user, customer, and society as a whole.

1.7. ABET Accreditation

- The ABET Criteria/Final Report Cross-Reference Table (follows Table of Tables) provides objective evidence that the team has demonstrated the required level of mastery for each ABET criterion.

1.8. Project Notebook

- A paragraph similar to the following should appear in this section:

The team organizes all its work into a team project notebook that is used throughout the project to document the work. The notebook contains detailed descriptions of all trade studies, analyses, tests, and team decisions. The final report is written as a comprehensive, standalone document. However, it refers to the notebook as needed to direct the reader to more detailed information regarding the design.

1.9. Report Organization

- The report organization has already been given in the Table of Contents. However, it is proper to also summarize the organization of the report in the introduction.
- A paragraph similar to the following should appear in this section:

The report is divided into 15 sections. Section 1 discusses the societal need, the project problem statement, and the project scope. Section 2 presents the solution to the problem statement, i.e., the final production design and the engineering prototype results. The next five sections describe the design/development process, i.e., Preconcept Design, Requirements and Constraints, Conceptual Design, Preliminary Design, and Detailed Design. Sections 8 through 12 describe the design, manufacturing, and testing of the engineering prototype. Section 13 describes how the engineering prototype results were integrated into the final production detailed design. Section 14 describes how the team performed, while Section 15 lists the lessons learned on the project. Section 16 provides the acknowledgements. This is followed by the appendices.

2. Final Design Description

2.1. Design Description Overview

- Introduce the final production unit design as the solution to the project problem statement.
- Provide the final CAD model isometric drawing of the *production unit* with key component labels and overall dimensions. Never start a section with a graphic. There must be text that introduces each graphic or table before it appears in the report.
- For the production unit, provide a table of key characteristics, such as size, weight, and performancec.
- Explain why this is an optimum design to meet the customer's need.
- Introduce the engineering prototype that was designed, built, and tested during detailed design.
- Provide a CAD model isometric drawing of the *engineering prototype* including labeled components and overall dimensions.

2.2. Method of Operation

- Explain how the design works, starting with the basic physics required. Show how these concepts are embodied in the design to meet the problem statement. Take the reader through the process of operation, such as start, stop, change performance levels, and maintenance. Use graphics in this section.

2.3. Key Features and Benefits

- Introduce a table of features and benefits. Provide text to introduce the table and discuss the various features and benefits. Remember, a feature is some physical aspect of the design that provides the customer with benefits in terms of cost, performance, etc.

2.4. Key Performance Results

- This is the place for a detailed component weight table, efficiency data, part-power performance, and reliability estimates. *Sensitivity* of the device's performance to other factors such as ambient temperature should also be covered here.

2.5. Production Unit Requirements/Validation Matrix

- Explain how the customer's need was transformed into measurable engineering requirements, with target values.
- Explain what standards were used as part of the requirements.
- Introduce the production unit requirements validation matrix. This table should have four columns: requirement, method of validation, validation status, and report reference page.
- In the validation status column, state the predicted requirement value based on analysis and/or testing. This includes testing of the engineering prototype.
- State that final validation of the production unit requirements will be accomplished with the Phase 5 production prototype, but this activity is beyond the scope of the current capstone project.

2.6. Cost Analyses

- Present the life cycle costs (LCC) of the product from the customer's perspective. Discuss how these costs (such as the sell price) were estimated.
- Present an LCC analysis from the producer's perspective. This should include a break-even analysis and internal rate of return (IRR) analysis.
- Provide a unit production cost table with component production costs.
- Discuss the costs associated with the engineering prototype.

2.7. Drawing Package Overview

- Describe how the design is documented. It may be a combination of formal drawings, hand drawings on a common format, and fly sheets from catalogs—give examples of each.

Provide a drawing tree or indented drawing list (bill of materials). *Explain that all the drawings are provided as an appendix to the report.*

2.8. Engineering Prototype

- Explain the purpose of the engineering prototype.
- Present the engineering prototype requirements validation matrix.
- Present the engineering prototype CAD model design (labeled with dimensions).
- Summarize how the engineering prototype was manufactured.
- Describe how the build book was used to document the engineering prototype configuration.
- Present labeled photographs of the completed engineering prototype.
- Describe the engineering prototype testing and the key results.

2.9. Intellectual Property Considerations

- Identify proprietary aspects of the production design. Determine which items should be trade secrets and which should be patented. Describe any patent searches conducted. Are there any features that can be patented? If so, what are they, and what steps have been taken to obtain a patent? What further steps should be taken? If no intellectual property exists, state that fact.

2.10. Commercialization Plan Summary

- Explain how the Goldsmith Commercialization Model was used to prepare the plan.
- Provide a Gantt chart for all six phases of the PPD process for the development and production of the unit. Include for each year of production the sales quantity and sell price.
- Discuss the marketing efforts done during the project and what else needs to be done during further development and production. Include market segments, competition, market entry strategy, and market exit strategy.
- Explain why the production unit design will be superior to potential offerings from other competitors.
- Identify the key remaining production issues that need to be addressed in the go-forward efforts.
- Present a brief business plan summary. Refer to the break-even and IRR analyses presented in Section 2.6.

3. Project Proposal for Phase 1: Preconcept

- This section presents the project proposal submitted to the sponsor.
- The purposes of the project proposal are to (1) convince the sponsor that they should fund the project and (2) provide the project team with a top-level plan for conducting the project.
- The project proposal is the deliverable for Phase 1: Preconcept Design of the PPD process. This phase uses a preconcept to prepare the proposal. The preconcept is only a representative concept that will be replaced by a better concept during conceptual design.

- Explain that the project plan is written in the future tense because it is the plan. The project's actual performance is presented in Section 14.

3.1. Selecting the Project Problem Statement

- Describe briefly how the team was formed and the problem statement was selected.

3.2. Preconcept Design

- To arrive at a preconceptual design, a preset of engineering requirements was developed. These should be provided in a table and discussed.
- Based on the prerequirements, a preconceptual design is defined. This will be used to envision a representative design and development program that can then be broken down into tasks and task descriptions.
- The preconcept design should be defined by a narrative description that includes a sketch or solid model with labeled components; a functional block diagram; a list of key parameters, such as size, weight, input requirements, and output values; and a table of features and their associated benefits.

3.3. Strategies to Address Key Issues

- Based on the preconcept, the team undergoes a visioning process for the entire design and development process. Once this overall program vision is made, the key issues associated with the program are listed.

3.4. Technical Approach

- This is a brief narrative that takes the reader through the technical steps that will be followed to design and develop the product. It will feature specific tasks that address the key issues identified in the previous subsection.

3.5. Project Management Approach

- This section describes in specifics how the team will manage the project to arrive at a product that meets all of the requirements while staying on time and within budget.
- The following should be addressed in specifics:
    - A narrative that covers the management functions of planning, organizing, and controlling
    - How the project plan will be used and updated as needed
    - When, where, and how team meetings will be used to manage the project, with a discussion of forms for meeting agendas and minutes
    - Project metrics
    - Method of addressing variances
    - Method of decision-making
    - Team problem-solving process
    - Teammate accountability

3.6. Risk Management Plan

- Describe how the team identified potential project (not technical) risks and prioritized them.
- Include a table listing the top five risks along with the plan to mitigate each risk.

3.7. Work Breakdown Structure (WBS) and WBS Dictionary

- Introduce and discuss the WBS diagram and WBS dictionary.

3.8. Project Schedules

- Introduce and discuss the project schedules. Discuss the key design review milestones and how the schedules accommodate the academic calendar.
- Explain how the schedules will be updated to show project progress (i.e., by filling in the bar an appropriate amount for each task) and showing task slips with dotted lines.

3.9. Labor Loading and Labor Budget

- Explain how the labor budget was determined. Discuss how the project was designed to evenly divide work among the team members.
- Present the project labor loading chart and explain how it will be updated during the project. Use the generic tables and figures of the handbook as templates for team's labor loading and labor budget.

3.10. Monetary Budget

- Present the monetary budget table and/or plot. Describe how the money is planned to be used as a function of calendar time.

3.11. Project Success Factors

- A good way to sum up the project plan is to list the five key factors that give the plan a high probability of meeting its goal. These factors should be well-reasoned and specific.

3.12. Proposal Presentation

- Describe what was presented to the sponsor and what resulted from the presentation, including any feedback.

3.13. Phase-Exit Criteria Checklist

- Provide a copy of the completed checklist and describe any additional actions required by the sponsor to exit this phase.

4. Phase 2 Conceptual Design Requirements and Constraints

- Have an opening paragraph that explains that Phase 2 starts with the determination of the engineering requirements and then proceeds into determining the final conceptual design that best meets the engineering requirements.

4.1. Refining the Project Plan for Phase 2

- Describe how the detailed Phase 2 project plan was prepared.
- Present the Phase 2 detailed schedule and labor loading table.

4.2. Needs to Requirements

- Describe how the voice of the customer (VOC) was accessed through research.
- Describe the process of converting the VOC into measurable engineering requirements.
- Provide a quality function deployment (QFD) graphic that shows the needs and also the requirements.
- Explain that the detailed requirements list is provided in Section 4.4.

4.3. Applicable Standards and Regulations

- Describe the process of identifying the applicable standards and regulations and list them. Explain what parts of these apply to this product and have been turned into requirements.

4.4. Validation Methods

- Describe the ways that a requirement can be validated at the end of the product development activities (i.e., inspection, demonstration, analysis, or test).
- Explain that each requirement listed in Section 4.4 will have to be validated by one of the methods described above.

4.5. Requirements/Validation Matrix

- Be clear that these requirements apply to the production unit design.
- Introduce the matrix table and explain how the status will be updated during the project. Remember, these updates represent work that is being done to predict how well the production unit will meet the requirements during Phase 5: Production Prototype Development and Validation.

- The requirements/validation matrix table should be in the following format:

Prototype Requirement	Method of Validation	Status

- Be sure each requirement is measurable. Leave the status column open. It will be filled in as the design progresses. For the "Method of Validation" column, be sure to include specific details on the type of validation. For example, rather than simply writing "analysis," you should state, for example, "finite element analysis". Another example would be the requirement "To operate down to 10 degrees Fahrenheit." Instead of just stating "test" in the "Method of Validation" column, be more specific, such as "use a cold soak chamber."

5. Conceptual Design

- Add a paragraph here that explains that this section describes what was done during the conceptual design phase to explore the entire design space and then attempt to find the best conceptual design. Summarize what will be presented in this section.

5.1. Functional Block Diagram

- This diagram shows the product functions and how they interrelate in order to meet all the engineering requirements.
- It is important to not include components in the functional block diagram. For example, instead of writing "motor," state "conversion of electrical energy into rotational motion."

5.2. Research of Prior Art

- Sponsors don't want to pay to reinvent the wheel. They expect the team to review all pertinent literature, be aware of state-of-the-art technology, benchmark other products, and be knowledgeable of the competition. Each year, U.S. taxpayers invests a huge amount of money in research and development (R&D) activities, mostly through the national labs and universities. Companies also invest in R&D and often report their findings in the open literature to demonstrate their core competencies.
- Be sure to include a description of any prior art and the team's efforts in researching the market for products related to the capstone problem statement.

5.3. Using Morphological Analysis to Identify Candidate Conceptual Designs

- Describe how the design space was explored through research, brainstorming, etc. Show representative sketches. Note: These sketches must be readable and labeled. Photos of drawings on a white board are often too poor in quality. In that case, redraw.
- Explain how for each function several component options were identified and placed on a morphological chart.
- Various combinations of components are then identified to provide three to five candidate conceptual designs.
- Show the resulting morphological chart.
- Describe the options in detail and include readable sketches and a descriptive narrative for each option. Be sure to label the sketches as *Option 1*, *Option 2*, etc., along with a descriptive title.

5.4. Method of Selecting Final Conceptual Design

- Explain how a Pugh matrix was used to select the final conceptual design.
- Describe the process of selecting evaluation criteria and scoring the options.
- Present the evaluation criteria, including the weighting. Provide the rationale.
- Describe what analyses and testing were used to obtain the necessary comparison information.

5.5. Final Selection Comparisons and Rationale

- Present the comparison trade study matrix table and explain the rationale for each rating.
- Identify the selected final design and refer the reader to Section 5.8 for a complete description of the final conceptual design.

5.6. Analyses

- Include an introductory paragraph here describing the content in Section 5.6.

5.6.1 Analysis Plan

- In most cases, the team will need to do some analyses to support the trade study done in Section 5.5. It is important to help the reader understand what analyses were done.
- Analysis is needed for four reasons: (1) to assess the feasibility of an approach; (2) to support trade studies that evaluate different concepts; (3) to optimize a particular design relative to a specific performance variable, such as weight and efficiency; and (4) to show that a particular design meets the requirements.
- Use a flowchart to show how all of the analyses performed are related to each other and to the project timeline. Indicate the analysis name. Put the requirement numbers that apply in parentheses.
- Have introductory text that refers to the flowchart and explains it.

5.6.2. *Example Analyses*

- Explain that two representative analyses are provided here and the rest are given in an appendix.
- For each analysis, use the reporting format provided in Appendix BA3 of the handbook.

5.7. Proof-of-Concept (POC) Testing

- Describe any POC testing that was accomplished in this phase.

5.8. Final Production Unit Conceptual Design

- Provide a well-constructed sketch or drawing of the final conceptual design. Be sure to label the parts. Discuss the key features and their benefits. Add tables of weights, performance, etc., as needed to define the design.
- Provide a requirements/validation matrix table and show in the status column how each requirement was addressed for the final conceptual design.

5.9. Commercialization

- Update the commercialization section provided in the prior phase to show the additional information that has been added.

5.10. Conceptual Design Review

- Describe what was presented in the design review and include any feedback received.

5.11. Conceptual Design Phase-Exit Checklist

- Provide a copy of the completed checklist and describe any additional actions required by the sponsor to exit this phase.

6. Preliminary Design

- Add a paragraph here that explains that this section describes what was done during preliminary design to refine the conceptual design. Explain that this refinement is broken down into work to create four increasingly improved versions of the product's CAD model.

6.1. Refining the Project Plan for Phase 3

- Describe how the detailed Phase 3 project plan was prepared.
- Present the Phase 3 detailed schedule and labor loading table.

6.2. Creating CAD Model Rev 0: Product Baseline Preliminary Design

6.2.1. Design Driver Analyses

- Identify the two or three key design drivers for this design and explain what is needed to arrive at a final design that meets all the engineering requirements and improves the goal function.
- Present the analyses accomplished to address each design driver and explain how the results changed the Phase 2 conceptual design.

6.2.2. Commercial-Off-the-Shelf (COTS) versus Fabrication Component Decisions

- Explain the process used to decide whether each component will be purchased from a supplier as a COTS item or fabricated either by the team or a supplier.

6.2.3. COTS Component Selections

- Explain the process used to select each COTS component and include the name of the supplier and their part number for the component. Show the analysis used to determine the technical specifications of the COTS items.

6.2.4. Fabricated Component Designs

- Explain the process used to design each fabricated component. Include, using neat sketching, the geometric dimensions and specifications of the fabricated parts. Also include the analysis used to arrive at the parts specifications.

6.2.5. Initial Assembly Process

- Describe how the product will be assembled. Include a table of steps.

6.2.6. CAD Model Rev 0

- Incorporate the above changes into this CAD model.

6.3. Creating CAD Model Rev 1: Updating the Requirements/Validation Matrix

6.3.1. Engineering Requirements Analyses

- These analyses ensure that the preliminary design, represented by CAD model Rev 0, will meet all the engineering requirements.
- Include an updated requirements/validation matrix table. Show the changes in the CAD model Rev 0 that result from the engineering requirements analysis.

6.3.2. CAD Model Rev 1

- Incorporate the above changes into this CAD model.

6.4. Creating CAD Model Rev 2: Reliability, Maintainability, and Safety (RMS) Analyses

- Explain that this section addresses RMS in the engineering requirements and that additional RMS analyses needed to ensure a robust design.

6.4.1. Reliability Analyses

- Describe the product's operational life and actions necessary to remove infant mortality and wear-out modes during the operational life.
- Provide a table of predicted component failure rates and explain how they were estimated.
- Discuss how the unit's predicted mean time between failures was determined from the above failure rates.
- Include what design changes are needed in CAD model Rev 1, arising from the reliability analysis, to meet the product's reliability requirements.

6.4.2. Maintainability and Logistics Analyses

- Describe the analyses that addressed scheduled and unscheduled maintenance.
- Describe features of the design that address logistics support.

6.4.3. Safety Analyses

- Describe the analyses that addressed system safety and include a hazards analysis table.

6.4.4. CAD Model Rev 2

- Incorporate the above changes into this CAD model.

6.5. Creating CAD Model Rev 3: FMEA, DFMA, and DTC

6.5.1. Failure Modes and Effects Analysis (FMEA)

- Briefly explain the purpose and method of the FMEA. Summarize the top five failure modes in a table. Follow this summary with the more detailed FMEA. If the FMEA is large, put it in the appendices and reference it here.
- Explain how the FMEA results were used to improve the preliminary design. Identify what changes in CAD model Rev 2 are required, based on the FMEA analysis.

6.5.2. Design for Manufacturing and Assembly (DFMA) Analysis

- Briefly explain the purpose and method of the DFMA. Summarize the team's efforts in improving the CAD model Rev 2 design using DFMA analysis. Identify what changes in CAD model Rev 2 are required as a result of DFMA analysis.

6.5.3. Design to Cost (DTC)

- Describe both the customer's and the producer's product life cycles and list the associated cost targets and rationale for selecting that target value.
- Describe how the product has been designed to meet each of these cost targets.
- Explain how the DTC results were used to improve the preliminary design.

6.5.4. CAD Model Rev 3

- Incorporate the changes that are dictated by FMEA, DTC, and DFMA analysis into the CAD model Rev 2. Show an isometric view of this CAD model that is labeled and has overall dimensions.

6.6. Defining the Final Preliminary Design

- Provide information that defines the design, such as narrative description, table of features and benefits, configuration block diagram, performance table, and requirements validation matrix with status.

6.7. Initial Engineering Prototype Planning

- Describe what engineering requirements will be addressed by the engineering prototype and how these requirements will be addressed.

6.8. Commercialization Plan Update

- Include Worksheet 5.

6.9. Preliminary Design Review

- Describe what was presented, the feedback provided, and how the team has acted upon this feedback.

6.10. Preliminary Design Phase-Exit Checklist

- Provide a copy of the completed checklist and describe any additional actions required by the sponsor to exit this phase.

7. Detailed Design Overview

- Add a paragraph here that explains that this section describes what was done during detailed design to arrive at a complete detailed drawing package of the production unit. Explain how the engineering prototype was used to improve the production unit detailed design.
- Describe how the detailed Phase 4 project plan was prepared.
- Present the Phase 4 detailed schedule and labor loading table.

- Present the top-level schedule and labor loading chart, including any slips in schedule, estimated time to complete (ETC) line, and completed tasks (percentage complete) and milestones.
- Explain how the team has managed the project to date and how they plan to manage Phase 4 to complete Phase 4 on schedule and within budget.

8. Engineering Prototype Preliminary Design

8.1. Scope and Engineering Requirements

- Explain the production requirements that are being addressed and how.

8.2. CAD Model

- Modify CAD model Rev 3 to produce the engineering prototype CAD model.
- Include exterior, interior, and exploded views that are labeled and have key dimensions.

8.3. Summary of Analyses and POC Testing

- Summarize analyses and testing used to establish status in Section 8.4.

8.4. Requirements Validation Matrix Status

- What was done, and what is the confidence level of successfully validating this requirement?

8.5. Long Lead Items

- What items were ordered early to remain on schedule?

8.6. Phase 3P Exit Checklist

- Provide a copy of the completed checklist and describe any additional actions required by the sponsor to exit this phase.

9. Engineering Prototype Detailed Design

9.1. Additional analyses and testing

- Was any additional work done before completing the prototype drawings?

9.2. Indented Drawing List and Bill of Materials

- Be sure to introduce these tables.

9.3. **Top Assembly Drawing**

9.4. **Example of Fabricated Part Drawing**

- Explain how tolerances and fabrication materials were selected.

9.5. **Example of Test Rig Drawing**

9.6. **Manufacturing Plan**

- Use the format given in Figure B-94.

9.7. **Test Plan**

- Use the format given in Figure B-95.

9.8. **Phases 3P and 4P Design Review (Design Review 4)**

- Summarize content, feedback, and team actions as a result of feedback.

9.9. **Phase 4P Exit Checklist**

- Provide a copy of the completed checklist and describe any additional actions required by the sponsor to exit this phase.

10. Engineering Prototype Manufacturing

10.1. **COTS Items**

- Use a table and emphasize a liaison engineer was assigned to each component.
- Discuss any issues encountered.

10.2. **Fabricated Items**

- Use a table and emphasize a liaison engineer was assigned to each component.
- Include photographs of items being fabricated.
- Discuss any issues encountered.

10.3. **Incoming Inspection**

- Describe the process and give an example.

10.4. **Manufacturing Troubleshooting**

- Describe any issues and how they were resolved.

10.5. Assembly

- Describe actual assembly process and include photographs.
- Describe the issues and how they were resolved.

10.6. Build Book

- Describe how the build book was organized.

10.7. First Article Inspection

- Describe this process, include the inspection sheet, and describe any issues found.

10.8. Test Readiness Review (Design Review 5)

- Summarize content, feedback, and team actions as a result of feedback.

11. Engineering Prototype Development

11.1. Updated Test Plan and Procedures

- Include any updates to the test plan.
- Provide test procedures.

11.2. Test Results and Development Issues

- For each test, describe how the data were collected, analyzed, and interpreted.
- Describe any issues encountered and how rework and/or repairs were accomplished.

11.3. Mid-Course Adjustments

- Describe any mid-course adjustments made to stay on schedule and budgets.

12. Engineering Prototype Validation

12.1. Updated Test Plan and Procedures

- Include any updates to the validation test plan.
- Provide test procedures.

12.2. Test Results

- For each test, describe how the data were collected, analyzed, and interpreted.
- Describe any issues encountered and how they affected the rest of the validation testing.

12.3. Test Results Presentation (Design Review 6)

- Summarize content, feedback, and team actions as a result of feedback.

13. Completing Detailed Design

13.1. Changes in Production Unit Due to Engineering Prototype Results

- List and describe the changes.

13.2. Production Unit Detailed Drawing Package

- Provide an overview of package and make the complete package an appendix.

13.3. Commercialization Plan Including Phases 5 and 6

- Show how the PPD process correlates with the Goldsmith Commercialization Model stages.
- Discuss how the project could continue beyond the capstone course.

14. Project Performance

- Provide the final project schedule, labor chart, and prototype budget with results.
- Assess how well the team managed the project.
- Assess how well the team demonstrated the entrepreneurial mindset indicators.

15. Lessons Learned

- List and describe at least five major lessons the team learned by conducting the project.

16. Acknowledgements

- Acknowledge the persons and organizations that provided support to the team during the project execution phases.

17. Appendices

Index

2-D, 188
3C's, 16
3-D, 260, 261, 262, 268, 366

A

accountable, 22, 36, 38
accreditation, 188, 323, 336
Accreditation Board for Engineering and Technology, 323, 336, 354, 371
accuracy, 268, 304, 375
aesthetics, 242, 243
affordable, 20, 45, 54
agreement, 36, 84, 85
AlliedSignal Inc., xxix, 5
alternatives, 66, 175, 192
American Marketing Association, 122
analyses, 84, 127, 131, 135
analysis plan, 116, 134
approach, 108, 117, 123, 127
approval, 94, 225, 235, 238
Arizona State University (ASU), xxix, xxxi, 2, 39
articles, 2, 88
ASME, 40, 188, 258
assembly, 104
assembly drawing, 144, 155, 194, 203
assembly plan, 139, 142, 144, 147
assessment, 237, 343, 364
assignment, 4, 18, 21
assumption, 21, 66, 116, 117, 120
author, 121, 183, 199
availability, 186, 187, 234

B

background, 5, 15, 192, 343, 378
back-of-the-envelope, 141, 168
back-ordered, 234
baseline, 51, 161, 164, 399
bathtub curve, 179, 180, 185
behavior, 14, 38, 91, 173
benefit, 121, 122, 125, 176
bill of materials (BOM), 232, 233, 337
block flow diagram, 242
Boeing 787 Dreamliner, 173
border patrol, 25
borescope, 186
brainstorming, 8, 77, 108, 397
brand, 66, 122, 123, 125
budget, 83, 85, 108, 133
build, 121, 122, 125, 234
build book, 237, 266, 274, 289
built, 234, 235, 295, 322
burn-in, 180, 185

business, 199, 200, 201, 209
business case, 19, 20
business model, 124
business venture, 121
buying power, 89

C

CAD model, 390, 392, 398, 399
CAD solid model, 161, 257
calculations, 116, 365, 366, 376
candidate, 82, 117, 118, 127, 135
candidate solutions, 21
capstone, 22, 25
capstone end-of-course deliverables, 340
capstone project activities, 233, 249, 274, 344
casting and molding processes, 244
catalogues, 151
cause, 193, 222
change management process, 295
changes, 297, 324, 326
characteristics, 242, 243, 275, 295
class session, 10, 43
clearance fit, 269
climate, 208, 209
commercial, 144, 148, 151
commercialization, xxviii, xxx, 207, 329
commercialization plan, 62, 65, 66
commerical off the shelf (COTS), 148
communicate, 61, 122, 162, 242
company, 20, 265, 380
competitor, 78, 123, 177, 223
complete package, 233, 405
complex problem, 21, 364
component, 366, 378, 384
computer, 23, 366, 370, 372
computer-aided design (CAD), 144, 151, 241, 388
computer model, 141, 166
concepts, 153, 175, 178, 185
conceptual design, 138, 199, 207
conclusions, 149, 313, 322, 327
conduct, 329, 337, 339
confidence, 319, 320, 321, 324
configuration block diagram, 248, 401
configuration control, 295
configuration management, 265, 295
connect, 16, 152, 166, 246
constraints, 172, 200, 201, 220
construction, 94, 162, 258, 337
consultants, 22, 85
contact wear, 180
contents, 257, 289, 291, 313
contract, 36, 84
contrarian, 16, 234

control drawings, 258, 337
corrosion, 40, 94, 180, 246, 385
cost, 186, 187
cost of electricity (COE), 177
create, 138, 141, 143, 144
create value, 16, 84
criteria, 5, 8, 13
cross-matrix, 92
cross-section, 254, 258, 262
curiosity, 4, 16, 17, 18, 234
curious, 4, 17, 18
current, 18, 38, 45, 66
customer, 39, 41, 42, 43, 46, 47, 48, 50
customer feedback, 9, 133, 228, 311
customer needs, xxx, 32, 64, 350
customer value proposition, 210, 221, 222

D

database, 185, 200, 265, 266
decision-making, 14, 175, 370, 393
decisions, 162, 182, 187, 196, 202
defects, 254, 380
defining equations, 41, 117, 366, 389
deliverables, 2, 388
demonstration, xxix, 20, 94, 104
Department of Homeland Security (DHS), 23
depot maintenance, 187
design, 187, 188, 189, 191
design analysis, 116, 315, 362, 364
design control, 94, 382
design drivers, 142, 148, 153, 399
design for assembly (DFA), 196
design for manufacturing and assembly (DFMA), 163, 196, 241
design for manufacturing (DFM), 196
Design I Course, 10
Design II Course, 12
design review, 67, 216, 225
design space, 34, 47, 53
design tasks, 141, 193
design team, 141, 146, 149, 155, 163
design to cost (DTC), 242, 254
design to unit production cost (DTUPC), 251
design to value, 199, 201, 228
destructive testing, 104
detail drawing, 257, 258, 337
detailed analyses, 233, 280, 341
detailed design, 281, 313, 328, 334
detailed design drawing package, 334, 337, 338, 340
detailed drawing, 242, 243, 257, 344
detail views, 263
development, 265, 266, 275, 279, 280
development test, 286, 316
development/validation, 313
deviations, 270, 290, 291, 295, 332
diagnose, 186, 187
diagram, 193, 208, 216, 227
diameter, 152, 155, 276
dimensioning, 267
dimensions, 267
distributor, 151
documentation, 173, 188, 214
dollars, 149, 150, 201, 221
double-pole-double-throw (DPDT), 157

down select, 47
drawing numbering system, 266
drawing tree, 266, 336, 392
drop test, 240, 250

E

effect, 330, 372, 375, 380
efficiency, 362, 391, 397
electric, 96, 144, 146, 157
electrical, 149, 150, 154, 190
embodiment design, 139
embrittlement, 180
EM@FSE 2.0 indicators, 234, 336
energy, 201, 243, 330
engineer, 236, 267, 315
engineering, 240, 242
engineering drawing format, 258
engineering judgement, 9, 121, 224
engineering prototype, 2, 9, 12, 13
Engineering Science Analysis Corporation (ESA), 23
entrepreneurial, xxix, 2, 14, 28
entrepreneurial mindset (EM), 75, 336, 354
environmental, 13, 17, 18
equation, 41, 100, 117, 182
errors, 182, 198, 297, 311
estimate, 253, 370, 375, 376
evaluation, 380, 381
example, 367, 368, 370, 388, 396
exit, 398, 401, 402, 403
exit criteria checklist, 76, 85, 120, 133, 394
experience, 111, 131, 145, 162, 168
experiment, 5, 372, 375
experts, 22, 121, 126, 127, 173
exploded drawing, 165
exploration, 40, 89, 91, 106, 108
exploratory testing, 316
exponential distribution, 180

F

fabricated, 139, 144, 147, 217
fabrication, 161, 198, 207
factory, 121, 180, 188, 381
faculty, 22, 234
failure, 239, 240
failure mode, 239, 335
failure mode and effects analysis (FMEA), 146, 192–193, 364
failure rates, 179, 181
fatigue, 162, 180, 186
fatigue crack, 186
feasibility, 5, 20, 385, 397
feedback, xxx, xxxi, 9, 17, 77
field data, 141, 182, 185
figure of merit, 201
filtering, 8, 42, 108, 111
final report, 111
finite element analysis (FEA), 364
first article inspection, 274, 301, 310, 404
fits, 38, 120, 270
fixtures, 163, 199
flowchart, 175, 273, 366, 397
focus groups, 89
format, 94, 95, 116, 148

forming and shaping, 244
friction fit, 269
functionality, 242, 243, 271, 272, 299

G

Gantt chart, 85, 141, 236, 336, 392
generic schedule, 145
geometric tolerances, 267, 275
goal function, 8, 93, 248, 362, 399
Goldsmith Commercialization Model, xxx, 62, 65
government regulations, 124, 234
graphic, 41, 92, 263
green run, 214, 215, 228
grinding, 197, 198, 245, 268
growth, 20, 22
guide, 2, 5
guidelines, xxx, 113, 196

H

handbook, 199, 201, 203, 204, 210, 215, 222
hard copy, 91, 106, 113, 314
hazards, 188, 400
hazards analysis (HA), 188
heat treatment, 198, 245
Herman E. Daly, 22
hole, 152, 155, 198, 259, 294
house of quality (HOQ), 92
human characteristics, 242
human factors, 242

I

incentive, 149, 150
incoming, 266, 274, 290, 291, 292
indented parts list, 266, 267
indicators, 336, 342, 354, 405
industrial design, 242, 243, 354
Industrial designer, 242, 243
industry, 265, 345, 348, 354, 373
in-house, 197
initiative, xxix, 2
inspection, 292, 293, 294, 301, 310
instructions, 68, 250, 258, 279
integrated product delivery and support (IPDS), xxix
interchangeability, 94, 197
interfaces, 94, 243, 246, 257, 337
internal rate of return (IRR), 125, 209, 391
International Organization for Standardization (ISO) brand standards, 122
internet, 66, 89, 111, 151
interviews, 25, 89
introduction, 20, 82, 138
investigate, 39, 41
isometric, 165, 390, 401
issues, 393, 403, 404
iterate, xxvii, 9

J

jet aircraft engine, 186
joining, 199, 245, 246
judgement, 9, 17, 77, 121, 342

K

Kern Entrepreneurial Engineering Network (KEEN), 354
kickoff meeting, 38
knowledge, 49, 55, 62, 77

L

labeling, 94, 294, 374
labor, 297, 302, 303
learning, 310, 314, 323, 325, 329
learning curve, 201
learning module, xxix, 10, 235, 354
lecture, 3, 10, 120, 339
legend, 2
lesson modules, 2, 3
level of mastery, 13, 14
library, 88, 193, 249
life, 179, 195, 217, 218
life cycle cost (LCC), 195
linear, 242, 249, 267
line fit, 269
line replaceable unit (LRU), 186
list, 262, 273, 281, 283
loading, 327, 365, 394, 395
locally made, 66, 131
logistics support, 174, 188, 400
long lead hardware, 241, 243

M

machining, 244, 245
maintainability, 239, 241, 242, 251, 254
maintenance, 246, 254, 391, 400
manufacturing, 390, 400, 403
manufacturing plan, 237, 247, 273, 282, 403
market, 330, 343, 392
marketing, 103, 121, 122
marketing mix, 123, 211, 222
marketing strategy, 123
market research, 84, 89, 95
marking, 94, 253, 274, 294
markups, 199, 213, 224
Martin Martinez, xxxi, 23
material review board (MRB), 94, 298
materials, 240, 243
mating parts, 269
matrix, 281, 303, 316
maturity, 20
maximize, 92, 150, 314, 362
maximum, 316, 317, 319, 320
McMaster-Carr, 151, 283
mean cost to repair (MCTR), 187
mean time between failure (MTBF), 181, 400
mean time between overhauls (MTBO), 187
mean time to repair (MTTR), 94, 187
mechanical, 182, 183, 190
meeting, 191, 195, 204, 207
methods, 140, 161, 162, 173
mid-course adjustments, 297, 299
mid-course corrections, 21, 195
milestone, 235, 299, 336, 379, 394, 402
military, 25, 95, 180, 182
Military Handbook 217, 182

military standards, 95, 362
MIL-SPECs, 95
MIL-STD 641A, 265
mindset, 234, 336, 342
minimize, 92, 93, 348, 355, 362
minimum, 92, 93, 111, 116, 122
minutes, 82, 86, 88, 91, 104, 113, 120
model, 116, 120, 121, 124
module, 82, 88, 120
module lesson sheet, 3
morphological chart, 84, 108, 109, 110

N

name plate, 94, 266
narratives, 106, 113
~National Aeronautics and Space Administration (NASA) Systems Engineering Handbook~, 173
~National Society of Professional Engineers (NSPE)~, 188
negotiation, 84
net present value (NPV), 125
nominal dimension, 161, 162, 164, 165
nonlinearity, 242
notebook, 248, 256, 280, 285

O

Occupational Safety and Health Administration (OSHA), 94-95
offshore, 155, 243
off the shelf, 148, 151
operation sheet, 276, 277
opportunities, 234, 315, 342, 343
optimization, 150, 177, 366, 367, 371
optimize, 150, 367, 397
orthogonal, 165
outcomes, 2, 10, 315, 324, 336
outline, xxix, 3, 4
outline drawing, 165, 257, 266, 337
overhaul, 186
overhead, 125, 199
overview, 204, 207, 214, 225

P

packaging, 147, 167, 173, 186
parallel, 157, 159, 182, 183, 184
parameter, 144, 152, 198, 201
part, 139, 198, 209, 220
peer evaluation, 67, 120
penalty, 149
perform, 141, 172, 188, 191, 199
Pew Research Center, 208, 209
Phase, 209, 210, 212, 214
Phase 1: Preconcept Design, 12, 32, 392
Phase 2: Conceptual Design, 46, 83, 134
Phase 3P: Engineering Prototype Preliminary Design, 242, 247, 279, 281
Phase 3: Preliminary Design, 121, 139, 227
Phase 4: Detailed Design, 243, 266, 341
Phase 5P: Engineering Prototype Development, 12, 257, 266, 279, 286
Phase 5: Product Prototype Development, 330, 348, 389, 395
Phase 6: Production and Support, 349
phased product development (PPD), 2, 3, 208, 348, 389
phase exit checklist, 142
phase-exit criteria, 133, 226, 340, 355, 394
plan, 8, 12, 13, 20, 24
planning, 33, 51, 57, 58
point estimate, 116, 364
police, 25
polishing, 198, 246
popular practice, 234
~Popular Science~, 24
post-lecture, 4
potential, 13, 17, 20
potential market, 48, 89
power, 66, 89, 104, 123
practicing engineers, 2, 354, 355, 365
precision, 116, 268
preconcept, 107, 234, 390, 392, 393
preface, 5
prelecture, 43, 46, 49
preliminary design, 53, 92, 121, 126, 127
presentation, 120, 127, 204, 214
press fit, 198, 269
price, 199, 203, 210, 211, 213
probability, 139, 168, 169, 221
probability distribution, 181
problems, 122, 126, 133, 196, 273
problem-solving, 21, 327, 336, 366, 393
problem statement, 8, 13, 34, 42
procedures, 274, 279, 280, 285
process, 289, 292, 295, 296, 297
process capability, 243
process planning, 274
process sheets, 258, 266, 337
producer, 315, 330, 336
producer value proposition, 330
product, 330, 335, 337, 339
product design and development, 2, 3, 339, 354, 355
production, 5, 6, 7, 9
production design, 53, 86, 205, 233, 242
production drawings, 9, 233
production prototype development, 234, 347, 348, 395
production rate, 132, 163, 221
production requirements, 236, 402
production unit design, 247, 254, 326
production unit drawing, 234, 333
production unit prototype, 315, 330, 345
production unit requirements, 236, 239, 249, 335
productivity, 35, 198, 199, 244
profilometer, 270
profit margin, 132, 199
program manager, 23, 25
project closure, 345
project management, 232, 234
proof-of-concept (POC), 5, 84, 139, 372, 398
proposal, 295, 374, 392, 394
prototype, 395, 401, 402, 403

Q

qualitative, 84, 92, 373
quality, 84, 92
quality function deployment (QFD), 84, 92, 395

quantitative, 92, 94, 373
questions, 89, 108, 116, 135
quiz, 4, 25, 28, 33

R

radiation, 94, 190, 361
random failures, 180, 181, 185
ratings, 99
rationale, 108, 114, 117, 125
reading, 138, 145, 153
reciprocal, 181
recommendations, 313, 324, 366, 373, 374
recycling, 22
redesign, xxvii, 92, 176, 184, 195
redundancy, 184, 185
relative weighting, 100
reliability, 94, 97, 101, 104
Reliability Analytics Toolkit, 182
reliability, maintainability, and safety (RMS), 9, 144, 172, 239
report, 2, 3, 5, 10, 12, 13
requirement, 6, 8, 9
requirements matrix, 9, 12, 98, 99
requirements validation matrix, 84, 119, 219
research, 86, 87, 88, 89
resistance, 157, 159, 246, 251, 317
resources, 233, 234, 236, 324, 343
responsibilities, 4, 13, 22, 85, 91
results, 93, 100, 111, 113, 116
return on investment (ROI), 124, 209
review, 126, 127, 133, 134, 136
revolving action item list (RAIL), 38, 236
rework, 53, 238, 286, 289, 290, 291
risk priority number (RPN), 193, 378, 385
roadblocks, 21
robust, 9, 53, 175, 400
roles, 36–37, 76, 85
routing sheet, 274, 275, 276, 278
rubric, 335

S

safety, 236, 239, 241, 242
safety factor, 94, 95, 162
sanity check, 116
scale, 89, 92, 96, 104, 123
scenario, 49, 55, 125, 149, 150
schedule, 50, 51, 55, 59
schematics, 151, 193, 378
scope, 144, 145, 147, 183, 201, 207
scrap, 197, 303
screw, 155, 156, 197, 198
search, 151, 152, 365
section view, 264
segmentation, 77, 88, 122
self-learning, 10
semester, 2, 3, 10
senior, 2, 22
sensitivity study, 116, 370
sequential, 141, 163, 179, 286
series, 157, 181, 182
shrink fit, 270
signature, 85, 313, 362, 388
similarity, 51

simulate, 154, 220
single-pole-single-throw (SPST), 159
size, 154, 161, 162, 169
sketches, 144, 161, 162, 165, 203
slides, 147, 214, 215
small business innovative research (SBIR), 23
societal needs, 17, 32, 39, 43
software, 116, 144, 161, 163, 164, 173
sole plate, 147, 154, 194, 208, 220
solid model, 161, 164, 165
SolidWorks, 144, 161, 163, 164
solution, 228, 236, 276, 327
sources, 16, 37, 246, 373
specifications, 84, 94, 144
SQUID, 24
stages, 35, 48, 200, 302
stakeholders, xxx, 122, 189
standard component, 197, 257, 258
standards, 243, 361, 362, 391, 395
statement, 388, 389, 390, 396
status, 391, 395, 398, 401, 402
strategies, xxvii, 2, 13, 21, 50, 393
strength, 133, 180, 181, 244, 250
stress, 181, 246, 253, 371
structural performance, 94
students, 91, 117, 153, 162
subassemblies, 145, 163, 165, 198, 203
subconscious, 108
submit, 151, 202, 225, 331
subproject, 233, 236, 333, 337, 341, 345
subsection, 10, 33, 116, 393
subteams, 148, 153, 161, 164, 169, 172
summary, 236, 241, 243, 247
supplier, 253, 258, 273
support, 281, 290, 335, 337, 342
surface finish, 243, 244, 245, 270, 275
surface treatment, 246
sustainability, 14, 22, 343, 354
syllabus, 4, 232
system, 14, 44, 66, 174, 266
systems engineering, 36, 172, 173, 174

T

target production cost, 162
targets, 199, 200, 201
target value, 100, 202, 335, 391, 401
tasks, 204, 206, 207, 214
team, 214, 215
team notebook, 138, 145, 148, 225
teamwork, 234, 358, 359
temperature, 95, 96, 319, 320, 365, 391
test, 84, 97, 104, 117, 134
test analyses, 323, 324
testing, 246, 248, 249, 253, 254
test plans, 237, 256, 257, 279
test readiness review, 301, 302, 303, 304, 310
test reporting, 323, 324
test rig, 239, 241, 328, 335, 340
textbook, 89, 258, 295, 364
thermostat, 130, 155, 166, 220
third angle projection, 258, 260, 262
Thomas, 151
threshold, 149, 176
title block, 258, 259, 262

tolerance, 259, 265, 267
tolerance stacking, 271, 272
tooling, 240, 244, 274, 275
tools, 245, 254, 274, 276
top assembly, 257, 262, 266, 283
total tolerance, 267
tradeoffs, xxvii, 22
trade study, 114, 115, 127, 397
training, 10, 173, 188, 274
transition fit, 269
transitioning, 4, 21
travel iron, 3, 69, 70, 71, 72
turbine blades, 186, 265
tutorials, 111
two-semester, 89, 90, 112, 121
types, 122, 126, 127, 130

U

Underwriters Laboratories (UL), 193, 239
unit production cost (UPC), 92, 175
university, 193, 197, 201, 249
unplanned events, 234, 342
unscheduled maintenance, 186, 400

V

validate, 2, 6, 9, 395
validation, xxix, 9, 119, 219, 284, 325
value proposition, 35, 46, 47, 48
values, 47, 122, 243
variability, 180
variable, 176, 187, 364, 365, 366
Vee model, 174
venture, 121, 123, 209
version, 161, 164, 172
views, 165, 206, 260, 262, 263–264
voltage, 283, 286, 315, 316, 317
voting system, 111

W

waivers, 94, 291
warnings, 94, 188, 254
warranty, 182, 187, 223
waste, 124, 136, 152, 197
watt, 149, 150, 218
webpage, 182
website, 15, 18, 24, 222
weight, 47, 93, 94, 97
weighting factors, 93, 114
wiring diagram, 159, 253
working drawing, 258, 265
worksheet, 18, 26, 203, 205, 210

About the Authors

Steven W. Trimble, PhD

For the past 15 years, Dr. Trimble has been a professor of practice in mechanical and aerospace engineering at Arizona State University (ASU). His research and teaching interests include systems engineering, energy, the engineering profession, servant leadership, and capstone course development, including multidisciplinary and interdisciplinary projects. He is also president of Trimble Consulting Inc. Prior to joining ASU, Dr. Trimble was vice president of engineering for Stirling Energy Systems, where he commercialized products for the solar energy market. He has over 45 years of industry experience in the design and development of products for the consumer, aerospace, defense, utility, automotive, industrial, and commercial markets. He has developed new products at  Texas Instruments, AlliedSignal, Honeywell, and Phoenix Controls. In addition, he has taught MBA and electrical engineering courses. Among his many honors are ASU Engineering's Top 5% Teaching Award and Honeywell's New Business Development Award. Dr. Trimble is past-president of the Tempe Rotary Club and past-chair of the AZ/NV Section of the Society of Automotive Engineers. He has authored a large number of technical papers, and he holds a patent in engine controls.

Abdelrahman N. Shuaib, PhD

For the past six years, Dr. Shuaib has been a professor of practice in mechanical and aerospace engineering at Arizona State University (ASU), where he specializes in the design and teaching of capstone product design and development courses. He is also co-investigator in the Industrial Assessment Center at ASU, which is funded by the DOE. Dr. Shuaib left King Fahd University of Petroleum and Minerals (KFUPM) as a full professor after 35 years of research and teaching. His areas of expertise include manufacturing, quality, materials, and product design and development. Dr. Shuaib has published over 90 journal and conference publications, one book chapter, and 13 technical reports. He is co-inventor of five issued US patents. At

KFUPM, he worked with the Massachusetts Institute of Technology (MIT)-KFUPM collaboration project to develop multidisciplinary capstone design projects. He also developed and conducted 28 industry continuing education courses. After graduating from University of Khartoum with a B.S. ME (first class honors), Dr. Shuaib started his career for five years in the appliance industry before he earned his doctorate in mechanical engineering from the University of Wisconsin–Madison.

Credits

IMG 0.1: Copyright © by ASU Enterprise Marketing Hub. Reprinted with permission.
IMG 0.2: Copyright © by ASU Enterprise Marketing Hub. Reprinted with permission.

CPSIA information can be obtained
at www.ICGtesting.com
Printed in the USA
LVHW020010260822
726797LV00005B/46